Feedstocks for Sustainable Biodiesel Production

Feedstocks for Sustainable Biodiesel Production

Characterization, Selection, and Optimization

Chinwe P. Okonkwo
Department of Pure and Industrial Chemistry
Nnamdi Azikiwe University
Awka
Nigeria

Modestus O. Okwu
Department of Mechanical Engineering
Federal University of Petroleum Resources
Effurun
Nigeria

WILEY

Published by John Wiley & Sons, Inc., Hoboken, New Jersey.
Published simultaneously in Canada.

Library of Congress Cataloging-in-Publication Data applied for:
Hardback: 9781394258079

Cover Design: Wiley
Cover Image: © krisanapong detraphiphat/Getty Images

Set in 9.5/12.5pt STIXTwoText by Straive, Pondicherry, India

Contents

Preface

Amid current energy challenges and ecological concerns, there is an urgent exploration for alternatives to conventional fossil fuels that are sustainable and environmentally friendly. The book titled *Feedstocks for Sustainable Biodiesel Production: Characterization, Selection, and Optimization* offers a comprehensive exploration into the critical role that novel and eco friendly feedstocks play in reshaping the scenery of the biodiesel industry. This book initiates an exploration that navigates through the varied aspects of sustainable biodiesel production, with specific emphasis on the invaluable contribution of biodegradable feedstocks. The book unfolds a narrative that spans from the cultivation and processing of these novel feedstocks to the intricate transformation of their oils into high-quality biodiesel, while also examining the optimization processes that enhance efficiency and effectiveness.

Against the backdrop of growing global concerns such as population growth, escalating energy demands, and pressing environmental issues, researchers and scientists are redirecting their efforts toward sustainable and renewable alternatives. Biodiesel, derived from both edible and nonedible sources, emerges as a symbol of hope in the twenty-first century due to its affordability, renewability, ecological friendliness, and minimal toxicity. Notably, this book underscores the immense potential lying within nonedible feedstocks, which offer sustainable alternatives without competing for valuable agricultural land or resources.

Fundamentally, this tome aims to shed light on the transformative potential of various feedstocks in progressing sustainable energy solutions and addressing environmental challenges. The aim of this study is to contribute to the knowledge foundation supporting a greener and more harmonious energy future by systematically examining both the scientific substructures and practical applications of these feedstocks. This comprehensive resource is intentionally crafted as an open text, tailored to cater to a broad spectrum of learners and researchers across various academic domains. It is methodically designed to be a valuable asset for undergraduate and graduate students as well as research fellows, covering a diverse array of fields such as professional and applied sciences (encompassing engineering, sciences, environmental studies, business, and agriculture, among others); natural sciences (including biology, chemistry, earth sciences, and related disciplines); economics; humanities; social sciences, and other relevant domains. Within its pages, this book provides a guiding inspiration for students and research fellows with a keen interest in energy studies, particularly focusing on the intricate process of biodiesel production utilizing novel feedstocks. The wealth of knowledge encapsulated in this resource is poised to empower learners and researchers alike, whether they are exploring the fields of sustainable energy, environmental studies, or interdisciplinary approaches. Whether the goal is to comprehend the scientific intricacies, inquire into economic considerations, or grasp the societal implications, this book serves as a guide for those navigating the multifaceted area of biodiesel production using innovative feedstocks. Whether for personal knowledge enhancement or as a foundation for more extensive research accomplishments, this open text stands as a comprehensive and accessible guide, fostering a deeper understanding of sustainable energy practices and biodiesel production for both domestic and industrial applications. It sheds light on the prospects and potentials of various biodegradable feedstocks including *Chrysobalamus icaco*, *Elaeis guineensis*, *Jansa*, waste cooking oils, *Moringa oleifera*, *Jatropha curcas*, *Hura crepitans*, water hyacinth, sunflower, and hemp, among others. Each feedstock receives dedicated chapters that comprehensively cover aspects ranging from sourcing and production to transesterification processes, biodiesel attributes, financial considerations, and environmental impacts and parametric studies.

The book is organized as follows: Chapter 1 provides a comprehensive overview and exploration of sustainable biofuel sources. It explores into the diverse areas of biofuels in terms of generation and types by investigating a range of feedstocks, their transformation processes, feasibility, environmental impact, and potential benefits of utilizing biofuels as an alternative to conventional fossil fuels. Chapter 2 offers an introduction to *Chrysobalanus icaco* as a sustainable and biodegradable seed feedstock for biodiesel production. Chapter 3 is about the utilization of *Elaeis guineensis* as an eco-friendly seed feedstock for biodiesel production. Chapter 4 deals with the improvement of biodiesel production process from waste cooking oil, by investigating the physical and chemical properties of a solid acid catalyst. Chapter 5 provides detailed information on the exploration of the viability of *Jansa* as a raw material for sustainable production of biodiesel, emphasizing the importance of oyster shell as a catalyst. Chapter 6 highlights the potential of *Hura crepitans* as a viable feedstock for the production of biodiesel. Chapter 7 provides a valuable feedstock: *Cyperus esculentus* (tiger nuts) as a biodegradable material for the production of biodiesel. Chapter 8 investigates the biodiesel potential of *Cucurbita pepo* seed oil, exploring its viability as a promising feedstock for biodiesel production. Chapter 9 investigates the sustainable production of biodiesel from sweet sorghum with detailed information on the farming practices, future prospects, and the production process. Chapter 10 offers a comprehensive overview of the current state of *Jatropha curcas* as a biodiesel feedstock, encompassing its botanical and agronomic traits, oil composition, and potential advantages and challenges. Chapter 11 covers various aspects of utilizing water hyacinth for biodiesel production, encompassing cultivation practices, harvesting methods, oil extraction techniques, and the biodiesel conversion process. Chapter 12 offers a comprehensive examination of the viability of sunflower as a promising feedstock for biodiesel production. Chapter 13 provides detailed exposition of *Moringa oleifera* as a potential and excellent feedstock for biofuel production due to its high oil content and economic and environmental benefits. Chapter 14 is an exposition of the potential of hemp as a promising and eco-friendly feedstock for biodiesel production. Chapter 15 focuses on the feasibility and potential of utilizing rapeseed oil as a primary feedstock for sustainable biodiesel production. Chapter 16 investigates the potential of Aidan seeds, a nonedible plant source, as a novel feedstock for sustainable biodiesel production. Chapter 17 investigates the viability of tobacco seeds as a sustainable feedstock for biodiesel production, employing a comprehensive parametric study facilitated by response surface methodology (RSM) and Grey Wolf Optimization (GWO) algorithm. Chapter 18 explores the utilization of natural heterogeneous catalysts, focusing on waste periwinkle shells, for biodiesel production from *Chrysobalanus Icaco*-seed oil and a comprehensive analysis by implementing fuzzy logic technique. Chapter 19 focused on the optimization of biodiesel production from waste cooking oil by implementing artificial neural network (ANN) models. Chapter 20 focuses on the analysis of results obtained based on biodiesel production from tobacco seed oil by implementing RSM and adaptive neuro-fuzzy inference system (ANFIS). Chapter 21 emphasizes the importance of innovation, efficiency, technological integration, global collaboration, policy support, and public awareness in navigating toward a greener and more sustainable energy frontier.

The holistic conclusion drawn from this collective exploration asserts that biodegradable feedstocks align with globally recognized standards, positioning them as pivotal players in the industrialization of biodiesel. As the intricacies of each feedstock is unraveled and investigation made into the unique contributions, it is evident that this book will serves as both a guide and an inspiration for students, researchers, practitioners, and enthusiasts alike, fostering a deeper understanding of the promise and potential inherent in advancing sustainable biodiesel production with biodegradable feedstocks.

1

Exploring Sustainable Biofuel Sources

A Comprehensive Overview

1.1 Introduction

The world's population is expected to increase to over nine billion people by 2050 [1], which will put enormous pressure on the world's energy reserves and accelerate their depletion [2, 3]. Fossil fuel combustion increases greenhouse gas (GHG) emissions, which has a negative impact on the environment [4]. Due to rapid industrialization and rising living standards, primary global energy consumption has increased over the past few decades [5]. For a decent standard of living, developing nations need a significant amount of energy. Currently, the world is made up of fossil fuels majorly gas, oil and coal, oil, and gas with the percentage of 80 approximately. The carbon emissions from these fuels account for approximately 98% [6, 7]. It is crucial to incorporate more renewable energy into the global energy mix to keep the surface temperature high to approximately 1.5 °C above preindustrial levels [8].

In the face of scarce resources, particularly water, and the ever-growing energy demand, alternative fuels emerge as the most feasible path forward [9]. Researchers are actively pursuing alternatives to meet this demand [10] aiming to address climate change impacts and fossil fuel dependence. Due to their potential to reduce GHG and the climate impacts related to transportation, biofuels are gaining more and more attention on national and international levels [11]. Utilizing bioenergy and/or biofuels offers a viable strategy for the production of renewable energy and low-cost synthesis [3]. Among biofuels, biodiesel has drawn significant interest due to its potential advantages over conventional fossil fuels and its adaptability in terms of feedstock.

Feedstocks for Sustainable Biodiesel Production: Characterization, Selection, and Optimization, First Edition. Chinwe P. Okonkwo and Modestus O. Okwu.

1.1.1 Comprehensive View of the State of Biodiesel, Feedstocks, and Production

Biodiesel has numerous advantages with attributes like sulfur-free content, adequate oxygen composition, ease of manufacturing, and reduced GHG emissions [12]. It is biodegradable [13], environmentally sustainable [14], nontoxic [15], efficient, and possesses low sulfur and aromatic content [16]. In addition, its higher flash point ensures safer transportation and storage compared to conventional diesel. By leveraging these valuable traits and the varied feedstock spectrum, ranging from waste frying oils to affordable nonedible sources, biodiesel holds substantial potential as an alternative fuel [17]. Positioned as an economical alternative, biodiesel offers promise in reducing global reliance on imported petroleum products. This chapter further delves into diverse biofuels and biomass resources that could potentially serve as biodiesel feedstocks. Supplementary chapters of this book will explore their properties, production procedures, factors influencing biodiesel synthesis, catalyst utilization, and other interesting area. Presentation of the comprehensive view of the state of biodiesel production and its associated prospects is further exposed.

1.2 Biofuels Generation and Types

Based on the sources and the materials used in their production, biofuels are divided into four generations. The traits of each generation of biofuels are presented below, with each generation representing a distinct method of biofuel production.

1.2.1 The First Generation

First-generation biofuels, which are regarded as conventional, are mainly made from two categories of edible feedstocks: sources based on starch (such as corn, wheat, barley, and potatoes) and sources based on sugar (such as sugarcane and sugar beetroot) [18]. Despite being easily accessible and convertible, these feedstocks are of great concern because they may result in a decrease in the world's food supply and their use them for agricultural purposes other than food production. Environmental concerns are exacerbated by the use of fertilizers and pesticides [11].

1.2.2 The Second Generation

The discussion surrounding the conflict between food and energy has sparked interest in second-generation biofuels using inedible sources like lignocellulosic biomass. Carbohydrates from various biomaterials, including trees, grasses, and agricultural waste, fall under this category [19]. These feedstocks have the advantage of requiring less agricultural land and edible plants [20]. Biomass exhibits the potential to stimulate the economy, improve energy security, and reduce GHG emissions [21].

1.2.3 The Third Generation

Waste oil and Algal biomass are the main sources of third-generation biofuels. These have benefits like faster growth rates, less reliance on farmland, more oil content, and less impact on food supplies. For the production of third-generation biodiesel, significant feedstocks include fish oil, microalgae, animal fat, and used cooking oil [22]. Utilizing algal biomass results in significantly higher yields of biofuel, but it is more expensive due to the processing requirements. Due to its quick cultivation time and high protein, carbohydrate, and lipid content, seaweed or macroalgae is equally gaining popularity [23].

1.2.4 The Fourth Generation

Fourth-generation biofuels incorporate genetic engineering, molecular biology, and other approaches to optimize biofuel yield. Genetically modified algae, boasting enhanced lipid and carbohydrate content, represent the core of fourth-generation biofuel production [24]. This approach targets increased biofuel production efficiency and yields through genetic manipulation. Cyanobacteria, micro-, and macroalgae form the foundation of fourth-generation feedstocks, demonstrating the potential for advanced biofuel synthesis [24].

Considering the generation of biofuels, it is important to note that the production of biodiesel is possible as considered in the available feedstock with oil content for the firstgeneration. The edible oils include palm, olive, mustard, soybean, and others. Figure 1.1 shows some vital feedstocks and their oil content. It has become more important to focus on identifying sustainable feedstocks for sustainable production of biodiesel in supplementary chapters.

1.3 Biomass Sources for the Production of Biodiesel

Fatty acid methyl ester (FAME) is a refined diesel fuel derived from a variety of biological sources, such as edible and nonedible oils, animal fats, and used cooking oil. FAME, which also stands for an oxygenated, organic, and environmentally friendly ester-based oil composed of fats and oils, and also contains long-chain fatty acid monoalkyl esters [2].

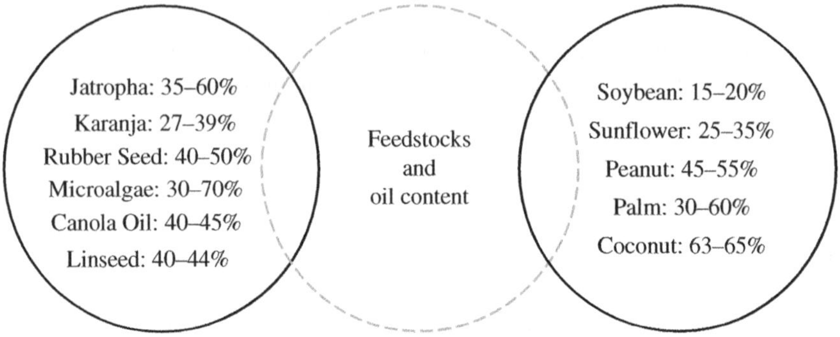

Figure 1.1 The oil content of feedstocks for biodiesel synthesis.

When considering factors like yield, price, composition, and purity, the choice of feedstock has a significant impact on the production of biodiesel. For example, while rapeseed and palm oil predominate in Europe and tropical areas, soybean dominates biodiesel production in the United States [25, 26]. This in-depth analysis highlights the significance of choosing the right feedstock by exploring the unique characteristics each source adds to biodiesel, particularly when combined with petroleum diesel. Exploring higher blends, even those exceeding 20%, becomes increasingly important as biodiesel's popularity grows [2]. In the United States, soybean and canola oil are the main sources of biodiesel, with soybean accounting for close to 50% of the feedstock input [27].

1.4 Feasibility of Using Biofuels in Place of Traditional Fossil Fuels

Several important factors must be taken into consideration for biofuels to be a viable replacement for traditional fossil fuels. One of the primary considerations is the availability of feedstock, which includes various biomass sources such as crops, agricultural residues, algae, and waste oils. The feasibility also depends on the technological advancements in biofuel production processes, which encompass methods such as transesterification, pyrolysis, and hydrothermal liquefaction [28]. Evaluating the scalability of these processes to meet the energy demands of modern society is crucial for determining the feasibility of biofuel adoption. In addition, factors such as energy efficiency, economic viability, and compatibility with existing infrastructure contribute to the overall feasibility assessment.

1.4.1 Environmental Impact of Biofuels

The environmental impact of biofuels compared to conventional fossil fuels is a central consideration in the transition toward sustainable energy sources. Biofuels have the potential to significantly reduce GHG emissions, contributing to the mitigation of climate change. While the combustion of biofuels releases carbon dioxide, the plants used as feedstock absorb CO_2 during their growth, forming a closed carbon cycle [28]. This contrasts with the open carbon cycle of fossil fuels, which release stored carbon into the atmosphere. However, the environmental impact of biofuels also depends on factors such as land use change, water consumption, and potential competition with food production [28]. Assessing the net environmental benefit requires a comprehensive analysis of these factors.

1.4.2 Potential Benefits of Biofuel Adoption

The adoption of biofuels offers a range of potential benefits that extend beyond environmental considerations. One of the most notable advantages is the reduction of dependence on fossil fuel reserves, which contributes to energy security. Biofuels can also enhance rural economies by creating opportunities for biomass cultivation and biofuel production in agricultural regions. In addition, the utilization of biofuels can diversify the energy mix, reducing the geopolitical tensions associated with oil-producing regions. From a technological standpoint, advancements in biofuel production can stimulate research and innovation, fostering economic growth in related industries. However, it is crucial to remember that while biofuels have many advantages, there are also drawbacks and trade-offs that must be carefully considered. These include the potential impact on food prices, land use conflicts, and the overall sustainability of biofuel production [29]. Striking a balance between the benefits and potential drawbacks is essential for ensuring a successful transition to biofuels as a viable and sustainable alternative to conventional fossil fuels.

In summary, exploring the feasibility, environmental impact, and potential benefits of utilizing biofuels as an alternative to conventional fossil fuels underscores the need for a comprehensive and interdisciplinary approach. It requires considering economic, environmental, social, and technological factors to determine the role that biofuels can play in addressing the challenges posed by climate change, energy security, and sustainable development.

1.5 Conclusion

This comprehensive book chapter is an exposition of the diverse generations of biofuels, emphasizing the escalating demand for renewable energy sources due to population growth and increased energy consumption. The pursuit of sustainable alternatives to traditional fossil fuels intensifies, drawing focus to the potential of biofuels. Other chapters will focus more on an exploration of novel assorted feedstocks for biodiesel production and further discuss biodiesel's multifaceted benefits and drawbacks. The aim is to provide a comprehensive understanding of the ongoing biodiesel revolution, underpinned by the potential of diverse feedstocks, to contribute to a more sustainable energy landscape. By analyzing the characteristics, production methods, catalysts, and environmental implications, this book will equip readers with a holistic view of the possibilities and challenges associated with biodiesel production, ultimately shaping the future of sustainable energy.

References

1 Okwu, M.O., Samuel, O.D., Otanocha, O.B., and Akporhonor, E. (2022). Production and characterization of briquettes produced from blend of rice husk and water-hyacinth. In: *Advances in Biofeedstocks and Biofuels, Production Technologies for Solid and Gaseous Fuels* (ed. L.K. Singh and G. Chaudhary). Wiley. https://doi.org/10.1002/9781119785842.ch12.

2 Maheshwari, P., Belal Haider, M., Yusuf, M. et al. (2022). A review on latest trends in cleaner biodiesel production: role of feedstock, production methods, and catalysts. *J. Clean. Prod.* 335: 131588.

3 Singh, A.R., Singh, S.K., and Jain, S. (2022). A review on bioenergy and biofuel production. *Mater. Today Proc.* 49: 510–516.

4 Alengebawy, A., Mohamed, B.A., Ghimire, N. et al. (2022). Understanding the environmental impacts of biogas utilization for energy production through life cycle assessment: an action towards reducing emissions. *Environ. Res.* 213: 113632.

5 Mbachu, V., Okwu, M., Nwabuto, C., and Tartibu, L. (2022). Biogas for electricity generation in Nigeria: asystematic review of the prospects, efforts and contemporary challenges. In: *Advances in Biofeedstocks and Biofuels, Production Technologies for Solid and Gaseous Fuels.* Wiley. https://doi.org/10.1002/9781119785842.ch4.

6 Seffati, K., Honarvar, B., Esmaeili, H., and Esfandiari, N. (2019). Enhanced biodiesel production from chicken fat using $CaO/CuFe_2O_4$ nanocatalyst and its combination with diesel to improve fuel properties. *Fuel* 235: 1238–1244.

7 Yusuf, M., Beg, M., Ubaidullah, M. et al. (2021). Kinetic studies for DRM over high-performance Ni–W/Al_2O_3–MgO catalyst. *Int. J. Hydrog. Energy* 47: 42150–42159.

8 Bacenetti, J., Negri, M., Fiala, M., and González-García, S. (2013). Anaerobic digestion of different feedstocks: impact on energetic and environmental balances of biogas process. *Sci. Total Environ.* 463: 541–551.

9 Iqbal, F., Abdullah, B., Oladipo, H. et al. (2021). Recent developments in photocatalytic irradiation from CO_2 to methanol. In: *Nanostructured Photocatalysts* (ed. V.-H. Nguyen, D.-V.N. Vo, and S. Nanda), 519–540. Amsterdam, The Netherlands: Elsevier.

10 Rosdin, R.D.b., Yusuf, M., and Abdullah, B. (2021). Dry reforming of methane over Ni-based catalysts: effect of ZrO_2 and MgO addition as support. *Mater. Lett. X* 12: 100095.

11 Jeswani, H.K., Chilvers, A., and Azapagic, A. (2020). Environmental sustainability of biofuels: a review. *Proc. R. Soc. A* 476: 20200351.

12 Tayari, S., Abedi, R., and Rahi, A. (2020). Comparative assessment of engine performance and emissions fueled with three different biodiesel generations. *Renew. Energy* 147: 1058–1069.

13 Okoye, P.U., Longoria, A., Sebastian, P.J. et al. (2020). A review on recent trends in reactor systems and azeotrope separation strategies for catalytic conversion of biodiesel-derived glycerol. *Sci. Total Environ.* 719: 134595.

14 Sun, S. and Li, K. (2020). Biodiesel production from phoenix tree seed oil catalyzed by liquid lipozyme TL100L. *Renew. Energy* 151: 152–160.

15 Garcia, R., Figueiredo, F., Brandão, M. et al. (2020). A meta-analysis of the life cycle greenhouse gas balances of microalgae biodiesel. *Int. J. Life Cycle Assess.* 25: 1737–1748.

16 Esmaeili, H. (2022). A critical review on the economic aspects and life cycle assessment of biodiesel production using heterogeneous nanocatalysts. *Fuel Process. Technol.* 230: 107224.

17 Singh, R., Singh, S., and Kumar, M. (2020). Impact of n-butanol as an additive with eucalyptus biodiesel-diesel blends on the performance and emission parameters of the diesel engine. *Fuel* 277: 118178.

18 Deora, P.S., Verma, Y., Muhal, R.A. et al. (2022). Biofuels: an alternative to conventional fuel and energy source. *Mater. Today Proc.* 48: 1178–1184.

19 Mat Aron, N.S., Khoo, K.S., Chew, K.W. et al. (2020). Sustainability of the four generations of biofuels–a review. *Int. J. Energy Res.* 44: 9266–9282.

20 Bhuiya, M., Rasul, M., Khan, M.M.K. et al. (2014). Second generation biodiesel: potential alternative to-edible oil-derived biodiesel. *Energy Procedia* 61: 1969–1972.

21 Okwu, M.O., Samuel, O.D., Otanocha, O. et al. (2020). Design and construction of bio-digester for gas production from hybrid waste. *World J. Eng.* https://doi.org/10.1108/WJE-07-2018-0249.

22 Alaswad, A., Dassisti, M., Prescott, T., and Olabi, A.G. (2015). Technologies and developments of third generation biofuel production. *Renew. Sust. Energ. Rev.* 51: 1446–1460.

23 Wirth, R., Lakatos, G., Böjti, T. et al. (2018). Anaerobic gaseous biofuel production using microalgal biomass–a review. *Anaerobe* 52: 1–8.

24 Abdullah, B., Muhammad, S.A.F.a.S., Shokravi, Z. et al. (2019). Fourth generation biofuel: a review on risks and mitigation strategies. *Renew. Sust. Energ. Rev.* 107: 37–50.

25 Kim, D.S., Hanifzadeh, M., and Kumar, A. (2018). Trend of biodiesel feedstock and its impact on biodiesel emission characteristics. *Environ. Prog. Sustain. Energy* 37: 7–19.

26 Umana, U.S., Ebong, M.S., and Godwin, E.O. (2020). Biomass production from oil palm and its value chain. *J. Hum. Earth Future* 1: 30–38.

27 Mahdavi, M., Abedini, E., and hosein Darabi, A. (2015). Biodiesel synthesis from oleic acid by nano-catalyst (ZrO_2/Al_2O_3) under high voltage conditions. *RSC Adv.* 5: 55027–55032.

28 Samuel, O.D., Okwu, M.O., Semiu, A., and Adeniran, S.A. (2020). Production of fatty acid ethyl esters from rubber seed oil in hydrodynamic cavitation reactor: study of reaction parameters and some fuel properties. *Ind. Crop. Prod.* 141: 1–13. https://doi.org/10.1016/j.indcrop.2019.111658.

29 Samuel, O.D. and Okwu, M.O. (2019). Comparison of response surface methodology (RSM) and artificial neural network (ANN) in modelling of waste coconut oil ethyl esthers production. *Energy Sources Part A*. https://doi.org/10.1080/15567036.2018.1539138.

2

Chrysobalanus icaco

A Sustainable and Biodegradable Feedstock for Biodiesel Production

2.1 Background of Study

The quest for an alternative energy source has been spurred by the vulnerabilities of fossil fuels focusing on their adverse ecological impacts when utilized in diesel engine systems. A feasible solution in this endeavor is the use of biofuels, which are both environmentally friendly and renewable [1]. According to ASTM, biofuels are composed of monoalkyl esters of long fatty acid chains made from renewable sources. They adhere to the ASTM D6751 standard. The need for supplementary water washing, the inability to recycle homogeneous alkali catalysts, and the production of waste are ongoing challenges for the biodiesel industry [2]. In order to lower production costs and waste generation, research on efficient and recyclable heterogeneous catalysts from inexpensive materials is ongoing [3].

The chemical process of transesterification involves the use of either homogeneous (acid or base) or heterogeneous (acid, base, or enzymatic) catalysts to produce biodiesel. Notably, the favored method is heterogeneous base-catalyzed transesterification due to its rapid product yield and short reaction times [4]. Nevertheless, variables like the water content of free fatty acids (FFAs), oils/fats, glyceride–alcohol molar ratio, catalysts, reaction time, and temperature can affect the transesterification reaction [5]. To achieve the proper methyl/ethyl ester transformation, FFAs must undergo an acid pretreatment before transesterification. Different feedstock-based triglycerides possess varying FFA levels influenced by factors like chemical composition, environment, and soil [6]. Economically viable feedstocks are rich in FFAs and moisture, which have great potential for making sustainable and competitive biodiesel [7].

C. icaco, also known by various names worldwide, such as cocoplum, gbafilo, fat-pork, icaco, and others, holds medicinal significance. It flourishes during the spring and summer, bearing immature and mature fruits on the same cluster that

Feedstocks for Sustainable Biodiesel Production: Characterization, Selection, and Optimization, First Edition. Chinwe P. Okonkwo and Modestus O. Okwu.

change color as they ripen [8]. Its presence spans diverse regions, from the United States to Central and South America, Africa, Asia, and Pacific island nations [9]. Traditionally, cocoplum has been used for treating conditions like leucorrhoea, bleeding, and chronic diarrhea and has been recognized for its diuretic, hypoglycemic, and anti-antigenic properties. Notably, its roots are used for diabetes treatment in Northern Brazil. Phytochemical analyses have identified myricetin and pomolic acid in its fruits, suggesting hypoglycemic potential.

This study is focused on the exposition of novel unconventional seed oil as alternative to biodiesel production. The scopeis on biodiesel production using a novel seed oil (*C. icaco*) with a natural heterogeneous catalyst and evaluating its performance. This research seeks to bridge the knowledge gap concerning the use of alternative feedstocks for biodiesel production.

2.2 Review of Literature

Okonkwo et al. [6] conducted a comprehensive study focusing on the kinetics and transesterification processes of oil derived from Cussonia bateri, commonly known as Jansa Seed, as a critical step in the production of biodiesel. Their research explored the potential of using a natural heterogeneous catalyst to enhance the efficiency and sustainability of biodiesel production. By employing such a catalyst, they aimed to address some of the common challenges associated with biodiesel production, such as the high costs and environmental impacts linked to conventional catalysts. The study carefully analyzed the reaction kinetics, which is fundamental in understanding the rate at which the transesterification process occurs. This analysis is crucial for optimizing the process parameters, such as temperature, reaction time, and catalyst concentration, to achieve maximum yield and efficiency. The use of a natural heterogeneous catalyst is particularly noteworthy, as it offers several advantages over traditional homogeneous catalysts, including easier separation from the biodiesel product, reusability, and reduced environmental footprint. Okonkwo et al.'s research contributes significantly to the existing body of knowledge on biodiesel production by demonstrating the feasibility of using non-edible oils and environmentally benign catalysts. Their findings indicate that Cussonia bateri oil, when subjected to transesterification using a natural heterogeneous catalyst, can produce biodiesel with desirable properties. This advancement not only promotes the use of sustainable feedstocks but also supports the development of greener production methodologies, aligning with global efforts to reduce reliance on fossil fuels and mitigate climate change.

Okonkwo et al. [10] focused their research on biodiesel production using *C. icaco*. They characterized the aforementioned seed oil using classical techniques such as Fourier transform infrared spectroscopy (FTIR) and gas chromatography–mass spectrophotometer (GC–MS). They also produced a catalyst from periwinkle shell ash (PSA) in heterogeneous solid form in three different states. Characterization was done using classical techniques like the FTIR and the scanning electron microscope (SEM). SEM analysis indicated that the calcined samples were preferable due to their larger surface area. The study showed a promising oil yield from the crop used to make commercial biodiesel. ANN toolbox from MATLAB was used to compare experimental and predicted values. Its potential as a source of biodiesel was demonstrated via transesterification of the seed oil with methanol. The catalyst used to accelerate the process is PSA.

Concerns about the availability of energy and the effects of global warming, which have drawn the attention of researchers all over the world, were addressed in the literature [11]. The study examined the improvement of several biodiesel blends and their uses in a variety of industries. They looked into the chemical and physical properties of various biodiesel types for a range of applications. While biodiesel usage positively impacts emissions reduction and pollution control, optimizing its efficiency and effectiveness both technically and nontechnically is essential. The study indicated that optimizing biodiesel's characteristics and parameters makes it a feasible and dependable renewable source of energy which the power generation and the automotive industry will benefit from in the future.

2.3 Materials and Method

All chemicals were procured from Sigma Aldrich and utilized as received. Extraction of oil from the raw seeds was carried out using a Soxhlet extractor, and the flashpoint was determined using a Pensky Martens flashpoint tester. Calcination was performed using a Muffle furnace.

A diesel engine with the model number TECHNO R175A was used to test the burning capacity. Also, a viscometer and an AP PAAR DMA 35 digital density meter were each used to measure the viscosities and densities of ester blends and seed oil, respectively. The seed oil and catalyst classical collections were examined using FTIR. An emission SEM and a spectrophotometer were used to examine the catalyst's morphology. The resulting ester's composition was determined using the GC–MS QP2010 Plus Shimadzu Japan. The setup is presented in Figure 2.1.

2.3.1 Sample Collection and Soil Extraction

Obtaining the sample of feedstock was made possible by accessing a local market called Ogbette main market in Enugu. An adequate quantity of cocoplum seed was obtained from this market. A taxonomist was able to identify the sample as *C. icaco*. After being transported to the site, the raw *C. icaco* was sundried and sample was ground in a commercial blender. A sieve with a mesh size of 50 μm was used to filter the samples.

Figure 2.1 Biodiesel production processes of *Chrysobalanus icaco*.

2.3.2 Extraction of Oil by Soxhlet Extractor

Samples of the *C. icaco* seeds were ground after drying and placed on a scale. The weighed material was placed within a thimble of a Soxhlet extractor in a semipermeable cotton material. The apparatus was linked to a round-bottom flask filled with *n*-hexane, and a condenser was attached to facilitate the process of extraction. The sample was subjected to reflux in the Soxhlet extraction setup until complete oil extraction was achieved. The defatted sample was discarded, and the oil extracted from the novel seeds was determined from the bottom layer under vacuum. The percentage yield of the obtained oil was then calculated.

2.3.3 Catalyst Preparation and Characterization

The snail shell sample collected from a local restaurant in Uwani, Enugu, Nigeria, underwent a process of cleaning, drying, and grinding to achieve a fine consistency. A quantity of 5 g of snail shell material was subjected to calcination at a temperature of 800 °C for two hours to produce calcium oxide. After calcination, the resulting material was washed with approximately 20 mL of distilled water until reaching a pH of 7. Subsequently, the samples were carefully dried in an oven before undergoing analysis as per the procedure outlined in Table 3.3, which presents the characteristics of the catalyst.

2.3.4 Biodiesel Production (Transesterification)

For each trial, a three-neck reactor (500 mL) was utilized to combine the catalyst (0.5 g) with ethanol, creating a total volume of 300 mL. After thorough stirring for a duration of 10 minutes, the mixture was completely dissolved. Subsequently, the oil (50 mL) was introduced to the solution, and the reaction mixture underwent refluxing at a temperature of 70 °C for a duration of two hours. To facilitate the separation of glycerol, the resulting reaction mixture was transferred to a separating funnel and allowed to settle for a period of 12 hours, forming distinct layers. The methyl ester (biodiesel) was then extracted using warm water, followed by its transfer to a 250 mL beaker. For moisture removal, the biodiesel was heated to 105 °C and subsequently cooled to room temperature. This procedure was repeated for snail shell ash (SSA) catalysts as well, following the approach outlined in a study [12].

2.3.5 Preparation of Biodiesel Blends

In the process of conducting characterization analyses, different biodiesel blends were formulated by combining them with fossil diesel in varying proportions of 20, 40, 60, 80, and 100%. These blends were designated as B20, B40, B60, B80, and B100, respectively. The evaluation of fuel properties encompassed parameters such as acid value, density, flash point, pour point, and viscosity at different temperature levels.

2.4 Results and Discussion

The results of the extracted seed sample's physicochemical properties are displayed in (Table 2.1).

The *C. icaco* crop exhibited a notably high oil yield percentage, rendering it suitable for commercial-scale biodiesel production. The oil obtained from this crop displayed a light brown hue. The majority of values aligned with the ASTM

Table 2.1 The extracted oils' physicochemical properties.

Properties	***Chrysobalanus icaco***
Color of material	Light brown
Yield (%)	60.80
Moisture (%)	3.20
Kinematic viscosity at 40 °C (mm^2/s)	22.54
Refractive index at 29 °C	1.375
Molecular weight (g/mol)	771.6
Acid value of oil (mgKOH/g)	1.18
Iodine value (g/100 g of oil)	105.0
Peroxide value (meq/kg)	0.91
Oil Saponification (mgKOH/kg)	185.0
Oil Density(g/mL)	0.928
Fire point of oil (°C)	254.0
Oil Pour point (°C)	4.0
Oil Flashpoints (°C)	220.0
Oil Cloud point (°C)	11.0
Energy value of oil (kJ/kg)	43 729

standards applicable to oils intended for biodiesel production, indicating that *C. icaco* seed oil holds promise as a valuable biodiesel source. Its relatively low acidity levels signify the feasibility of employing the base-catalyzed transesterification technique for biodiesel conversion without the need for additional treatment [13]. It is worth noting that excessive FFAs in oil can lead to catalyst deactivation, reducing biodiesel yield and resulting in soap formation. In the iodine test, *C. icaco* seed oil exhibited an iodine value of 105, classifying it as a nondrying oil. Oils with higher iodine content and greater unsaturation tend to undergo quicker oxidation and polymerization. The level of fatty acid unsaturation is directly associated with enhanced polymerization tendencies [14].

Consequently, biodiesel with high iodine content may show decreased stability at elevated temperatures.

The composition and ester content of the feedstocks used in the production of biodiesel has a significant impact on their iodine value. Fuel atomization and combustion quality are directly influenced by kinematic viscosity. Due to the oil's fatty acid composition and higher molecular weight, it had a higher kinematic viscosity than regular diesel fuel. Notably, according to Knothe and Steidley [15], the oil's kinematic viscosity is within the acceptable range for ASTM (Table 2.1). Table 2.2 is a view of the *C. icaco* seed oil fatty acid profile.

Table 2.2 *Chrysobalanus icaco* seed oil fatty acid profile.

S/no	Popular name	Scientific name	Shorthand	R. A (%)
1	Palmitic acid	Hexadecanoic acid	C16:0	5
2	Stearic acid	Octadecanoic acid	C18:0	25
3	Oleic acid	cis-9-Octadeaoic acid	C18 : 1	20
4	Linoleic acid	cis-9,12,15-Otacatrienoic acid	C18:3	11
5	α-Parinaric	Octadeca-9,11,13,15-tetraenoic acid	C18:2	11
6	α-Licanic	4-Oxooctadeca-9,11,13-trienoic acid	C18:3	10
7	Almonic acid	4-Oxo-octadeca-cis-9-trans-11-trans-13-cis-15-tetraenoic acid	C18:4	18
8	Arachidic acid	5,8,11,14-Eicosatetraenoic acid	C20:1	Trace

Table 2.2 outlines the presence of saturated, monounsaturated, and polyunsaturated fatty acids in *C. icaco*. The oil's relatively low linolenic acid content (13%) and the notable oxidative stability (20%) of oleic acid contribute to its prolonged storage potential. These characteristics position it as a promising feedstock option for biodiesel production, ensuring viability over an extended duration.

The catalyst's suitability for the necessary biodiesel production processes is evident from the characterization outcomes of the SSA catalyst, as detailed in Table 2.3. The calcined catalyst exhibited a surface area measurement of 842.00 m^2/g, indicating superior attributes attributed to its significant surface area. This attribute is associated with enhanced qualities and performance, as observed in related studies [16].

Table 2.3 Catalysts characterization of SSA.

Parameter	Calcined
Moisture (%)	0.04
Bulk density (g/cm^3)	0.542
Ph	5.68
Surface area (m^2/g)	842
Carbon (%)	1.86
Organic matter (%)	5.32
Loss on ignition (%)	0.52
Particle density ($g/cm^{3)}$	0.32
Total porousity (%)	23.54
Ash (%)	0.57

2.4.1 Surface Morphology of the Catalyst

The structure of the catalysts was examined using SEM, as presented in Figure 2.2. SEM images demonstrated that the duration of carbonization influenced the pore dimensions within the carbon structures of SSA catalysts; these dimensions expanded initially before collapsing. In contrast, calcined catalysts displayed well-defined and sizable pores. The substantial pore diameters contributed to an increased surface area of the catalyst, promoting the effective conversion of bio-oil. The process of producing the SSA-based catalyst involved a carbonization period exceeding 60 minutes at 800 °C, leading to the development of significant pore diameters. Furthermore, a prolonged one-hour carbonization into graphite could potentially lead to the disintegration of the carbon framework, thereby impacting the biomass conversion process [17].

The sample that had been calcined showed the diverse effects of the catalyst's larger surface area pores as shown in Figure 2.2.

2.4.1.1 Effect of Process Parameters (Snail Shell Ash) Using Ethanol (*Chrysobalanus icaco*)

The effect of catalyst concentration in production of biodiesel from *C. icaco* seed oil is further presented in Table 2.4.

It was found that using less catalyst produced higher yields. Larger molecules tend to react more slowly than smaller ones because they have a higher activation energy [18].

Table 2.5 shows the effect of time in the production of biodiesel from *C. icaco* seed oil.

SSA Calcined

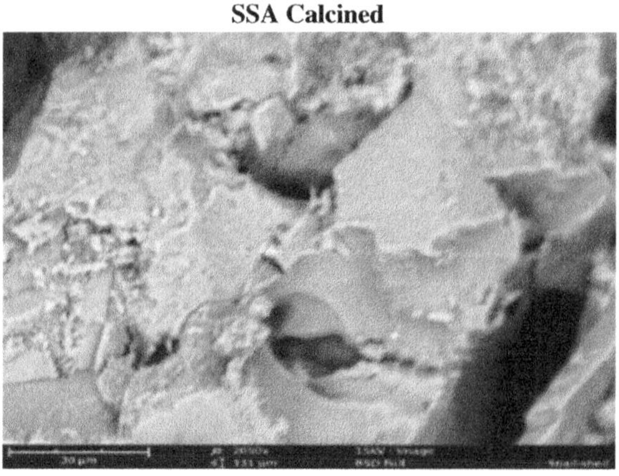

Figure 2.2 SEM images of SSA catalyst obtained from calcined forms.

Table 2.4 Effect of catalyst concentration (C. Snail shell with ethanol).

Mole ratio = 10:1, Speed = 300 rpm, Temperature = 55 °C, Time = 3 hours		
Runs	**Catalyst conc. (wt. % of Oil)**	**Yield of FAME *C. icaco* (vol. %) (calcined)**
1	0.75	59.00
2	1.50	67.00
3	2.25	66.00
4	3.00	59.00
5	3.75	58.00

Table 2.5 Effect of time (minutes).

Runs	Time (min.)	Yield of FAME *C. icaco* (vol. %) (calcined)
1	30	43.00
2	60	53.00
3	90	63.00
4	120	68.00
5	150	68.00

Maximum yield was reached at 120 and 150 seconds as the reaction's rate accelerated over time. The collision theory of reaction suggests that this is due to the variations in activation energy [19].

Table 2.6 shows the effect of temperature in the production of biodiesel from *C. icaco* seed oil.

Table 2.6 Effect of temperature (celsius).

Runs	Temperature (°C)	Yield of FAME *C. icaco* (vol. %) (calcined)
1	30	34.00
2	40	40.00
3	50	48.00
4	60	58.00
5	70	60.00

The sample yield increased as the temperature rose, reaching its maximum yield at 70 °C. This demonstrates that the reaction could occur at room temperature, but it would take a lot of time to produce an appreciable yield and conversion [20].

The effect of ethanol/oil in the production of biodiesel from *C. icaco* seed oil is presented in Table 2.7.

Ethanol's lower viscosity creates space for an efficient reaction that results in a higher yield and overall conversion. The yield increases with increasing alcohol ratio. The effect of stirring speed in production of biodiesel from *C. icaco* seed oil is presented in Table 2.8.

With the augmentation of stirring speed, the yield displayed an incremental trend, reaching its peak at 500 rpm. The outcomes indicate that optimal yield required heightened agitation. This observation underscores the necessity for intensified mixing between the two liquid phases, particularly as viscosity escalates [18].

The outcomes of the process parameter analysis revealed that the yield of the calcined catalyst was notably elevated. This can be attributed to the fact that

Table 2.7 Effect of ethanol/oil ratio.

Runs	Ethanol/oil ratio	Yield of FAME *C. icaco* (vol. %) (calcined)
1	6:1	47.00
2	8:1	61.00
3	10:1	73.00
4	12:1	74.00
5	14:1	71.00

Table 2.8 Effect of stirring speed (revolution per minute).

Runs	Speed (rpm)	Yield of FAME *C. icaco* (vol. %) (calcined)
1	100	44.00
2	200	64.00
3	300	73.00
4	400	79.00
5	500	59.00

Table 2.9 Regeneration of SSA catalyst.

Regeneration runs	Quantity of catalyst remains after each use (g)	Conversion yield *C. icaco* (vol. %) (calcined)
1	2.25	88.90
2	1.78	85.26
3	1.20	82.48
4	0.75	80.24
5	0.25	77.25

calcination leads to an increase in surface area, thereby facilitating the conversion of bulky molecules such as bio-oils. Table 2.9 presents a comprehensive overview of the *C. icaco* regeneration process utilizing the SSA catalyst.

During the initial regeneration cycles, using a residual catalyst mass of 2.25 g, the calcined SSA catalyst yielded the highest conversion rate at 88.90%, as indicated by the findings. However, as the subsequent regeneration cycles progressed, the conversion yield experienced a decline. The catalyst's ability to be reused was tested while monitoring process variables closely. Notably, the calcined SSA catalyst exhibited a notably elevated conversion efficiency due to its carbon based composition. Nonetheless, a reduction in catalytic activity emerged by the fourth cycle, presumably due to impurities leading to the deactivation of some active sites.

2.5 Conclusion

The exploration of alternative fuels through biodiesel development is captivating. The incorporation of *C. icaco* offers cost reductions in biodiesel production. The advantages of *C. icaco*, such as affordability and widespread availability, come to the forefront. Our study underscores the potential for future commercial biodiesel production from *C. icaco*, facilitated by a catalyst derived from calcined PSA. The research demonstrates that the oil yield from the crops is sufficient for commercial biodiesel production (60.8%). The catalyst's performance varied based on its hydrolytic stability and pore dimensions. The SEM image of the catalyst's surface revealed the expansive nature of the calcined version. The catalyst's reusability was investigated under identical operating conditions, revealing that the biofuel yield remained substantial even after five consecutive runs. This approach offers a means of conducting efficient and cost-effective experiments, circumventing the need for resource-intensive and time-consuming experimental inquiries.

References

1 Dash, S.K. and Lingfa, P. (2018, June). An overview of biodiesel production and its utilization in diesel engines. *IOP Conference Series: Materials Science and Engineering* (Vol. 377, No. 1, p. 012006). IOP Publishing.

2 Gholami, A., Pourfayaz, F., and Maleki, A. (2020). Recent advances of biodiesel production using ionic liquids supported on nanoporous materials as catalysts: a review. *Front. Energy Res.* 8: 144.

3 Ogbu, I.M., Ajiwe, V.I.E., and Okoli, C.P. (2018). Performance evaluation of carbon-based heterogeneous acid catalyst derived from Hura crepitans seed pod for esterification of high FFA vegetable oil. *Bioenergy Res.* 11 (4): 772–783.

4 Kara, K., Ouanji, F., Lotfi, E.M. et al. (2018). Biodiesel production from waste fish oil with high free fatty acid content from Moroccan fish-processing industries. *Egyptian J. Petroleum* 27 (2): 249–255.

5 Laskar, I.B., Rajkumari, K., Gupta, R. et al. (2018). Waste snail shell derived heterogeneous catalyst for biodiesel production by the transesterification of soybean oil. *RSC Adv.* 8 (36): 20131–20142.

6 Okonkwo, C.P., Ajiwe, V.I.E., Obiadi, M.C. et al. (2021). Kineticsand transesterification of the oil obtained from Cussonia bateri (Jansa Seed) as a step in biodiesel production using natural heterogeneous catalyst. *Am. J. Appl. Chem.* 9 (5): 164–171.

7 Pikula, K., Zakharenko, A., Stratidakis, A. et al. (2020). The advances and limitations in biodiesel production: feedstocks, oil extraction methods, production, and environmental life cycle assessment. *Green Chem. Lett. Rev.* 13 (4): 275–294.

8 Brown, S.H. and Frank, M.S. (2018). *Cocoplum (Chrysobalanus icaco L.) Identificationand Uses.* University of Florida, IFAS Extension, ENH1289, pp. 1–9.

9 Davies, R.M. and Zibokere, D.S. (2011). Some physical properties of gbafilo (*Chrysobalanus icaco*) fruits and kernels preparatory to primary processing. *Int. J. Agric. Res.* 6 (12): 848–855.

10 Okonkwo, C.P., Ajiwe, V.I.E., Mmaduakor, E.C., and Nwankwo, N.V. (2022). Modelling and optimization of biodiesel production process parameters from Jansa seed oil (Cussonia bateri) using artificial neural network. *Am. J. Appl. Chem.* 10 (1): 7–14.

11 Sentanuhady, J., Hasan, W.H., and Muflikhun, M.A. (2022). Recent progress on the implementation of renewable biodiesel fuel for automotive and power plants: raw materials perspective. *Adv. Mater. Sci. Eng. 2022*: 1–19.

12 Okonkwo, C.P., Ajiwe, V.I.E., Obiadi, M.C., and Okwu, M. (2021). A systematic review of biodiesel production with sustainable feedstock using assorted catalyst. *Am. J. Appl. Chem.* 9 (5): 154–163.

13 Thiruvengadaravi, K.V., Nandagopal, J., Sathya, S.B.V. et al. (2009). Kinetic study of the esterification of free fatty acids in non edible pongamiapinnatails using acid catalyst. *Indian J. Sci. Technol.* 2 (12): 20–24.

14 Barabas, I. and Todorut, I.-A. (2011). Biodiesel-Quality, Emissions and By-Products, pp. 3–28.

15 Knothe, G. and Steidley, K.R. (2018). The effect of metals and metal oxides on biodiesel oxidative stability from promotion to inhibition. *Fuel Process. Technol.* 177: 75–80.

16 Ogbu, I.M., Ajiwe, V.I.E., and Okoli, C.P. (2018). Performance evaluation of carbon-based heterogeneous acid catalyst derived from Hura crepitans seed pod for esterification of high FFA vegetable oil. *Bioenergy Res.* 11 (4): 772–783.

17 Wang, L., Dong, X., Jiang, H. et al. (2014). Preparation of a Novel carbon-based solid acid from cassava stillage residue and its use for the esterification of free fatty acids in waste cooking oil. *Bio Resour. Technol.* 9 (3): 283–287.

18 Ogbu, I.M. and Ajiwe, V.I.E. (2016). Fuel properties and their correlations with fatty acids structures of methyl-and butyl-esters of Afzelia africana, Cucurbita pepo and Hura crepitans seed oils. *Waste Biomass Valorization* 7. 373–381.

19 Moser, B.R. (2009). Biodiesel production, properties and feed stocks. *In Vitro Cell Develop. Biol. Plant* 45: 229–266.

20 Ogbu, I.M., Ajiwe, V.I.E., and Okoli, C.P. (2018). Performance evaluation of carbon-based heterogeneous acid catalyst derived from Hura crepitans seed pod for esterification of high FFA vegetable oil. *Bioenergy Res.* 11 (4): 772–783.

3

Utilization of *Elaeis guineensis* as an Eco-friendly Feedstock for Biodiesel Production

3.1 Introduction

A substantial portion of research endeavors have been directed toward environmentally conscious and more sustainable technologies that harness indigenous resources as a means of mitigating the detrimental impacts linked with fossil fuels, thereby promoting sustainable development [1]. In the pursuit of sustainable energy sources and the preservation of the environment, global attention has converged on renewable alternatives to conventional fossil fuels. The evolution of renewable and inventive energy avenues has been profoundly influenced by the exploration, production, and utilization of biodiesel in diesel engines [2]. Biodiesel, emanating from natural feedstocks and recognized for its low environmental impact, has emerged as a promising remedy. Among the plethora of potential biodiesel sources, *Elaeis guineensis* (*E. guineensis*), commonly referred to as the oil palm, stands out for its distinctive attributes [3].

This segment sets the foundation for delving into "*Elaeis guineensis*: Leveraging the Biodegradable Seed Feedstock for Biodiesel Production," an inquiry into the significance of this tropical plant and its potential to revolutionize the biodiesel landscape.

3.1.1 Imperative for Biodiesel

In an era marked by pressing concerns encompassing climate change, air quality degradation, and the diminishing reserves of traditional fossil fuels, the quest for eco-friendlier and enduring energy alternatives has taken center stage. Biodiesel, an alternative biofuel, emerges as a promising avenue for curbing greenhouse gas emissions and reducing dependency on finite energy sources. Sourced from

Feedstocks for Sustainable Biodiesel Production: Characterization, Selection, and Optimization, First Edition. Chinwe P. Okonkwo and Modestus O. Okwu.

natural origins such as plant oils, animal fats, and even algae, biodiesel presents a pragmatic solution for realizing a greener energy future.

3.1.2 The Advantages of Oil Palm (*E. guineensis*)

Prominently situated among the potential candidates for biodiesel feedstocks is *E. guineensis*, the oil palm. Indigenous to West Africa and now cultivated across tropical regions, this tree species is primarily renowned for its fruit, yielding palm oil [4]. Palm oil finds versatile applications across diverse industries, ranging from the culinary realm to cosmetics. However, the focal point of this study lies in uncovering the potential of *E. guineensis* as a sustainable and biodegradable resource for biodiesel generation.

3.1.3 *E. guineensis*: A Multifaceted Asset

The oil palm's distinct advantage as a biodiesel source stems from its elevated oil content. The oil obtained from its seeds is replete with fatty acids, rendering it an exemplary contender for biodiesel synthesis [3]. Furthermore, the oil palm tree's growth demands significantly less land compared to other oilseed crops, rendering it a compelling choice for biodiesel feedstock without contributing to deforestation.

3.1.4 Sustainability and Eco-friendliness

In the pursuit of authentically sustainable energy solutions, the environmental ramifications of feedstock cultivation and processing merit close scrutiny. *E. guineensis* aligns favorably with these concerns. The plant's capacity to sequester carbon dioxide during growth and its efficient land utilization underscore its sustainability. Notably, recent research has focused on alternative diesel replacements to reduce dependence on fuels based on petroleum and mitigate environmental pollutants [5]. The ongoing degradation of our ecosystem due to pollutants and emissions from fossil fuel usage underscores the urgency of reducing greenhouse gas emissions, as underscored by the recent Marrakesh UN climate conference [6]. It is imperative to remain vigilant against the escalating trend of environmental contamination [7]. In light of fossil fuel depletion, environmental concerns, and the persistent escalation in fossil fuel costs, scientists are actively exploring alternative fuel sources [6].

3.2 Review of Literature

The escalating demand for biodiesel is rooted in its perceived viability as a practical substitute for conventional diesel in the realm of transportation. Biodiesels stand out for their oxygenated and nontoxic attributes, alongside their inherent biodegradability. Furthermore, the utilization of biodiesel in engines requires no modifications, a facet underlined by Zhang et al. [8]. Notably, the European Union (EU) led global biodiesel production in 2014 with a share of 37.1%, trailed by the United States (16.7%), Brazil (11.9%), and Argentina (10.7%) [9]. It has been established that biodiesel can be blended with regular diesel for use in transportation engines.

It should be noted that the US diesel standard (ASTM D975) allows the incorporation of approximately 5% biodiesel without classifying the fuel as a biodiesel blend [10]. The high oxygen content that biodiesel naturally contains is essential for effective combustion. Additives, such as diethylene glycol methyl ether (DGM), diethyl ether, methanol, dimethyl ether (DME), ethanol, pentanol, butanol, diglyme, and others have been combined with the produced triglycerides to further enhance this oxygenation characteristic [5, 11].

The economic feasibility of biodiesel production is significantly influenced by the availability of feedstocks, with a total cost of approximately 80%. Endeavors to identify cost-effective year-round feedstocks have been numerous. Notably, vegetable oils exhibit substantial variations in fatty acid composition. Moreover, the selection of the catalyst contributes to the vegetable oil's final composition [11, 12]. The African oil palm (*E. guineensis*) is characterized as a single-stemmed palm tree with an adventitious root system that attains heights of 20–30 m [13].

It encompasses approximately 87.92% saturated fatty acid and the remaining percentage of 12.08% unsaturated fatty acid [14]. When juxtaposed with diesel fuel, vegetable oils exhibit lower volatility and elevated viscosity. However, the direct utilization of vegetable oils in diesel engines is precluded due to incomplete combustion and fuel deposition issues [15]. The process of transesterification typically ameliorates the properties of vegetable oils to render them suitable as biodiesel or green fuel in diesel engines [15, 16].

E. guineensis confers additional fuel property advantages, according to published research, it emits only a small amount of harmful exhaust gases such as nitric oxide, smoke, unburned hydrocarbons, and carbon monoxide. This positions *E. guineensis* as an appealing fuel option for biodiesel production. The challenges impeding commercial biodiesel ventures chiefly encompass production costs, fuel yield, and oil availability. Consequently, *E. guineensis* emerges as a potential solution to address these availability concerns [3, 4].

In comparison to acid catalyst-based yields, the employment of alkaline catalysts yields the highest conversion rate in biodiesel production, with 98% efficacy at low temperatures and pressure levels during the process. This aspect holds significance in determining the nature of the reaction [5]. Importantly, the selection of a catalyst hinges on the specific oil employed in the biodiesel production process [17, 18]. Transesterification, facilitated by a suitable catalyst, constitutes the principal methodology for biodiesel synthesis. Notably, heterogeneous catalysts are favored for their repeatability, product purity, ease of recovery, and production cost. Varied reactor designs are being explored to render biodiesel production commercially viable and cost-effective [17]. The adoption of heterogeneous catalysts from waste materials like seashells, bones, eggshells, and carbon-based biomass has garnered considerable attention due to their potential for sustainable and viable biodiesel production via transesterification reactions [19].

Natural heterogeneous catalysts such as snails and eggshells embody bioceramic composites infused with mineral components, comprising primarily of calcite aggregates permeated by an organic matrix, imparting robust mechanical properties [20]. Predominantly composed of calcium carbonate, supplemented by minor quantities of magnesium carbonate, phosphate, sodium, potassium, carbon, zinc, manganese, iron, and copper salts, these shells exhibit effective biodiesel conversion through transesterification and esterification reactions [21–23]. Notably, they outperform numerous heterogeneous catalysts due to their hydrophobicity facilitate active retention and reaction, thermal stability, and their cost-effectiveness [24]. This study is focused on investigating the potential for effective natural solid catalysts in the production of biodiesel alongside a comprehensive evaluation of a novel seed oil (*E. guineensis*) as a biodiesel feedstock.

3.3 Materials and Method

3.3.1 Materials

This section is focused on the materials used for the study. Every single reagent was bought from Sigma Aldrich. The flashpoint of the *E. guineensis* seeds was determined using a Pensky Martens flashpoint tester to extract the oil. The calcination process was carried out in a Muffle furnace. The TECHNO R175A diesel engine was used to assess the biodiesel's combustion potential. A viscometer and the digital density meter with the standard model of AP PAAR DMA 35 were implemented for effective measurement of the viscosities and densities of the seed oil and ester blends. Fourier-transform infrared spectroscopy (FTIR) was used to analyze the catalyst and seed oil functional group.

An Emission Scanning Electron Microscope (JSM-670IF) and a standard spectrophotometer (8400SSHIMADZU) were used to examine the catalyst's morphology. GC–MS QP2010 Plus Shimadzu Japan was used to determine the produced ester components. In order to achieve maximum effectiveness and efficiency of biodiesel in terms of production, specific steps must be implemented to help reduce emissions and pollutants. Conversely, the improvement of the characteristics and parameters of biodiesel demonstrate that this alternative energy is practical and can be used as a renewable fuel source for the power sector and the automotive industry now and in the future.

3.3.2 Oil Extraction and Sample Collection

E. guineensis were purchased as raw seeds from a neighborhood shopkeeper in Enugu State of Nigeria. The obtained products were identified by a plant taxonomist. It was later sun-dried and pulverized in an industrial blender. It was sieved with an 80 m mesh size. After sieving, the sample in dry form was weighed into a thimble-sized Soxhlet extractor's semipermeable cotton material. The condenser that the Soxhlet was connected to was fixed to a flask with a flat bottom that contained *n*-hexane (400 mL). After the sample had been fully extracted from available oil, the Soxhlet extraction system was refluxed at 700 °C (approximately six hours). The solvent mixture and solvent from the oil was removed in a vacuum. The solvent in the oil was left to be distilled at 500 °C. The weight of the extracted oil as well as the yield percentage were calculated.

3.3.3 The Preparation of Catalyst and Characterization Process

The cleaned, dried, and powdered snail shell was used to produce the snail shell ash (SSA). 5 g of SSA were calcined at 8000 °C for three hours, and 10 g were acidified with conc. Sulfuric acid (10 mL). Up until the pH was 7, SSA was repeatedly rinsed with distilled water after being acidified. The samples underwent thorough oven drying before being analyzed by SEM and FTIR.

3.3.4 Biodiesel Production (Transesterification)

Through transesterification reactions, the methyl esters were produced. For each experiment, the amounts of methanol, oil (6:1–14:1), and the modified catalyst (0.75–3.75 g) were added, and to ensure proper dissolution of the catalyst, the blend was pre-stirred for approximately 10 minutes. After reaching the proper temperature, the reaction mixture was blended with 50 mL of fatty acid-rich oil. The estimated time for each reaction is within the range of 30–150 minutes given to stir the process at a specific temperature (30–700 °C). A separating funnel was used to contain the reaction mixture, and it was left there for 12 hours so that the glycerol could separate. The methyl ester (biodiesel) was washed and transferred into a beaker using hot water (250 mL). It was heated to 105 °C, and to get rid of the moisture, the biodiesel was allowed to cool to room temperature.

3.3.5 Effect of Process Parameters

The effects of process variables such as time, catalyst amount, methanol/oil ratio, temperature, and agitation speed were examined using standard procedures.

3.3.6 Engine Performance Test

The performance of the produced biodiesel samples in terms of fuel consumption rate, emissions level, and generating capacity was evaluated by comparing it with a diesel engine (TECHNO R175A). Each sample (50 mL) was repeatedly run to determine the consumption rate through the system. The exhaust from the emission test samples from the engine performance tests was run through a conical flask containing 25 mL of a 1 M sodium hydroxide solution. The final product was titrated against a solution of 1 M hydrochloric acid. The vapor's acidity was then determined using the formula below:

$$\text{Acidity of fumes} = (\text{a} - \text{b}) \times (\text{molar mass of NaOH}) \times (\text{molarity of NaOH}) \tag{3.1}$$

The formula for calculating the fuel consumption rate is presented in Eq. (3.2):

$$\text{Rate of fuel consumption} = \frac{\text{volume of fuel consumed}}{\text{time taken}} \tag{3.2}$$

3.4 Results and Discussion

3.4.1 Physicochemical Properties

The physicochemical characteristics of the extracted seed oil (*E. guineensis*) are shown in Table 3.1. Acidic compounds can be found in biodiesel from a variety of sources, including byproducts of the oxidation of other organic acids, residual free fatty acids from ester hydrolysis, and residual mineral acids from the production process [25].

The determined acid value (AV) of *E. guineensis* was 0.60 mg KOH/g. This AV, falling within the range of 0.5–0.8, satisfied the standard outlined in the ASTM D6751 standard. Consequently, the oil could be directly subjected to base-catalyzed transesterification for biodiesel conversion [26]. Notably, elevated concentrations of free fatty acids deactivate base catalysts, culminating in soap formation and reduced biodiesel yield [27].

The iodine value, reliant on the type of feedstock's and content of ester used in the production of biodiesel, elucidates the extent of unsaturation in fatty acids. A heightened iodine value signifies amplified unsaturation in fatty acids. As a result of this increase in unsaturation, fats are more susceptible to lipid peroxidation, which antioxidants can prevent. Biodiesel with elevated iodine content (115–120) tends to polymerize, causing deposits to form on ring grooves, piston rings, and injector nozzles. Owing to their greater unsaturation, unsaturated fatty acids are more inclined to polymerize, as a result of this, high-iodine biodiesel is less stable

Table 3.1 Extracted oil's physicochemical properties.

Properties	*E. guineensis*
Color	Pale yellow
Yield (%)	58.87
Moisture (%)	5.50
Kinematic viscosity at 40 °C (mm^2/s)	14.96
Refractive index at 29 °C	1.56
Energy value (kJ/kg)	44 147
Peroxide value (meq/kg)	1.21
Saponification (mgKOH/kg)	248.00
Molecular weight (g/mol)	840.54
Iodine value (g/100 g of oil)	52.00
Flashpoint (°C)	247.00
Fire point (°C)	256.00
Pour point (°C)	5.00
Cloud point (°C)	10.50
Density (g/mL)	0.91
Acid value (mgKOH/g)	0.60

at high temperatures [28]. In this context, *E. guineensis* demonstrated an iodine value of 52.0 (g/100 g), diverging from the standards of 120 and 115 for European and German specifications, correspondingly, as stipulated in DIN 5160 and EN 14214 [29].

The kinematic viscosity wields a direct influence on fuel combustion, atomization, and overall performance. A biodiesel sample in commercial form (methyl soyate) manifests a kinematic viscosity of 4.09 mm^2/s, whereas the value for gasoline–diesel fuels stands at 1.54 mm^2/s [30]. The oils' kinematic viscosities (2.5–3.2 mm^2/s at 40 °C) surpass those of gasoline and diesel due to their higher molecular weight and fatty acid composition [31].

3.4.2 Characterization of Snail Shell Ash Catalysts

The outcomes of the characterization of the SSA catalyst demonstrate its potential as a valuable resource for biodiesel synthesis. This is presented in Table 3.2.

The scanning electron microscope (SEM) images of the calcined SSA sample depicted diverse impacts arising from the catalyst's enlarged surface area and pores.

Table 3.2 SSA catalyst characterization.

Parameter	Calcined SSA
Moisture (%)	0.03
Bulk density (g/cm^3)	0.46
Ph	5.87
Surface area (m^2/g)	832
Carbon (%)	1.86
Organic matter (%)	5.32
Loss on ignition (%)	0.62
Particle density (g/cm^3)	0.42
Total porosity (%)	25.54
Ash (%)	0.67

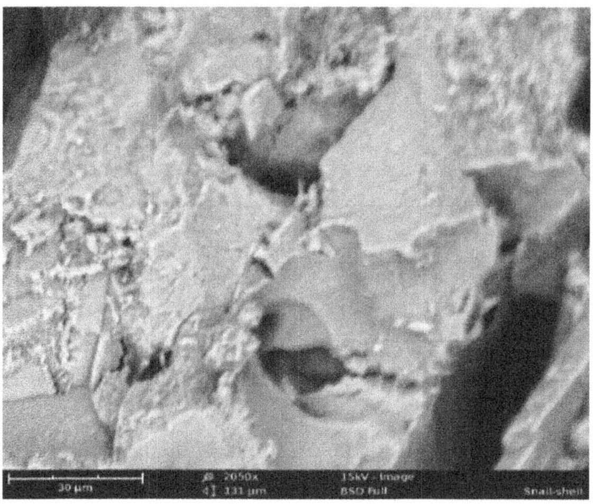

SSA Calcined

Figure 3.1 SEM images of the calcined form of the SSA catalyst.

The calcination process generates more room for the attachment of active sites, thereby enhancing both the transesterification reaction and the yield of biodiesel [32]. Figure 3.1 represents SEM images for calcined SSA which shows the wide surface area for effective reaction.

3.4.3 Effect of Process Parameters

Tables 3.3–3.7 show the effects of the process variables, which include temperature, catalyst concentration, methanol-to-oil ratio, stirring speed, and duration on the production of biodiesel from seed oil (*E. guineensis*) using methanol through transesterification and a catalyst made of SSA.

3.4.3.1 Effect of Quantity of Catalyst

The effect of catalyst concentration in the production of biodiesel from *E. guineensis* is presented in Table 3.3.

Table 3.3 Effect of catalyst concentration (E. Snail shell with methanol).

Runs	Catalyst conc. (wt. % of oil)	Yield of FAME *E. guineensis* (vol. %) (calcined)
1	0.75	63.00
2	1.50	70.00
3	2.25	68.00
4	3.00	64.00
5	3.75	47.00

Table 3.3 illustrates that a reduced amount of catalyst resulted in a higher yield. Utilizing *E. guineensis*, a substantial yield of 70.00% was achieved by employing calcined SSA and methanol for transesterification at a catalyst concentration of 0.75. This outcome can be attributed to the fact that larger molecules exhibit slower reactions and elevated activation energies compared to smaller molecules [33].

3.4.3.2 Effect of Time

Table 3.4 displays the effect of time on the production of biodiesel from *E. guineensis* (ESO).

Table 3.4 Effect of time (minutes).

Runs	Time (min.)	Yield of FAME *E. guineensis* (vol. %) (calcined)
1	30	47.00
2	60	54.00
3	90	68.00
4	120	73.00
5	150	75.00

Time has a noticeable impact on biofuel production, as observed from the data. The reaction rate progressively rose with time, culminating in the highest yield between the 120- and 150-minute intervals (refer to Table 3.4). Methanol, in general, yielded a significantly favorable outcome. This phenomenon can be elucidated by the collision theory of reactions, which posits that discrepancies in activation energy play a pivotal role [34]. Remarkably, a notable yield of 75.00% was attained at the 150-minute mark from *E. guineensis* through the application of calcined SSA and methanol for transesterification (as depicted in Table 3.4).

3.4.3.3 Effect of Temperature

The effect of temperature in the production of biodiesel from *E. guineensis* (ESO) is shown in Table 3.5.

Table 3.5 Effect of temperature (Celsius).

Runs	Temperature (°C)	Yield of FAME *E. guineensis* (vol. %) (calcined)
1	30	36.00
2	40	41.00
3	50	49.00
4	60	62.00
5	70	73.00

The influence of temperature on biofuel yield exhibited an ascending trend as the temperature rose, ultimately peaking at 60 and 70 °C for the respective samples (refer to Table 3.5). This signifies that while the reaction could take place at room temperature, achieving a substantial yield and conversion would demand a prolonged duration [35]. For *E. guineensis*, the utilization of calcined SSA and methanol for transesterification yielded commendable results, particularly at 70 °C, where a substantial yield of 73.00% was attained (as indicated in Table 3.5).

3.4.3.4 Effect of Alcohol to Oil Ratio

The effect of the methanol/oil ratio in the production of biodiesel from *E.guineensis* seed oil (ESO) is presented in Table 3.6.

Table 3.6 Effect of methanol/oil ratio.

Runs	Methanol/oil ratio	Yield of FAME *E. guineensis* (vol. %) (calcined)
1	6:1	52.00
2	8:1	71.00
3	10:1	78.00
4	12:1	75.00
5	14:1	74.00

The impact of the alcohol-to-oil ratio on biofuel generation is outlined in Table 3.6. Notably, the employment of calcined snail shell catalyst resulted in a noteworthy yield of 78.00 when the methanol-to-oil ratio was 10:1, after which the yield began to decline. This heightened yield attributed to methanol can be attributed to its low viscosity, which contributes to enhancing the overall conversion process.

3.4.3.5 Effect of Stirring Rate

Table 3.7 shows the effect of stirring speed in the production of biodiesel from *E. guineensis* seed oil (ESO).

Table 3.7 Effect of stirring speed (revolution per minute).

Runs	Speed (rpm)	Yield of FAME *E. guineensis* (vol. %) (calcined)
1	100	46.00
2	200	69.00
3	300	79.00
4	400	82.00
5	500	82.00

The yield exhibited an upward trend with the increasing mixing speed, culminating at 500 rpm, as indicated in Table 3.7. Notably, a significant yield of 82.00 was observed when the agitation speed was set at 400 and 500 rpm in the presence of the calcined catalyst. This high yield was notably influenced by the augmentation of agitation. The findings underscore the need for heightened agitation to achieve maximum yield. Given the variance in their molecular weights and viscosities, methanol also displayed a promising yield. It is noteworthy that greater viscosity necessitates increased agitation to ensure optimal mixing between the two liquid phases, a concept emphasized in previous studies [35].

Collectively, the outcomes from the process parameter analysis demonstrated the attainment of a substantial yield, which is likely attributed to the effect of calcination enhancing the surface area of the catalyst and promoting the transformation of larger molecules like bio-oils. For the industry to improve production efficiency while simultaneously curbing production costs, optimization of response parameters is imperative. This involves optimizing factors such as catalyst concentration, time, reaction temperature, methanol-to-oil molar ratio, and agitation rate [32] (Table 3.8).

Table 3.8 SSA catalyst regeneration.

Runs	Catalyst amount remaining after each use (g)	Conversion yield of *the Seed Oil* for the calcined sample (vol. %)
1	2.25	90.65
2	1.78	87.26
3	1.20	81.58
4	0.75	75.24
5	0.25	72.75

During the initial cycles of regeneration, employing a catalyst residue of 2.25 g, the calcined SSA catalyst demonstrated a notable peak conversion yield of 90.65%, as deduced from the findings. However, as subsequent cycles of regeneration ensued, the yield experienced a decline. The reusability of the catalyst was methodically examined while closely monitoring the process variables. Intriguingly, as the cycles progressed, a noticeable reduction in catalytic activity emerged, likely due to the deactivation of specific active sites caused by impurities. This phenomenon became evident by the fourth cycle of reusability.

3.4.4 Seed Oils Fuel Properties and their Derivatives

The examined sample exhibited favorable characteristics suitable for use as fuel, aligning with the ASTM specifications for standard diesel, as indicated by the outcomes of the fuel sample analysis. Specifically, Sample X demonstrated commendable fuel properties in terms of viscosity following the transesterification process involving *E. guineensis* and methanol, facilitated by the use of a calcined SSA catalyst. Furthermore, when subjected to blending at elevated temperatures (100 °C), the viscosity of the sample notably decreased to 0.40 mm^2/s. This reduction in viscosity highlights the beneficial impact of incorporating biodiesel, which enhances compatibility with modern diesel engines featuring intricate fuel injection systems that are sensitive to alterations in viscosity, as highlighted in prior studies [36] (Table 3.9).

3.4.5 Fume's Acidity Concentration Produced by Burning the Diesel and Methyl Ester Samples

Table 3.10 provides a comprehensive presentation of the combustion outcomes for biodiesel and diesel samples, incorporating details about the acidity concentration within the emitted fumes. The concentration of acidic fumes of the samples (methyl esters) from the engine performance tests is presented in the table.

The power capability assessment conducted on the Techno R175A diesel engine revealed that all samples exhibited comparable engine performance. Notably, when compared to diesel fuel emissions, which appeared darker, the samples demonstrated higher emissions with a milder odor. By adopting biofuels as an alternative, the deterioration of air quality associated with the combustion of fossil fuels can be mitigated, as highlighted by the EPA in 2002. The findings from the study are an indication that the concentration of acid in the exhaust of the engine was lower compared to those observed in hydrocarbon diesel emissions. Interestingly, as the proportion of diesel in the biodiesel blend increased, acid levels in the exhaust emissions showed an upward trend. The highest increase in emissions was observed with B100, the blend with the highest biodiesel content. This behavior may be attributed to the elevated oxygen content in biodiesel, promoting more comprehensive combustion and subsequently reducing carbon

Table 3.9 Seed oil fuel properties by implementing the SSA catalyst and its derivatives.

Sample		Density at 28 °C	Flashpoint (°C)	Calorific value (J/kg)	Pour point (°C)	Viscosity at 40 °C (mm^2/s)	Viscosity at 70 °C (mm^2/s)	Viscosity at 100 °C (mm^2/s)	Acid value (mgKOH/g)
Sample 1/SSA/Calcined/ Methyl ester	B_{100}	0.896	174	44 357	6.0	7.5	3.8	3.6	0.7
	B_{80}	0.886	152	44 938	5.5	6.8	2.5	3.2	0.6
	B_{60}	0.874	138	44 938	4.5	5.5	1.7	2.2	0.5
	B_{40}	0.85	127	45 334	3.5	4.75	1.2	0.9	0.4
	B_{20}	0.845	114	45 334	2.0	3.2	0.7	0.4	0.3
ASTM Limit		0.8–0.9	100 (min.)	NA	16 (max.)	1.9–6.0	NA	NA	0.80 (max.)

Max. = maximum, Min. = minimum, NA = Not Available.

Table 3.10 Combustion outcomes for biodiesel and diesel samples.

	Elaeis guineensis (g/dm^3)
Diesel	0.0043
B_{100}	0.0012
B_{80}	0.0014
B_{60}	0.0017
B_{40}	0.0020
B_{20}	0.0025

monoxide and hydrocarbon emissions. Therefore, the inclusion of more transesterified vegetable oils in the blend has the potential to enhance the performance of diesel engines. The outcomes of this study align with the research conducted by Ajjiwe et al. ([37, 38], respectively, as well as Ezekwe and Ajiwe [31]).

3.5 Conclusion

The intriguing investigation into biodiesel production centers on the development of alternative fuel sources. Elaeis guineensis emerges as an economically viable candidate for biodiesel production due to its cost-effectiveness and wide availability. The study's results indicate that *E. guineensis* can yield commercial biodiesel with a crop oil yield of 56.87%. Variations in catalyst activity were observed in relation to pore sizes. Scanning electron microscope (SEM) images of the catalyst revealed a significant surface area in the calcined version. The research also established a sustained high biofuel yield after five consecutive runs, demonstrating the catalyst's reusability under consistent operating conditions. The suitability of biodiesel and its blends was determined by analyzing fuel properties, including pour point, acid value, density, and flash point, all of which complied with ASTM D6751 biodiesel standard limits. Encouraging large-scale biodiesel production is recommended. Additionally, the by-product of the process, glycerol, holds potential for various industrial applications that could foster job growth and contribute to the country's economic advancement.

References

1 Krishna, Y., Aslfattahi, N., Saidur, R. et al. (2020). Fatty acid/metal ion composite as thermal energy storage materials. *SN Appl. Sci.* 2 (5): 1–10.

2 Ogunkunle, O. and Ahmed, N.A. (2019). A review of global current scenario of biodiesel adoption and combustion in vehicular diesel engines. *Energy Rep.* 5: 1560–1579.

3 Rulli, M.C., Casirati, S., Dell'Angelo, J. et al. (2019). Interdependencies 580 and telecoupling of oil palm expansion at the expense of Indonesian rainforest. *Renew. Sust. Energ Rev.* 105: 499–512. https://doi.org/10.1016/j.rser.2018.12.050.

4 Kolakoti, A., Setiyo, M., and Rochman, M.L. (2022). A green heterogeneous catalyst production and 541 characterization for biodiesel production using RSM and ANN approach. *Int. J. Renewable Energy Dev.* 11 (3): 703–712. https://doi.org/10.14710/ijred.2022.43627.

5 Dey, S., Reang, N., Das, P., and Deb, M. (2021). A comprehensive study on prospects of economy, environment, and efficiency of palm oil biodiesel as a renewable fuel. *J. Clean. Prod.* 286: 124981.

6 Shote, A.S., Betiku, E., and Asere, A.A. (2019). Characteristics of CO and NOx emissions from combustion of transmethylated palm kernel oil-based biodiesel blends in a compression ignition engine. *J. King Saud Univ. Eng. Sci.* 31 (2): 178–183.

7 Eloka-Eboka, A.C., Igbum, G.O., and Inambao, F. (2017). Biodiesel methyl ester production and testing from selected African tropical seed oil feedstocks. *Energy Procedia* 142: 755–767.

8 Zhang, M., Gao, Z., Zheng, T. et al. (2018). A bibliometric analysis of biodiesel research during 1991–2015. *J. Mater. Cycles Waste Manage.* 20 (1): 10–18.

9 Kass, M., Kaul, B., Armstrong, B. et al. (2022). Stability, rheological and combustion properties of biodiesel blends with a very-low sulfur fuel oil (VLSFO). *Fuel* 316: 123365.

10 Janaun, J. and Ellis, N. (2010). Perspectives on biodiesel as a sustainable fuel. *Renew. Sust. Energ. Rev.* 14 (4): 1312–1320.

11 Chozhavendhan, S., Singh, M.V.P., Fransila, B. et al. (2020). A review on influencing parameters of biodiesel production and purification processes. *Curr. Res. Green Sustainable Chem.* 1: 1–6.

12 Nursal, R.S., Khalid, A., Abdullah, I.S. et al. (2021). Autoignition behavior and emission of biodiesel from palm oil, waste cooking oil, tyre pyrolysis oil, algae and jatropha. *Fuel* 306: 121695.

13 Atasie, V. and Akinhanmi, T. (2009). Extraction, compositional studies and physico-chemical characteristics of palm kernel oil. *Pak. J. Nutr.* 8 (6): 800–803.

14 Paulin, Y.A. and Irène, K.A. (2019). Physico-chemical characterization of palm kernel oil extracted from the seeds of two varieties of oil palm (*Elaeis guineensis* Jacq.) for possible use in feed or food. *Eur. J. Nutr. Food Saf.* 9 (4): 341–353.

15 Almasi, S., Najafi, G., Ghobadian, B., and Jalili, S. (2021). Biodiesel production from sour cherry kernel oil as novel feedstock using potassium hydroxide catalyst: optimization using response surface methodology. *Biocatal. Agric. Biotechnol.* 35: 102089.

16 Franjo, M., Šalić, A., and Zelić, B. (2018). Microstructured devices for biodiesel production by transesterification. *Biomass Convers. Biorefinery* 8 (4): 1005–1020.

17 Brahma, S., Nath, B., Basumatary, B. et al. (2022). Biodiesel production from mixed oils: a sustainable approach towards industrial biofuel production. *Chem. Eng. J. Adv.* 10: 100284.

18 Ejeromedoghene, O. (2021). Acid-catalyzed transesterification of palm kernel oil (PKO) to biodiesel. *Mater. Today Proc.* 47: 1580–1583.

19 Quah, R.V., Tan, Y.H., Mubarak, N. et al. (2020). Magnetic biochar derived from waste palm kernel shell for biodiesel production via sulfonation. *Waste Manag.* 118: 626–636.

20 Boro, J., Deka, D., and Thakur, A.J. (2012). A review on solid oxide derived from waste shells as catalyst for biodiesel production. *Renew. Sust. Energ. Rev.* 16 (1): 904–910.

21 Huang, Y.-B. and Fu, Y. (2013). Hydrolysis of cellulose to glucose by solid acid catalysts. *Green Chem.* 15 (5): 1095–1111.

22 Lee, J.H., Lee, J.H., Kim, D.S. et al. (2019). Biodiesel production by lipases co-immobilized on the functionalized activated carbon. *Bioresour. Technol. Rep.* 7: 100248.

23 Zong, M.-H., Duan, Z.-Q., Lou, W.-Y. et al. (2007). Preparation of a sugar catalyst and its use for highly efficient production of biodiesel. *Green Chem.* 9 (5): 434–437.

24 Liu, T., Li, Z., Li, W. et al. (2013). Preparation and characterization of biomass carbon-based solid acid catalyst for the esterification of oleic acid with methanol. *Bioresour. Technol.* 133: 618–621.

25 Berthiaume, D. and Tremblay, A. (2006). Study of the Rancimat test method in measuring the oxidation stability of biodiesel ester and blends. *Oleotek Inc., NRCan project CO414 CETC-327.*

26 Thiruvengadaravi, K., Nandagopal, J., Bala, V.S.S. et al. (2009). Kinetic study of the esterification of free fatty acids in non-edible Pongamia pinnata oil using acid catalyst. *Indian J. Sci. Technol.* 2 (12): 20–24.

27 Lam, M.K., Lee, K.T., and Mohamed, A.R. (2010). Homogeneous, heterogeneous and enzymatic catalysis for transesterification of high free fatty acid oil (waste cooking oil) to biodiesel: a review. *Biotechnol. Adv.* 28 (4): 500–518.

28 Barabás, I. and Todoruţ, I.-A. (2011). Biodiesel quality, standards and properties. In: *Biodiesel-Quality, Emissions and by-Products*, IntechOpern, 3–28.

29 Calais, P. and Clark, A. (2004). *Waste Vegetable Oil as a Diesel Replacement Fuel.* Western Australia: Murdoch University and Western Australian Renewable Fuels Association.

30 Knothe, G. and Steidley, K.R. (2005). Kinematic viscosity of biodiesel fuel components and related compounds. Influence of compound structure and comparison to petrodiesel fuel components. *Fuel* 84 (9): 1059–1065.

31 Ezekwe, C.C. and Ajiwe, V. (2014). The effects of temperature on the viscosities of biofuels from thevetia, gourd and bush melon. *Recent Sci. Res.* 5 (3): 611–617.

32 Abdullah, R.F., Rashid, U., Ibrahim, M.L. et al. (2021). Bifunctional nano-catalyst produced from palm kernel shell via hydrothermal-assisted carbonization for biodiesel production from waste cooking oil. *Renew. Sust. Energ. Rev.* 137: 110638.

33 Ogbu, I. and Ajiwe, V. (2016). FTIR studies of thermal stability of the oils and methyl esters from Afzelia africana and Hura crepitans seeds. *Renew. Energy* 96: 203–208.

34 Moser, B.R. (2011). Biodiesel production, properties, and feedstocks. In: *Biofuels*, 285–347. Springer.

35 Ogbu, I.M., Ajiwe, V.I.E., and Okoli, C.P. (2018). Performance evaluation of carbon-based heterogeneous acid catalyst derived from Hura crepitans seed pod for esterification of high FFA vegetable oil. *BioEnergy Res.* 11 (4): 772–783.

36 Jaichandar, S. and Annamalai, K. (2011). The status of biodiesel as an alternative fuel for diesel engine – an overview. *J. Sustain. Energy Environ.* 2 (2): 71–75.

37 Ajiwe, V., Ajibola, V., and Martins, C. (2001). Possible vegetable diesels from rubber seed oil: methyl ester and ester-diesel blends. *Afr. J. Sci.* 3 (2): 277–289.

38 Ajiwe, V., Mbonu, S., and Enukorah, E. (2006). Biofuel from *Irvinga gabonesis* seed oil and its methylesters, *Proceedings of the NASEF'06 International Conference*, Awka. Solar Energy Society of Nigeria, 247–252.

4

Waste Cooking Oils for Biodiesel Production Using a Pyrolyzed Corn Cob

4.1 Background of Study

Biodiesel is gaining popularity as a renewable source of energy, providing a sustainable and environmentally beneficial fuel option. It is obtained from organic or biodegradable feedstocks such as plant oils, animal fats, or recycled cooking oils and is a more environmentally friendly alternative to traditional petroleum-based fuels. According to Yildiz [1], fossil energy has been the primary source propelling global industrial, technological, social, and economic progress. Despite its good contributions to progress, it has also been a major source of air pollution, aggravating the greenhouse effect and contributing to the global climate problem, as Perera [2] emphasizes. The global concerns about decreasing fossil fuel resources and the urgent need to solve the climate issue have prompted environmental campaigners in the European Union to recommend a transition away from crop-based fuels and toward second-generation biofuels. According to Keating [3], these biofuels made from waste oils and algae are a more sustainable alternative to first-generation biofuels produced from raw feedstock. In the pursuit of alternative energy solutions, biodiesel derived from nonfossil fuel sources has received substantial interest as a potential substitute for petroleum diesel. Several studies [4, 5] highlight biodiesel's potential due to remarkable characteristics such as high biodegradability, low toxicity, and lower combustion emissions when compared to petroleum diesel. Furthermore, biodiesel functions within a closed carbon cycle, which reduces its impact on global warming. Its interoperability with existing diesel engines is intriguing, requiring minimal modification and suffering just negligible performance losses.

Biodiesel not only provides an alternative fuel option for energy sustainability but it also contributes to the creation of a more ecologically sensitive and

Feedstocks for Sustainable Biodiesel Production: Characterization, Selection, and Optimization, First Edition. Chinwe P. Okonkwo and Modestus O. Okwu.

resilient energy system. Kumar et al. [6] emphasize biodiesel's sustainability by emphasizing its position as a renewable, cost-effective fuel that may handle any petroleum shortages. Nunez [7] predicts that growing biodiesel use will have a good economic and environmental impact in the future. In essence, biodiesel research and implementation constitute a possible path toward a more sustainable and ecologically friendly energy system. The synthesis of biodiesel requires the reaction of triglyceride and alcohol, with triglyceride serving as the principal raw material sourced primarily from plants and animals. Plants are divided into edible oils (such as coconut oil, soybean oil, peanut oil, palm oil, and rapeseed) and nonedible sources (such as algae, Karanja, sea mangoes, jatropha, and halophytes) for biodiesel generation. However, utilizing these resources poses obstacles, such as the food vs fuel dilemma, as well as severe environmental issues, such as soil resource depletion, deforestation, and the occupation of a considerable amount of available arable land. Furthermore, some of these sources may fail to supply global energy demands properly and perform poorly in colder climates.

Animal fats, which are predominantly generated from tallow, yellow grease, chicken fat, and byproducts of fish oil, are another type of triglyceride source. However, as Bitang [8] points out, getting certain types of animal fats can be difficult due to their high level of saturated fatty acids. The complexity connected with both plant- and animal-based triglyceride sources highlights the necessity for a full review of the environmental, economic, and logistical issues in biodiesel production methods. Improper waste cooking oil (WCO) disposal, combined with hazardous and solid waste mismanagement, has become an increasingly serious problem in many urbanized areas [8]. WCOs are a mixture of oils and fats often used in cooking in food manufacturing businesses, fast-food restaurants, and residences [9]. The widespread use of oils for deep-frying, notably at fast-food restaurants for items such as French fries and fried chicken, is a significant contributor to the production of WCO [10]. According to Xu [11], one of the earliest ways of food preparation is frying, which is done at temperatures ranging from 160 to 200 °C. This complex cooking process requires a plethora of physical and chemical events, such as heat and mass transfer. Cooking oils should ideally be used at temperatures ranging from 35 to 180 °C, according to Anisah [12]. Cooking at lower temperatures, as Brown [13] point out, increases fat uptake in fried meals.

The development of food outlets, fueled by increased human consumption, has resulted in a significant increase in the output of WCO. Improper waste disposal practices may result in uncontrolled waste release, providing a considerable environmental risk [14, 15]. In response to this environmental concern, recycling appears as a viable solution for used cooking oil. Rather than being thrown, WCO can be recycled and used to make animal feed, soap, and biodegradable lubricants [16]. This recycling technique not only reduces environmental impact but also contributes to the creation of value-added products, encouraging a more sustainable and eco-friendly waste management plan.

4.2 Review of Literature

The global focus on biodiesel as a potential answer shows a shared understanding of the need for more sustainable energy practices. Governments, companies, and researchers around the world are aggressively supporting and developing biodiesel technology to fulfill rising energy demands while also addressing environmental issues. As a result, it is critical to evaluate what Phan [17] reported on the Alkali-catalyzed transesterification of WCOs collected in Ho Chi Minh City, Vietnam, with methanol in a laboratory scale reactor. The influence of methanol/WCO ratio, potassium hydroxide concentration, and temperature on biodiesel conversion was investigated. According to the research, biodiesel has a higher but significantly narrower boiling range than normal diesel. This implies that with modest adjustments, they may be employed in engines.

Zhang [18] disclosed a solid acid catalyst made by sulfonating pyrolyzed rice husk with concentrated sulfuric acid, and the catalyst's physical and chemical characteristics were comprehensively characterized. The catalyst was then utilized to catalyze both esterification and transesterification, resulting in biodiesel made from WCO. Ayoola [19]. presented the usage of response surface methodology (RSM) and artificial neural network (ANN) in the investigation of crude palm kernel oil (CPKO) biodiesel synthesis using KOH and NaOH catalysts. The four process parameters addressed during the manufacturing and modeling stages were the mole ratio of methanol/oil, catalyst concentration, reaction temperature, and minutes of reaction time. The regression coefficients for RSM models were 0.9324 for KOH catalyzed processes and 0.8935 for NaOH catalyzed processes, whereas the overall correlation coefficients for ANN models were 0.82921 for KOH catalyzed processes and 0.89396 for NaOH catalyzed processes.

According to Buasri [20], rising human use of plant and animal oils has led to an increase in WCO production. Instead of disposing leftover cooking oil, new technological breakthroughs have made it possible to use WCO as a sustainable feedstock for biofuels production, maximizing the value of biowastes through energy recovery while also addressing the disposal issue. While WCO production is restricted, it has the potential to provide waste diversification if issues like carbon chain length, bio-jet fuel quality requirements, harsh reaction conditions, and the performance of selected catalyst support systems can be addressed. WCO is a biowaste with great potential for advanced transportation fuel generation in the ground and aviation industries, according to these studies. Technological advancements in fuel production and appropriate policies would speed the use of sustainable WCO biofuels.

4.3 Experimental Analysis

4.3.1 Materials

Stirrer, conical flask, fume cupboard, magnetic stirrer, density bottle, Pensky–Martens flash point tester, pour point tester, samples (WCO, esters, ester/diesel blends, and diesel), separator funnel, Soxhlet extractor, calcinator, muffle furnace, oven, digital magnetic heater, digital weighing balance are all required for the experiment. TECHNO R175A diesel engine, viscosity bath and viscometer, digital density meter model AP PAAR DMA, FTIR, Spectrophotometer (8400SHIMADZU), Emission Scanning Electron Microscope (JSM-670IF), Shimadzu Japan GC–MS QP2010 Plus.

4.3.2 Reagents

Reagents of analytical grade were utilized. Potassium hydroxide (BDH), hydrochloric acid (Sigma-Aldrich), concentrated sulfuric acid (Sigma-Aldrich), chloroform (BDH), distilled water, glacial acetic acid (BDH), iodine monochloride, methanol and ethanol (BDH), phenolphthalein, indicator, potassium iodide, sodium hydroxide, sodium (vi).

4.3.3 Sample Preparation and Collection

The WCO was obtained from Nnamdi Azikiwe University's campus cafeteria in Awka, Nigeria. After filtering out all insoluble contaminants, the oil was heated to 100 °C to eliminate moisture. All of the analytical grade compounds were given by Sigma-Aldrich. corn cob char (CCC) was pulverized to powder (particle size 50.24 nm), and a 5 g sample was sulfonated for 30 minutes in a 250 mL flask regulated at 1000 °C in an oil bath using 60 mL concentrated sulfuric acid (90–95%). The mixture was added to distilled water, mixed, and then filtrated after cooling to room temperature. The precipitate was then washed with hot distilled water several times until the filtrate was free of sulfate ions. Following filtering, the sample was dried in an air-drying oven at 800 °C for 24 hours.

4.3.4 Pretreatment of WCO Feedstock

The collected oil is pretreated to remove solid particles/residues (filtering) and excess moisture (heating) before being tested for quality and free fatty acid (FFA) content. To avoid alcohol vaporization, a three-neck reactor (250 mL) equipped with a reflux condenser was employed for treatment. The procedure uses 50 mL WCO, 10 mL methanol, and 0.2% acid catalyst. The reaction mixture was injected into a batch reactor, and the experiment was run at 50 °C for six hours. Mineral acids (HCl, H_2SO_4, and H_3PO_4) were utilized as pretreatment. After the specified time interval, the samples were withdrawn and centrifuged. The methanol layer was removed, and the WCO was recovered and rinsed three times with deionized water. The water content was removed by vacuum evaporation, and the FFAs values were determined. The influencing parameters were optimized (methanol to oil ratio, catalyst dose, reaction time, and temperature), and the FFA was estimated using Eq. (4.1).

$$\text{Free fatty acid conversion}\ \% = \frac{A_i - A_t}{A_i} \times 100 \tag{4.1}$$

where A_i and A_t denote acidity (at zero and time t).

4.3.5 Biodiesel Production

The methyl ester transesterification processes were carried out in a 500 mL round bottom, glass spherical, three-neck reactor. Before adding 50 mL of the fatty acid-containing WCO to the experimental setup, the predicted amount of methanol with pyrolyzed maize cob was added in the amounts indicated for each experiment and prestirred for 10 minutes to guarantee appropriate catalyst dissolution. In the reactor's mid-neck, a reflux condenser with cold water circulating at the outer jacket was placed. The right neck of the reactor was supplied with a mercury thermometer housed in a plastic bung. A capsular stirring nob was put into the reactor via the reactor's left neck and secured with a plastic bung. The stirring system was also turned on and kept at a constant speed and temperature, with this point serving as the reaction's time zero. After the transesterification reaction, the product was allowed to stand in a separating funnel for 12 hours to allow for glycerol separation. The methyl ester (biodiesel) was left behind after the crude glycerol was removed through a funnel tap. The biodiesel was poured into a 250 mL beaker after being washed with hot water. It was heated to 105 °C to remove water molecules from the biodiesel. After cooling, the biodiesel was stored in calibrated specimen bottles.

4.3.6 Results and Discussions

The following characteristics determine whether WCO can be utilized as a biodiesel feedstock: wa, acid value (AV), AV, water content, and saponification value of WCO indicate whether it may be used as feed for biodiesel [21, 22] (Table 4.1). The waste oil's low AV of 2.6 could be due to its FFA level. A high FFA

Table 4.1 Physical and chemical properties of WCO.

Property	Units	Value
Palmitic acid	wt. %	7.5
Stearic acid	wt. %	2.9
Oleic acid	wt. %	23.4
Linoleic acid	wt. %	52.5
Linolenic acid	wt. %	6.2
Water content	wt. %	1.7
Density	cm^3/g	0.89
Viscosity	mm^2/s	46.04
Kinematic viscosity (40 °C)	mm^2/s	3.1
Saponification value	mgKOH/g	204
Acid value	mgKOH/g	2.6
Iodine number	gI2100 g	1.73
Sodium content	mg/kg	7.1
Peroxide value	mg/kg	28

concentration promotes soap production, limiting transesterification efficiency [23, 24]. For successful transesterification, a maximum FFA content of roughly 3% is required. The oil includes 1.7% water, which aids in the creation of soap [25]. The saponification value, which is 204, is another property used to determine the quality of WCO. This is the quantity of base (NaOH/KOH) required in milligrams to transform 1 g of WCO. To reduce the high FFA content, WCO can be acid catalyzed esterified prior to transesterification, pretreated with an adsorbent, and passed through an acid ion-exchange resin [26, 27].

$$\underset{\text{Triglycerid}}{C_3H_5\left(OCOR\right)_3} + \underset{\text{Water}}{H_2O} \leftrightarrow \underset{\text{Diglyceride}}{C_3H_5\left(OH\right)\left(OCOR\right)_2} + \underset{\text{Ester}}{RCOOH} \tag{4.2}$$

$$\underset{\text{FFA}}{RCOOH} + \underset{\text{Base}}{NaOH} \leftrightarrow \underset{\text{Soap}}{RCOONa} + H_2O \tag{4.3}$$

4.3.7 Catalyst Characterization

These mesopores allow macromolecules like oleic acid, triolein, and methyl oleate to diffuse in and out of the catalyst, improving catalytic activity [28]. It is worth noting, however, that pretreatment of CCC with NaOH solution [29] had almost no effect on the surface area of the catalyst. This was most likely due to silicon dioxide. The FTIR spectrum of CCC-SO_3H revealed a band at 1700 cm^{-1}, which corresponds to the ASO_3H stretching mode [30]. The presence of AOH in CCC-SO_3H is attributed to the OAH stretching modes of the phenolic AOH and ACOOH, resulting in broadband at 3700 cm^{-1}. The band assigned to the ACOOH at around 1700 cm^{-1} was not observed because it could have been overlapping with the ASO_3H band [29]. The morphology was examined by scanning electron microscopy (SEM, Sirion 200). The framework vibration was investigated using Fourier transform infrared spectroscopy (FTIR, Nicolet 6700, Thermo Scientific Instrument). According to the SEM image, the catalyst was composed of irregular particles with micrometer grain sizes and a few distinct pores on the surface. The adsorption isotherm confirmed the existence of the undeveloped porous structure's textural feature. The surface area of CCC-SO_3H was only 3 m^2/g, and the average pore size was 6.7 nm, indicating the presence of some mesopores.

4.3.8 Catalytic Activity in the Biodiesel Production

Because transesterification is reversible and endothermic, more methanol and a higher reaction temperature are necessary to boost the reaction rate and favor forward reaction; consequently, all tests were conducted at a molar ratio of methanol to WCO of 10:1 at a temperature of 1000 °C (Table 4.2). In addition, a modest stirring velocity of 350 revolutions per minute and a small amount of catalyst were utilized. When compared to previous published solid acid catalysts, CCC-SO_3H displayed greater activity in biodiesel generation from WCO. As a result, greater FFA content feedstocks are more suitable for biodiesel production using CCC-SO_3H as a solid acid catalyst. The strong catalytic performance of CCC-SO_3H could be attributable to the three variables described below. For starters, CCC-SO_3H contains a significant concentration of extremely acidic ASO_3H groups. Second, because ASO_3H is hydrophilic, it can integrate huge amounts of water into the carbon bulk. Finally,

Table 4.2 Effect of catalyst concentration.

Runs	Catalyst conc. (wt.% of oil)	Yield of FAME of Raw CCC	Yield of FAME CCC-SO_3H (vol. %) (acid activated)
1	0.75	72.00	82.00
2	1.50	70.00	81.00
3	2.25	68.00	76.00
4	3.00	66.00	74.00
5	3.75	63.00	73.00

Table 4.3 Stability of CCC catalyst.

Runs	Quantity of catalyst remains after each use (g)	FFA conversion of WCO	FAME yield
1	2.25	98.23	89.65
2	1.78	96.78	87.26
3	1.20	94.90	81.58
4	0.75	92. 25	75.24
5	0.25	90.75	70.50

unlike Amberlyst-15, the produced catalyst contains roughly 1.6 mmol/g of AOH, which can adsorb the hydrophilic components of the reactants via hydrogen bonding [31]. As a result, methanol's strong affinity for the catalyst leads to its high catalytic activity in the generation of biodiesel from WCO (Table 4.3).

The stability of CCC-SO_3H was studied for 15 hours under reaction circumstances using the method proposed by Park [32]. In each run, the FFA conversion was assessed after three hours, and the FAME yield was determined after the reaction. After five cycles, the FFA conversion and FAME yield remained at 90.75% and 70.50%, respectively, confirming the catalyst's high stability. The recovered catalyst after five cycles was characterized by ultimate analysis. The results showed that the content reduced when compared to fresh catalyst, showing that the minor drop-in activity is attributable to the leaching of certain ASO_3H groups [33].

Table 4.4 shows the biodiesel characterization results, demonstrating the viability of WCOs oil biodiesel as a fuel. The outcomes were consistent with the biodiesel standard.

Table 4.4 Characterization of the biodiesel from WCOs.

Fuel properties	Unit	Value	Biodiesel standard
Density at 15 °C	kg/m^3	790	878
Viscosity at 40 °C	mm^2/s	2.9	1.9–6.0
Flash point	°C	> 130	100–170
Moisture content	ppm	Trace	0.05% max
Cetane number	—	—	48–65
Acid value	mg KOH/g	0.5	0.5
Polyunsaturated (≥4double) methyl esters	% m/m	0.0	0.0
Methanol content	% m/m	0.12	0.2

4.4 Environmental and Economic Implications

4.4.1 Sustainability of Using WCOs and Pyrolyzed Corn Cob in Biodiesel Production

4.4.1.1 Reducing Waste and Environmental Impact

WCOs: Using WCOs as a feedstock for biodiesel production greatly reduces waste. WCOs are frequently disposed of incorrectly, resulting in degradation of the environment. The environmental impact of inappropriate disposal is reduced by reusing these oils. WCOs aid in addressing the issues connected with an increasing number of culinary waste, particularly in urban areas with a high concentration of food enterprises.

Pyrolyzed corn cob: As a catalyst, pyrolyzed corn cob offers a sustainable option obtained from agricultural waste. The usage of pyrolyzed maize cob eliminates additional strain on resources and increases the utilization of agricultural byproducts.

4.4.1.2 Biodiesel Production Efficiency

WCOs: WCOs provide a valuable and sustainable feedstock for biodiesel production. The variety of sources, which includes both edible and nonedible oils, enables flexibility and adaptation in the manufacturing process. The efficiency of WCO biodiesel production coincides with the circular economy concept, which promotes resource reuse in a closed-loop system.

4.4.1.3 Economic Viability and Cost-Effectiveness

WCOs: The availability of WCOs from various sources, including household cooking and fast-food establishments, makes them a cost-effective feedstock. This accessibility contributes to the economic viability of biodiesel production. WCOs, when the purposed for biodiesel, can potentially reduce the demand for traditional fossil fuels, leading to economic savings and a more sustainable energy model.

Pyrolyzed corn cob: The use of pyrolyzed maize cob as a catalyst improves transesterification efficiency. It functions as a solid acid catalyst, offering a more environmentally friendly alternative to standard chemical catalysts. The usage of pyrolyzed maize cob adheres to green chemistry principles, reducing the use of harsh chemicals and supporting a more ecologically friendly manufacturing process.

4.4.1.4 Reducing Dependence on Conventional Feedstocks

The use of WCOs and pyrolyzed maize cob minimizes the need for conventional feedstocks like edible oils and chemical catalysts. By reducing vulnerability to fluctuations in the availability and cost of traditional resources, this diversification enhances resilience in the biodiesel manufacturing business.

4.4.1.5 Addressing Food vs Fuel Dilemma

WCOs: The article addresses the longstanding controversy over "food vs fuel" by repurposing WCOs for biodiesel production. WCOs derived from discarded cooking oils do not compete directly with edible oils in food manufacturing.

Pyrolyzed maize cob: Using pyrolyzed maize cob as a catalyst alleviates further concerns about the food vs fuel conundrum. Rather than relying on chemical catalysts generated from food sources, pyrolyzed maize cob makes use of an agricultural by-product.

4.4.1.6 Potential for Localized and Small-Scale Production

The proposed approach's sustainability extends to its potential for localized and small-scale biodiesel production. WCOs are widely available in local communities, and using pyrolyzed maize cob as a catalyst enables decentralized and community-based activities.

4.4.1.7 Closing the Loop in Agricultural Practices

The use of pyrolyzed maize cob in biodiesel synthesis completes the agricultural cycle. The approach conforms with circular economy concepts by repurposing maize cobs, a byproduct of corn farming, minimizing waste, and maximizing resource efficiency.

The use of WCOs and pyrolyzed maize cob in biodiesel synthesis demonstrates a green and sustainable strategy. This methodology solves waste management challenges, increases economic viability, and helps to make the biodiesel sector more robust and diverse. The utilization of these alternative feedstocks not only coincides with sustainability goals but also highlights the possibility for novel solutions in the search for renewable energy sources.

4.4.2 Evaluation of the Environmental Impact and Reduction in Waste

Environmental Impact and Waste Reduction in Biodiesel Production Using WCOs and Pyrolyzed Corn Cob.

4.4.2.1 Mitigation of Improper Disposal

WCOs: One of the key environmental benefits is the reduction of incorrect WCO disposal. When WCOs are not maintained properly, they frequently end up in drainage systems, causing clogs and pollution. This problem is addressed by repurposing WCOs for biodiesel production, which provides a sustainable and controlled disposal approach.

Pyrolyzed corn cob: The usage of pyrolyzed corn cob helps to reduce agricultural waste. Corn cobs, which would otherwise be discarded, are used in the biodiesel synthesis process. This decreases the environmental effect associated with agricultural by-product disposal.

4.4.2.2 Waste Cooking Oils

Reduced Reliance on Landfills, WCO biodiesel generation reduces the reliance on landfills for the disposal of spent cooking oils. When these oils are incorrectly disposed of, they can contribute to soil contamination and the release of dangerous compounds. The requirement for landfill space is reduced by using them for biodiesel.

Pyrolyzed corn cob: The incorporation of pyrolyzed corn cob as a catalyst further contributes to the reduction in landfill waste. Rather than discarding corn cobs, which can be challenging to decompose, their utilization in the biodiesel production process minimizes the environmental burden associated with agricultural waste.

4.4.2.3 Prevention of Soil Resource Loss and Deforestation

WCOs: Using WCOs in biodiesel production helps to avoid soil resource loss associated with traditional agricultural practices. It minimizes the demand for extending agricultural fields for the cultivation of oilseeds, hence reducing the risk of deforestation.

Pyrolyzed corn cob: Including pyrolyzed corn cob in the process improves environmental sustainability by eliminating the requirement for additional agricultural practices that could result in deforestation. The usage of a byproduct, maize cobs, contributes to a more circular and resource.

4.4.2.4 Contribution to Circular Economy

A circular economy paradigm is exemplified by the use of WCOs and pyrolyzed maize cob in biodiesel manufacturing. It entails repurposing waste materials (WCOs) and by-products (corn cobs) from various sectors, reducing linear resource flow, and encouraging a more sustainable and closed-loop system.

4.4.2.5 Minimization of Greenhouse Gas Emissions

WCOs: Repurposing WCOs for biodiesel production helps to reduce greenhouse gas emissions. When waste oils are poorly disposed of, they degrade and emit methane, a strong greenhouse gas. This environmental impact is reduced by converting them into biodiesel.

Pyrolyzed corn cob: Using pyrolyzed corn cob as a catalyst eliminates the requirement for chemical catalysts, which may contribute to greenhouse gas emissions during the manufacturing process. The use of a renewable and agricultural by-product is consistent with the larger goal of lowering the carbon footprint associated with biodiesel production.

4.4.2.6 Promotion of Sustainable Resource Use

The integrated technique of employing WCOs and pyrolyzed maize cob increases resource sustainability. It adheres to the concepts of using renewable resources, reducing waste, and encouraging appropriate farming practices, all of which contribute to the overall sustainability of the biodiesel manufacturing process.

In conclusion, the assessment of the environmental impact and waste reduction in biodiesel synthesis using WCOs and pyrolyzed maize cob highlights the approach's potential to address major environmental concerns. Not only does it decrease waste and pollution but it also adds to a more sustainable and circular approach to biodiesel production, coinciding with worldwide efforts to move to cleaner and greener energy options.

4.4.3 Potential Economic Benefits and Cost-Effectiveness of Biodiesel Production Using WCOs and Pyrolyzed Corn Cob

4.4.3.1 WCOs Economic Viability as Feedstock

Feedstock at a low cost: WCOs are frequently more cost-effective as a feedstock for biodiesel production than standard edible oils. WCOs are a widely available and commercially effective resource due to their availability from numerous sources, including families and food outlets. Furthermore, using waste materials to produce biodiesel reduces the demand for virgin feedstocks, potentially leading to cost savings in the total production process.

4.4.3.2 Costs of Biodiesel Production Could be Reduced

WCOs: The usage of WCOs in biodiesel production can help to reduce production costs. Because WCOs are a by-product of many cooking operations, their acquisition cost may be lower than that of standard feedstocks, improving the overall economics of biodiesel production.

Pyrolyzed corn cob: As a catalyst, pyrolyzed corn cob has the potential to save money. It represents a resource that, as an agricultural by-product, maybe more economically practical than chemical catalysts. When compared to the creation of synthetic catalysts, the pyrolysis method may take less energy, which contributes to cost-effectiveness.

4.4.3.3 Agricultural By-Product Utilization

WCOs and pyrolyzed maize cob: Using WCOs and pyrolyzed maize cob in biodiesel manufacturing encourages the use of agricultural by-products. This not only eliminates waste but also delivers an economic benefit by extracting value from items that would otherwise be discarded.

4.4.3.4 Reduced Reliance on Traditional Fossil Fuels

Cost savings from reduction in reliance: Biodiesel generation from used cooking oils and pyrolyzed maize cob reduces reliance on traditional fossil fuels. This reduction may result in cost savings for biodiesel-producing companies and businesses, resulting in a more economically resilient energy sector.

4.4.3.5 Economic Advantages of Agricultural Practices

Pyrolyzed corn cob: The use of pyrolyzed corn cob provides agricultural economic benefits. Farmers may find an additional source of money from what would otherwise be considered waste by recycling maize cobs. This economic incentive can help to promote more sustainable and profitable agricultural practices.

4.4.3.6 Local Economic Impact and Job Creation

The development of biodiesel manufacturing facilities employing WCOs and pyrolyzed maize cob has the potential to create jobs and stimulate local economies. This localized influence may extend beyond the biodiesel generation process, influencing garbage collection, agricultural, and catalyst producing businesses.

4.4.3.7 Diversification Promotes Economic Resilience

Reduced reliance on traditional feedstocks: Increasing economic resilience by diversifying biodiesel production with alternative feedstocks such as WCOs and pyrolyzed maize cob. Reduced reliance on traditional and sometimes volatile edible oil markets can provide biodiesel producers with a more stable economic footing.

4.4.3.8 Green Technology Investment

Attracting investment: Biodiesel production employing sustainable feedstocks and catalysts may entice environmentally aware investors to invest. Investors seeking both financial gains and environmental impact may find the linkage with green technologies and the possibility for economic benefits in a rising renewable energy sector enticing.

4.4.3.9 Subsidies and Incentives

Government Assistance: Governments and regulatory agencies may provide subsidies or incentives for ecologically friendly and sustainable biodiesel manufacturing technologies. The use of WCOs and pyrolyzed maize cob corresponds with green energy goals, making the biodiesel production process financially viable.

Finally, taking into account potential economic benefits and cost-effectiveness in biodiesel production using WCOs and pyrolyzed corn cob reflects the twin goal of developing a sustainable energy source while also contributing to economic viability. This strategy not only offers cost savings opportunities but also corresponds with the increased emphasis in the renewable energy sector on green and economically resilient practices.

4.5 Biodiesel Production Prospects Using WCOs and Pyrolyzed Corn Cob

4.5.1 Investigating Potential Advancements and Improvements

4.5.1.1 Catalyst Optimization

Further research into optimizing the pyrolyzed maize cob catalyst is critical for increasing the efficiency of biodiesel synthesis. Experimenting with different pyrolysis settings and treatments can result in catalysts with increased catalytic activity, stability, and reusability.

4.5.1.2 Process Integration and Automation

Advances in process integration and automation can streamline biodiesel synthesis using WCOs and pyrolyzed maize cob in the Industry 4.0 era. Implementing modern control systems and monitoring technologies has the potential to:

1) Increase system process efficiency;
2) Reduce resource consumption;
3) Increase overall productivity in biodiesel manufacturing.

4.5.1.3 Investigating Innovative Combintaions of WCOs

Investigating innovative combinations of WCOs with other nontraditional biodegradable feedstocks can broaden the spectrum of available resources. This may entail combining WCOs with other waste oils, such as water hyacinth, or looking into unusual sources to diversify the feedstock assortment, which will also improve biodiesel production.

4.5.1.4 Advanced Reaction Engineering

It is also critical to consider alternate methods of advancing reaction engineering techniques in order to improve transesterification processes. This is achieved by experimenting with different reaction conditions, such as pressure and temperature changes, which can improve biodiesel yield and quality.

4.5.1.5 Comprehensive Life Cycle Assessments

Comprehensive life cycle assessments can give a holistic picture of the environmental impact of the full biodiesel production process. This can direct changes at key phases, resulting in a more sustainable and ecologically friendly overall process.

4.5.2 Suggestions for Additional Field Research and Development

4.5.2.1 Feedstock Availability and Collection Strategies

It is critical to explore research on efficient methods of collecting and processing WCOs. Understanding geographical differences in waste oil availability and creating optimized collection mechanisms will help to make large-scale manufacturing feasible.

4.5.2.2 Economic Feasibility Studies

Conducting extensive economic feasibility studies is critical for determining the economic viability of producing biodiesel from used cooking oils and pyrolyzed maize cob. This includes assessing the proposed method's manufacturing costs, possible revenue streams, and overall economic sustainability.

4.5.2.3 Techno-Economic Analysis

A thorough techno-economic study will aid in identifying cost drivers and potential cost savings. This analysis should take into account equipment costs, energy consumption, and catalyst production costs.

4.5.2.4 Market Acceptance and Consumer Perception Studies

It is critical to assess market adoption and consumer perception of biodiesel produced using this technology. Consumer preferences, willingness to adopt, and perceptions of environmental impact can all be used to inform marketing efforts and encourage greater adoption.

4.5.3 Considering the Scalability and Practicality of the Proposed Method

Scalability evaluations: Scalability evaluations are critical for establishing the feasibility of scaling up the biodiesel production process. Feedstock availability, catalyst production scalability, and process adaptability to bigger volumes all require careful consideration.

Integration with existing infrastructure: Assessing the compatibility of the proposed method with existing biodiesel production infrastructure is essential. Compatibility ensures a smoother transition for existing facilities looking to adopt this method and contributes to the overall practicality of implementation.

Recognizing the importance of regional variations, the proposed strategy should be adaptable to varied geographical and industrial situations. Adapting the method to regional trash supplies, climate conditions, and legal frameworks improves its practicality and broad applicability.

Community engagement and stakeholder collaboration: It is critical for successful implementation to engage with local communities and stakeholders. Collaborativeefforts including local communities, waste management agencies, and biodiesel producers can build support, address logistical issues, and ensure the suggested method's feasibility on a community level.

Continuous improvement and iterative development: Adopting a mentality of continuous improvement and iterative development supports the suggested method's continual practicability and relevance. Regularly incorporating comments, monitoring technological improvements, and reacting to emerging issues will help to ensure the long-term success of biodiesel synthesis utilizingWCOs and pyrolyzed maize cob.

In conclusion, the future prospects for biodiesel production using WCOs and pyrolyzed maize cob are encouraging, with areas for advancement, additional study, and scalability and practical application issues. The company's continued dedication to innovation and sustainability will be critical in influencing the future of this ecologically conscious biodiesel production process.

4.6 Conclusion

The results of the inquiry into WCO biodiesel manufacturing, notably the use of a solid acid catalyst obtained from CCC, represent a milestone in the field of sustainable and efficient biodiesel synthesis. The solid acid catalyst, derived from pyrolyzed maize cob, has demonstrated exceptional catalytic activity and stability, making it a suitable candidate for expanding biodiesel manufacturing procedures. The catalyst's capacity to simultaneously catalyze the esterification of FFAs is an important feature. The quick and significant conversion of FFAs throughout the esterification process demonstrated the efficacy of this catalyst. Within three hours in the presence of the catalyst, the FFA conversion exceeded an astonishing 88%, setting the framework for a quick and successful conversion of feedstocks with high FFA content. Furthermore, as the esterification process continued for an extended length of time, specifically 15 hours, the cumulative fatty acid methyl ester (FAME) yield approached a classic level of roughly 90%. This consistent and high yield demonstrates the catalyst's long-term durability and ability to allow a lengthy biodiesel synthesis process. The importance of these findings extends beyond the laboratory and into real-world applications. The CCC catalyst appears as a critical component in solving the issues faced by high FFA content feedstocks such as WCO, soap stock and inedible oil. The prospective applications of this catalyst in biodiesel production show promise for providing companies wrestling with the management and transformation of these various feedstocks with economically feasible and environmentally benign solutions. The catalytic power of CCC, as demonstrated in this study, not only allows for the efficient conversion of WCO but also opens the door to the valorization of other difficult feedstocks that were previously deemed unsuitable for biodiesel production due to their high FFA content. The catalyst contributes to the optimization of biodiesel production processes and encourages the utilization of resources that would otherwise be underutilized by dramatically decreasing FFA levels in these feedstocks.

The reported efficacy of the CCC-derived catalyst in WCO biodiesel production has enormous potential in the broader context of the global push toward sustainable and renewable energy sources. It is consistent with the goals of developing cleaner, greener alternatives to traditional fossil fuels, as well as contributing to ongoing efforts to reduce environmental impact and battle climate change.

References

1 Yildiz, I. (2018). Fossil fuels. *Comprehensive Energy Systems* 1: 521–567. https://doi.org/10.1016/B978-0-12-809597-3.00111-5.

2 Perera, F. (2017). Pollution from fossil-fuel combustion is the leading environmental threat to global pediatric health and equity: solutions exist. *International Journal of Environmental Research and Public Health* 15 (1): 16. https://doi.org/10.3390/ijerph15010016.

3 Keating, D. (2019). EU labels biofuel from palm oil as unsustainable, banssubsidies. Forbes (5 July). https://www. forbes.com/sites/davekeating/2019/03/14/eu-labels-biofuelfrom-palm-oil-as-unsustainable-bans-subsidies/.(accessed 5 July 2019).

4 Suzihaque, M.U.H., Alwi, H., Ibrahim, U.K.et al. (2022). Biodiesel production from waste cooking oil: a brief review. *Materials Today: Proceedings*.

5 Ahmad, A.L., Yasin, N.H.M., Derek, C.J.C., and Lim, J.K. (2011). Microalgae as a sustainable energy source for biodiesel production: a review. *Renew. Sustain. Energy Rev.* 15 (1): 584–593. https://doi.org/10.1016/j.rser.2010.09.018.

6 Kumar, N., Sonthalia, A., Pali, H.S., and Sidharth. (2020). Alternative fuels for diesel engines: new frontiers. *Diesel and Gasoline Engines*. https://doi.org/10.5772/intechopen.80614.

7 Núñez-Regueiro, M.M., Siddiqui, S.F., and Fletcher, R.J. Jr. (2021). Effects of bioenergy on biodiversity arising from land-use change and crop type. *Conservation Biology* 35 (1): 77–87.

8 Bitang, J. F. H., Cabeza, J. M. B., Pacot, E. R. S., &Vitor, K. C. A. (2021). Waste cooking oil as feedstock for small-scale biodiesel production. *Journal of umindanao, University of Mindano.*

9 Raqeeb, M.A. and Bhargavi, R. (2015). Biodiesel production from waste cooking oil. *Journal of Chemical and Pharmaceutical Research* 7 (12): 670–681.

10 Zhang, Q., Saleh, A.S.M., Chen, J., and Shen, Q. (2012). Chemical alterations taken place during deep-fat frying based on certain reaction products: a review. *Chemistry and Physics of Lipids* 165 (6): 662–681. https://doi.org/10.1016/j.chemphyslip.2012.07.002.

11 Xu, Z., Leong, S.Y., Farid, M., Silcock, P., Bremer, P., and Oey, I. (2020). Understanding the Frying Process of Plant-Based Foods Pretreated with Pulsed Electric Fields Using Frying Models. *Foods* 9: 949. https://doi.org/10.3390/foods9070949

12 Anisah, P. M., & Agustian, E. (2019, March). Effect of transesterification on the result of waste cooking oil conversion to biodiesel. *Journal of Physics: Conference Series* 1170 (1): 012067.

13 Brown, H., Iverson, B., Anslyn, E., and Foote, C. (2017). *Organic Chemistry*, 8e. States: Cengage Learning.

14 De Feo, G., Di Domenico, A., Ferrara, C. et al. (2020). Evolution of waste cooking oil collection in an area with long-standing waste management problems. *Sustainability* 12 (20): 8578.

15 Goh, B.H.H., Chong, C.T., Ge, Y. et al. (2020). Progress in utilisation of waste cooking oil for sustainable biodiesel and biojet fuel production. *Energy Conversion and Management* 223: 113296.

16 Panadare, D.C. (2015). Applications of waste cooking oil other than biodiesel: a review. *Iranian Journal of Chemical Engineering (IJChE)* 12 (3): 55–76.

17 Phan, A.N. and Phan, T.M. (2008). Biodiesel production from waste cooking oils. *Fuel* 87 (17–18): 3490–3496.

18 Zhang, N., Li, Y., Wen, S. et al. (2021). Analytical methods for determining the peroxide value of edible oils: a mini-review. *Food Chemistry* 358: 129834.

19 Ayoola, A.A., Hymore, F.K.Omonhinmin, C.A.Babalola, P.O.Fayomi, O.S.I.Olawole, O.C.Olawepo, A.V and Babalola, A. (2020): Response surface methodology and Artificia Neural Network analysis of crude palm kernel oil biodiesel production, *Chemical Data Collections*. https://doi.org/10.1016/j.cdc.2020.100478.

20 Buasri, A., Sirikoom, P., Pattane, S. et al. (2023). RSM process optimization of biodiesel production from waste cooking palm oil in the presence of SO3H-PSC catalyst. *Eng. Proc.* 37: 73.

21 Yaakob, Z., Mohammad, M., Alherbawi, M. et al. (2013). Overview of the production of biodiesel from waste cooking oil. *Renew. Sustain. Energy Rev.* 18: 184–193.

22 Refaat, A.A. (2011). Biodiesel production using solid metal oxide catalysts. *International Journal of Environmental Science Technology* 8 (1): 203–221.

23 Sahar, Sadaf, S., Iqbal, J. et al. (2018). Biodiesel production from waste cooking oil: an efficient technique to convert waste into biodiesel. *Sustain. Cities Soc.* 41: 220–226.

24 Thoai, D.N., Hang, P.T.L., and Lan, D.T. (2019). Pre-treatment of waste cooking oil with high free fatty acids content for biodiesel production: an optimization study via response surface methodology. *Vietnam J. Chem.* 57: 568–573.

25 Shalaby, E.A. and El-Gendy, N.S. (2012). Two steps alkaline transesterification of waste cooking oil and quality assessment of produced biodiesel. *Int. J. Chem. Biochem. Sci.* 1: 30–35.

26 Asri, N.P., Sari, D.A.P., Poedjojono, B., and Suprapto. (2015). Pre-treatment of waste frying oils for biodiesel production. *Mod. Appl. Sci.* 9: 99–106.

27 Omotoso, M.A. and Akinsanoye, O.A. (2014). A review of biodiesel generation from non-edible seed oils crop using non-conventional heterogeneous catalysts. *Journal of Petroleum Technology and Alternative Fuels* 6 (1): 1–12.

28 Shu, Q., Zhang, Q., Xu, G.H. et al. (2009). Synthesis of biodiesel from cottonseed oil and methanol using a carbon-based solid acid catalyst. *Fuel Process. Technol.* 90 (7–8): 1002–1008.

29 Yingqun, M. and Yu, L. (2019). *Biodiesel Production: Status and Perspectives*, 503–537. Elsevier Inc., Academic Press, Chapter 21.

30 Dehkhoda, A.M., West, A.H., and Ellis, N. (2010). Biochar based solid acid catalyst for biodiesel production. *Appl. Catal., A* 382 (2): 197–204.

31 Hara, M. (2009). Environmentally benign production of biodiesel using heterogeneous catalysts. *Chemsuschem* 2 (2): 129–135.

32 Park, J., Kim, D., Wang, Z. et al. (2008). Production and characterisation of biodiesel from Tung oil. *Applied Biochemistry and Biotechnology* 148: 109–117.

33 Fonseca, J.M., Teleken, J.G., de Cinque Almeida, V., and da Silva, C. (2019). Biodiesel from waste frying oils: methods of production and purification. *Energy Convers. Manag.* 184: 205–218.

5

Exploring Jansa Tree Seed Oil as a Biodegradable Feedstock for Sustainable Biodiesel Production

5.1 Introduction

Biodiesel is often considered an ideal alternative to traditional diesel fuel owing to its various potential benefits. It is environmentally friendly, renewable, simple to produce, contributes to improved air quality, enhances energy security, and offers safety advantages. This has generated significant enthusiasm for biodiesel as a promising alternative for fossil fuel in terms of diesel which is quite common, primarily because fossil fuels are nonrenewable and environmentally harmful. The large-scale production of biodiesel is cost-effective and simple; the process involves utilizing alkali-catalyzed transesterification on freshly extracted vegetable oils [1–3]. *Cussonia bateri*, is known by various names such as Shigo in Yoruba, Jansa in Cameroon, Takandagiwa in Hausa, Ugbaokwe in Igbo, Bumarlahi in Fulani, and, is a prevalent tree in Northern Nigeria characterized by its twisted savanna form and thick corky bark. The greenish-white flowers and whitish fruits of this tree have an obovate shape with lateral nerves, while the leaves are brittle and oborate. Its seeds, which emit a pleasant scent, contain a semi-drying oil suitable for manufacturing varnishes, oil paints, confections, cosmetics, pharmaceuticals, and food supplements [4]. These properties, coupled with its abundance in nature without competing with the food chain supply, have piqued interest in exploring the composition of the plant seed [5].

Recent findings suggest the existence of a stable and efficient recyclable catalyst derived from naturally occurring heterogeneous carbonaceous materials for biodiesel production [6–8]. The transesterification process necessitates a catalyst with a porous structure, numerous pores, high basicity or acidity, a dense concentration of active species, and effective dispersion of active sites. These attributes

Feedstocks for Sustainable Biodiesel Production: Characterization, Selection, and Optimization, First Edition. Chinwe P. Okonkwo and Modestus O. Okwu.

enhance the surface area of the active site in contact with reactants, thereby improving mass transfer, diffusion, and catalytic activity [9].

The surface acid level or basicity of the catalyst is directly related to its catalytic effectiveness, as it plays an important role in facilitating the transesterification process, usually by activating carbonyl groups in triglyceride molecules.

The improper disposal of oyster shells contributes to environmental pollution. However, utilizing these waste shells for the preparation of solid catalysts not only offers an economical solution for biodiesel conversion but also serves as an effective waste disposal strategy [10].

Oysters, being saltwater bivalves, hold significant nutritional and medicinal value, making them widely consumed worldwide. Apart from their high nutritional content, oysters serve as a low-calorie, low-cholesterol protein source and a remarkable supplier of zinc, known for its immune system-strengthening properties. In addition, oysters are rich in bioactive compounds, showcasing diverse biological activities. The wide array of biological compounds found in oyster shells, along with their chemical constituents, finds applications across the food, pharmaceutical, and medical industries [11].

Oyster shell powder (OSP), comprising 5% organic matter, 95% calcium carbonate, emerges as a promising material with excellent flame-retardant potential. Beyond its flame-retardant properties, OSP also proves to be a valuable source of basic catalyst for heterogeneous transesterification in the biodiesel production process [12]. By repurposing oyster shells in this manner, environmental concerns related to waste management will be addressed as well as unlocking a plethora of applications in various industries, ranging from nutrition and medicine to sustainable energy production (Figures 5.1 and 5.2).

Figure 5.1 Sample of *Cussonia bateri*.

Figure 5.2 Sample of Oyster shell.

5.2 Review of Literature

The phytochemical and physicochemical analysis representing the seed oil and seed marc of *Cussonia bateri* was presented in a study conducted by Nwokonkwo et al. [4]. The presented result is a demonstration that the oil can be used as a vegetable oil, cosmetics, paints, resin, pharmaceutical industrial blend, and other products. The presence of flavonoid, alkaloid, tannin, glycoside, saponin, and phenol in the seeds' is an indication that they have medicinal value, this is evidence via phytochemical analysis.

Okonkwo et al. conducted research focused on the production of clean biodiesel using Jansa seed oil as a feedstock and introducing heterogeneous catalysts that appear in natural form [1]. The study did not end at the level of production of the biodiesel, the research further explored the prediction of other process variables using artificial neural networks. The results demonstrate the strength of the ANN model in terms of parametric studies to enhance performance prediction, particularly when the stochastic dataset is required, specifically when considering biodiesel transesterification.

Lin et al. [12] reported the utilization of calcium oxide derived from tropical oyster shells to develop a shell waste product. The following preparations were employed in the study. According to the experimental findings, the efficiency of the process increased with the rise in the methanol/oil molar ratio, reaching its peak at 12:1 (MeOH/oil) with a 5% catalyst concentration. The oyster shell catalyst demonstrated remarkable resilience, as it could be recovered and reused in the transesterification reaction up to seven times without a loss of activity. Incorporating oyster shell waste in catalyst preparation not only has the potential to reduce the catalyst's cost but also diminishes the presence of contaminants. This approach contributes to making biodiesel more affordable and environmentally sustainable.

5.2.1 Transesterification Process

The transesterification process stands as a pivotal step in the utilization of oils as substitutes for fossil diesel, resulting in the production of biodiesel characterized by a closely aligned cetane index and viscosity [13]. One of the earliest methods employed for biodiesel production is the catalyst-assisted approach, which can manifest in either homogeneous or heterogeneous processes. The homogeneous catalysis method is characterized by its speed, simplicity, high yield, few limitations, and the use of potent catalytic agents [14].

In contrast, the heterogeneous catalysis approach for biodiesel production holds particular intrigue due to its selectivity and environmentally friendly attributes [15]. Transesterification involving various oils can generate esters and glycerin by introducing ethanol or methanol as the prevailing alcohol [16]. To facilitate the conversion of triglycerides into diglycerides, addition of catalyst is required, the product is then further converted into monoglycerides, eventually leading to the production of glycerin. Alkali, acid, and enzymatic catalysts emerge as the most commonly employed catalysts in biodiesel production [16]. This underscores the significance of employing catalysts that not only facilitate chemical transformation but also align with eco-friendly principles in the pursuit of sustainable energy alternatives.

The adaptability of biodiesel production methods is enhanced by the versatility of transesterification, its compatibility with diverse oils, and the availability of various catalyst options. This multifaceted approach not only ensures the effective conversion of triglycerides into biodiesel but also enables the strategic utilization of different catalysts tailored to specific production requirements. Consequently, this enhances the overall sustainability and feasibility of biodiesel as a viable alternative fuel source.

5.2.2 Chemical Reactions of Transesterification

The transesterification of oil plays a crucial role in the esterification of fatty acids in the biodiesel production process. In this procedure, both oils and fatty acids can undergo transesterification and esterification when exposed to alcohol and a catalyst. The reaction process outlined below offers a stoichiometric insight into this phenomenon. This technique commonly employs ethanol and methanol as short-chain alcohols and catalysts encompassing chemical catalysts, biocatalysts, and nonenzymatic heterogeneous catalysts, which are frequently utilized in the process.

$$\begin{array}{lllll} CH_2-COOR^1 & & & R^1COOCH_3 & CH_2-OH \\ CH-COOR^2 & + \; 3CH_3OH & \leftrightarrow & R^2COOCH_3 \; + & CH-OH \\ CH_2-COOR^3 & & & R^3COOCH_3 & CH_2-OH \\ \text{Triglyceride} & \text{Methanol} & & \text{Methyl ester} & \text{Glycerol} \end{array}$$

5.2.2.1 Triglyceride Transesterification

When a catalyst is present, the unutilized free fatty acids (FFAs) may lead to the formation of soap, potentially reducing the overall yield. The primary drawback in conducting oil alcoholysis is associated with minimizing the presence of water and FFAs in the oils.

$$\begin{array}{llll} R_1-COOH + & NaOH & \leftrightarrow R_1COONa + & H_2O \\ \text{FFA} & \text{Sodium Hydroxide} & \text{Soap} & \text{Water} \end{array}$$

5.2.2.2 Soap Formation/Saponification

This has an impact on the quality of the end product. Water presence in the reaction system hampers the transesterification rate by hydrolyzing triglycerides, thereby elevating the content of the FFA in the system.

$$\begin{array}{lllll} CH_2-O-CO-R_1 & & & CH_2-OH & \\ CH-O-CO-R_2 & + & H_2O \leftrightarrow & CH-O-CO-R_2 & + \; R_1-COOH \\ CH_2-O-CO-R_3 & & & CH_2-O-CO-R_3 & \\ \text{Triglyceride} & & \text{Water} & \text{Diglyceride} & \text{FFA} \end{array}$$

The breakdown of triglycerides leads to the formation of acid molecules, which are subsequently employed in the production of soap. The interaction between the FFA in the acid-catalyzed system and alcohol can give rise to esters. Consequently, a two-stage catalyzed process emerges. An acid-catalyzed treatment is required for oils with FFA levels exceeding 3.0%.

5.3 Materials and Method

It is important to list all the required materials, reagents used in this study, and the method necessary to achieve the stated objectives.

5.3.1 Materials

Pensky-martens flash point tester, conical flask, powdered seeds, fume cupboard, pour point tester, material samples (esters, oil, diesel, and ester/diesel blends), soxhlet extractor, separatory funnel, muffle furnace, calcinator, oven, digital weighing balance, digital magnetic heater, viscosity bath and viscometer, TECHNO R175A diesel engine, digital density Emission Scanning Electron Microscope (JSM-670IF), GC–MS QP2010 Plus Shimadzu Japan.

5.3.2 Reagents

A list of reagents that are of analytical grande is presented thus: Magnesium hydroxide, Hydrochloric acid, chloroform (BDH), glacial acetic acid (BDH), concentrated sulfuric acid, methanol and ethanol (BDH), iodine monochloride, distilled water, sodium hydroxide, phenolphthalein, indicator, *n*-hexane (BDH), sodium (vi), potassium iodide.

5.3.3 Collection of Sample and Extraction of Oil

The collection of feedstock in the form of *Cussonia bateri* raw oil seeds was obtained from a neighborhood main market. The acquired feedstock was dried under the hot sun, and the samples were crushed with the help of an industrial blender. The sieving process was carried out. Petroleum diesel was purchased from the NNPC Mega Station in New Haven within the state.

5.3.4 Oil Extraction

Soxhlet extraction was employed to extract oil from the sample. A semipermeable cotton material, along with the sample's weight, was positioned within the thimble of a 500 mL Soxhlet extractor. To initiate the process, a 500 mL round-bottom flask was filled with 400 mL of *n*-hexane.

The condenser was connected to the Soxhlet apparatus via the flask containing *n*-hexane, with the extraction thimble containing the sample enveloped in a semipermeable membrane.

The Soxhlet extraction system was heated on a hot plate while water circulated through the condenser's outer jacket. While water circulated through the condenser's outer jacket, the extraction system of Soxhlet with the help of a heating element was heated further. Once all of the oil had been extracted from the sample, the extraction was terminated. The de-fatted sample was discarded, and the mixture of oil and *n*-hexane in the round-bottom flask was separated using a rotary evaporator apparatus. The yield percentage was then computed.

5.3.5 Catalyst Preparation and Characterization

5.3.5.1 Preparation of Catalysts

Oyster shells, considered waste, were gathered and subjected to a cleaning process with water to eliminate dust and impurities. Subsequently, the cleansed shells underwent drying in an oven at 70 °C for a duration of 10 hours. After drying, the oyster shells were crushed and sifted through screens ranging from 35 to 78 μm. The solid catalyst was produced through a calcination process, wherein spent shells were calcined in an air environment for three hours at 700 °C, with a heating rate of 12 °C/min [17]. The resulting CaO catalyst appeared as a white powder. Both the raw and calcined samples were stored in a sealed container before utilization to prevent reactivity with CO_2 and moisture in the air [18].

5.3.6 Oil Characterization

Oil from the feedstock, *Cussonia bateri* seeds was characterized to determine their moisture content, density, saponification value, refractive index, viscosity, peroxide value, acid value (AV), and iodine value.

5.3.7 Preparation of the Methyl and Ethyl Esters

Dissolving 47.5 g of sodium hydroxide pellets in 150 mL of distilled water yielded a sodium hydroxide solution. This sodium hydroxide solution was mixed vigorously with a 1000 mL beaker containing 30 mL of the oil sample until soap formation occurred. After resolving the soap in water, a super-saturated sodium chloride solution was added while vigorously stirring at 200 rpm and heating. The spent lye (a mixture of water, glycerol, and salt) settled to the bottom of the solution, causing pure soap to precipitate.

This precipitate was separated using filter paper, and then excess brine was added to the separated soap to remove any remaining glycerol and impurities, enhancing the color and facilitating drying. In a 100 cm^3 beaker with 250 mL of distilled water, the precipitated soap was redissolved and heated until a homogeneous solution formed. Litmus paper testing confirmed the alkalinity of the solution. To neutralize the solution to the indicator's neutral point, concentrated sulfuric acid was diluted with water in a 1:3 ratio (concentrated sulfuric acid:water). The oily layer was separated from the aqueous layer using a separatory funnel, and this oily layer contained a mixture of fatty acids contained a mixture of fatty acids.

5.3.8 Transesterification Process for Biodiesel Production

The methyl ester transesterification reaction was carried out in a 500 mL round-bottom glass reactor with three necks. According to the experiment's specification, a methanol predetermined amount was added, this is in addition to the modified catalyst which is according to the specification of the experiment. With the mixture in place, prestirring process took place for approximately 10 minutes, this is to ensure appropriate dissolution of catalyst, after which oil of 50 mL containing fatty acids was introduced into the experimental setup. With the experimental setup in place, a reflux condenser containing cold water circulating in the outer jacket was fitted into the middle neck of the reactor. A mercury thermometer was secured with a plastic bung in the right neck, while a capsular stirring knob was inserted through the left neck and secured with another plastic bung. The setup was turned on and placed on a magnetic heating mantle, which activated the stirring system at a constant speed and temperature, indicating the start of the reaction. Each reaction was allowed to run for the time and temperature specified.

The product was left in a separating funnel for 12 hours after the transesterification reaction so as to facilitate glycerol separation. When the crude glycerol was drained through the funnel tap, the biodiesel (methyl ester) remained. After washing with hot water, the biodiesel was transferred to a 250 mL beaker and heated to 105 °C to remove water molecules. The biodiesel was stored in calibrated specimen bottles after cooling.

5.4 Results and Discussion

5.4.1 Extracted Oils Physicochemical Properties

It is important to discuss the outcomes of the analysis based on the physicochemical properties of the extracted seed sample, which are outlined in Table 5.1. The high percentage of oil yield from the crop indicates a promising potential for substantial biodiesel production.

The acid number or neutralization number which is also known as the AV, signifies the potassium hydroxide (KOH) quantity with unit represented in milligrams that is required to neutralize the components of acid within 1 g of the sample. For the presented feedstock, *C. bateri*, the AV was measured at 1.25. The low AV suggests that the oil can be directly converted into biodiesel through the base-catalyzed transesterification method [19]. Elevated levels of FFAs in oil can deactivate the base catalyst, leading to soap formation and a reduction in biodiesel yield. The acidic compounds in biodiesel may arise from residual mineral acids in the manufacturing process, residual FFAs from the hydrolysis process or post-hydrolysis process of esters, and oxidation of by-products into other organic acids [20].

Table 5.2 shows the fatty acid profile of *C. bateri* in their various proportions. *C. bateri* exhibited an iodine value of 121.00, falling within the semi-drying oil (>120) range. Since the iodine value reflects the degree of unsaturation in fatty acids, *C. bateri* is prone to oxidize and polymerize at higher temperatures. A higher iodine value indicates greater oil unsaturation, resulting in faster oxidation and polymerization [21]. The kinematic viscosity of the oil exceeded that of petrol

Table 5.1 Physicochemical properties.

Properties	*C. bateri*
Color of extracted oil	Golden yellow
Percentage yield	69.84%
Percentage moisture	3.00%
Refractive index	1.46 at 29 °C
Kinematic viscosity	10.77 at 40 °C(mm^2/s)
Acid-value	1.25(mgKOH/g)
Energy value	44 384 (kJ/kg)
Peroxide value	0.650 (meq/kg)
Saponification	110.00 (mgKOH/kg)
Molecular	815.64 weight (g/mol)
Iodine value	121.00 (g/100 g of oil)
Cloud point (°C)	12.50
Flashpoint (°C)	260.00
Fire point (°C)	265.00
Pour point (°C)	7.00
Density (g/mL)	0.85

Table 5.2 *C. bateri* seed oil fatty acid profile [25].

S/no	Common name	Systematic name	Shorthand	R.A (%)
1	Linoleic acid	cis-9,12,15-Otacatrienoic acid	18:2	63
2	Oleic acid	cis-9-Octadeaoic acid	18:1	12
3	Palmitic acid	Hexadecanoic acid	16:0	14
4	Stearic acid	Octadecanoic acid	18:0	10
5	Myristic acid	Tetradecanoic acid	14:0	Trace

diesel due to differences in fatty acid composition and molecular weight [22]. Kinematic viscosity directly influences fuel performance, affecting atomization and combustion quality.

Reporting the results, *C. bateri* comprises 10% stearic acid, 12% oleic acid, 63% linoleic acid, 14% palmitic acid, and trace amounts of myristic acid. The composition reveals a higher presence of polyunsaturated fats, a modest quantity of unsaturated fatty acids, and a small proportion of saturated fatty acids, specifically palmitic, stearic, and myristic acids. The notable concentration of linoleic acid suggests that the oil possesses limited oxidative stability, implying a requirement for antioxidants if stored for an extended duration.

5.4.2 Catalyst Characterization

Based on the catalyst characterization results presented in Table 5.3, it is evident that Oyster shell exhibits promising features for biodiesel production. Notably, the calcined form of the catalyst displays superior attributes compared to the raw form, particularly in terms of its significantly larger surface area. The raw Oyster shell exhibits a surface area value of 720.00 m^2/g, while the calcined form demonstrates an enhanced surface area of 840.00 m^2/g. This increased surface area in the calcined state suggests improved catalytic potential and underscores the efficacy of the Oyster shell as a favorable resource for enhancing biodiesel production processes.

5.4.3 Effect of Process Parameters on the Yield

Table 5.4 presents the impact of catalyst concentration on the production of biodiesel derived from *C. bateri* seed oil (CSO). The C. Oyster shell, in conjunction with methanol, was subjected to a temperature of 70 °C over a two-hour duration, employing a methanol to oil ratio of 12:1 and maintaining an agitation speed of 400 rpm.

Table 5.3 Catalyst characterization of oyster shell.

Parameter	Raw	Calcined
Moisture (%)	0.05	0.04
Bulk density (g/cm^3)	0.73	0.54
pH	6.74	6.07
Surface area (m^2/g)	720.00	840.00
Carbon (%)	14.08	12.17
Organic matter (%)	16.09	12.44
Loss on ignition (%)	9.34	0.72
Particle density ($g/cm^{3)}$	1.52	0.83
Total porosity (%)	52.04	35.06
Ash (%)	3.22	1.54

Table 5.4 Effect of catalyst concentration considering C. Oyster shell with methanol.

Runs	Catalyst conc. (wt.% of oil)	Yield of FAME *C. bateri* (vol. %) (raw)	Yield of FAME *C. bateri* (vol. %) (calcined)
1	0.75	90.00	96.00
2	1.50	84.00	94.00
3	2.25	82.00	92.00
4	3.00	78.00	89.00
5	3.75	75.00	85.00

5.4.4 Effects of Process Parameters on Biodiesel Synthesis

It was observed that a smaller quantity of the catalyst led to a higher yield, which can be attributed to the fact that larger molecules possess higher activation energy and consequently react at a slower rate compared to smaller molecules [22]. The reaction rate exhibited a gradual increase with time. Moreover, the yield obtained from the calcined oyster shell surpassed that from the raw oyster shell, a difference that can be explained by variances in activation energy, aligning with the collision theory of reactions [23]. Generally, the sample yield demonstrated an upward trend as the temperature increased. This indicates that while the reaction could theoretically occur at room temperature, achieving a substantial yield and conversion would be a time-consuming process [23]. Notably, owing to methanol's lower viscosity compared to other alcohols, it facilitates a more efficient reaction, resulting in a higher yield and complete conversion.

The yield demonstrated an upward trend with increasing stirring speed, reaching its peak at 400 rpm. The findings suggest that a higher degree of agitation is essential for optimizing the yield. This observation may be attributed to variations in molecular weight and viscosity. Increased viscosity necessitates heightened agitation to attain effective mixing between the two liquid phases [24]. As indicated by the outcomes of the process parameters, the calcined yield exhibited the highest average yield for the seed oil. This outcome is attributable to the fact that calcination enhances the surface area of the catalyst, thereby facilitating the conversion of larger molecules such as bio-oils [25].

5.5 Conclusion

In this study, the evaluation of the efficacy of the transesterification process for biodiesel production using *C. bateri* seed oil, catalyzed with both raw and calcined forms of oyster shell was evaluated. Utilizing calcined oyster shells as a catalyst was chosen due to their ready availability and noncorrosive nature. The results indicate that the oyster shell catalyst can be efficiently recovered and reused in transesterification reactions without observable loss of activity. Biodiesel properties, as per international standards, surpassed those of conventional diesel, signifying its potential utility in biodiesel transesterification processes. Furthermore, the study revealed that calcined oyster shells yielded a higher quantity of biodiesel. The calcination process, characterized by the removal of volatile substances and an increase in friability, increased the number of basic sites available at the active phase in the transesterification process. This enhancement resulted in elevated biodiesel conversion rates, demonstrating the positive impact of calcination on the efficiency of biodiesel production.

References

1 Okonkwo, C.P., Ajiwe, V.I.E., Obiadi, M.C. et al. (2021). Kinetics and transesterification of the oil obtained from *Cussonia bateri* (Jansa seed) as a step in biodiesel production using natural heterogeneous catalyst. *Am. J. Appl. Chem.* 9 (5): 164–171.

2 Gebremariam, S.N. and Marchetti, J.M. (2018). Economics of biodiesel production: review. *Energy Convers. Manag.* 168: 74–84. https://doi.org/10.1016/j.enconman.2018.05.002.

3 Mansir, N., Teo, S.H., Rashid, U. et al. (2018). Modified waste egg shell derived bifunctional catalyst for biodiesel production from high FFA waste cooking oil: a review. *Renew. Sustain. Energy Rev.* 82: 3645–3655. https://doi.org/10.1016/j.rser.2017.10.098.

4 Nwokonkwo, D.C., Nwaolisa, S.U., and Odo, H.A. (2016). Comparative physicochemical study of the oils from the seeds of *Cussonia bateri* (JANSA) and Brasssica Juncea (MUSTARD). *Int. J. Appl. Chem.* 12 (1): 51–58.

5 Knott, G.D. (2015). *Chemical Kinetics: Simple Binding: Heritage Park Circle Silver*, 20906. Spring MD.

6 Ayoola, A.A., Hymore, F.K., and Omonhinmin, C.A. (2016). Optimization of biodiesel production from selected waste oils using response surface methodology. *Biotechnology* 16 (1): 1–9.

7 Ayoola, A.A., Babalola, R., Ojewumi, M.E., and Fayomi, O.S.I. (2018). Assessment of KOH and NaOH catalysed biodiesel from melon seed oil. *Mat. Focus* 7 (5): 650–656.

8 Ayoola, A.A. (2015). Production and life cycle assessment of biodiesel produced from three waste oils.Ph.D. thesis.Chemical Engineering Department, Covenant University, Nigeria.

9 Ali, G., Fathollah, P., and Akbar, M. (2020). Recent advances of biodiesel production using ionic liquids supported on nanoporous materials as catalysts: areview front. *Energy Res.* https://doi.org/10.3389/fenrg.2020.00144.

10 Ikbal, B.L., Kalyani, R., Rajat, G. et al. (2018). Waste snail shell derived heterogeneous catalyst for biodiesel production by the transesterification of soybean oil. *R. Soc. Chem. Adv.* 8: 20131–20142. https://doi.org/10.1039/c8ra02397b.

11 Ulagesan, S., Krishnan, S., Nam, T.J., and Choi, Y.H. (2022). A review of bioactive compounds in oyster shell and tissues. *Front. Bioeng. Biotechnol.* 10. https://doi.org/10.3389/fbioe.2022.913839.

12 Lin, Y.C., Amesho, K.T., Chen, C.E. et al. (2020). A cleaner process for green biodiesel synthesis from waste cooking oil using recycled waste oyster shells as a sustainable base heterogeneous catalyst under the microwave heating system. *Sustain. Chem. Pharm.* 17: 100310.

13 Giakoumis, E.G. and Sarakatsanis, C. (2019). A comparative assessment of biodiesel cetane number predictive correlations based on fatty acid composition. *Energies* 12: 422.

14 Farobie, O. and Matsumura, Y. (2017). State of the art of biodiesel production under supercritical conditions. *Prog. Energy Combust. Sci.* 63: 173–203.

15 Fonseca, J.M., Teleken, J.G., de Cinque Almeida, V., and da Silva, C. (2019). Biodiesel from waste frying oils: methods of production and purification. *Energy Convers. Manag.* 184: 205–218.

16 Quah, R.V., Tan, Y.H., Mubarak, N.M. et al. (2019). An overview of biodiesel production using recyclable biomass and non-biomass derived magnetic catalysts. *J. Environ. Chem. Eng.* 4: 103219.

17 Khemthong, P., Luadthong, C., Nualpaeng, W. et al. (2012). Industrial eggshell wastes as the heterogeneous catalysts for microwave-assisted biodiesel production. *Catal. Today* 190: 112–116.

18 Buasri, A., Chaiyut, N., Loryuenyong, V. et al. (2013). Calcium oxide derived from waste shells of mussel, cockle, and scallop as the heterogeneous catalyst for biodiesel production. *Sci. World J.* 2013: 460923.

19 Sivanesan, S. (2009). Kinetic study of the esterification of free fatty acids in non- edible Pongamia pinnata oil using acid catalyst. *Indian J. Sci. Technol.* 2 (12): 20–24.

20 Berthiaume, D. and Tremblay, A. (2006). Study of the Rancimat Test Method in Measuring the Oxidation Stability of Biodiesel Ester and Blends NRCan project No.CO414CET-327O. LEOTEK Inc, Quebec, Canada. http://www.tehttp://www.technopolethetford.ca/Industrialolechemistry/info_observatoiredel eochimie_etudes-etrecherches_187-ang.cfmchnopolethetford.ca/Industrialolechemistry/info_observatoiredel eochimie_etudes-etrecherches_187-ang.cfm (accessed 16 February 2016).

21 Thomas, A. (2000). Fats and fatty oils. In: *Ullmann's Encyclopedia of Industrial Chemistry*. Weinheim: Wiley VCH. https://doi.org/10.1002/14356007.a10_173.

22 Ajiwe, V.I.E. and Ogbu, M. (2016). FTIR studies of thermal stability of the oils and methyl esters from Afelezia Africana and Hura crepitans seeds. *Renew. Energy* 96: 203–208.

23 Moser, B.R. (2009). Biodiesel production, properties and feed stocks. *In Vitro Cell Dev. Biol. Plant* 45: 229–266.

24 Anekwe, O.J. and Ajiwe, V.I.E. (2015). Comparative analysis of the fuel properties of ethylester biodiesels from *Cyperus esculentus, Seasamum indicum* and *Colocynthus vulgais* seed oils. *Energy Nat. Resour.* 4 (3): 40–44.

25 Okonkwo, C.P., Ajiwe, V.I.E., Mmaduakor, E.C., and Nwankwo, N.V. (2022). Modelling and optimization of biodiesel production process parameters from Jansa seed oil (Cussonia bateri) using artificial neural network. *Am. J. Appl. Chem.* 10 (1): 7–14.

6

Production of Biodiesel by Implementing *Hura crepitans* as Feedstock

6.1 Background of Study

Society's dependence on the finite reservoir of petroleum-derived fuels, including diesel, natural gas, and coal, has spurred ongoing efforts to explore alternative energy sources to fulfill global energy requirements [1]. The heightened utilization of fossil fuels has adversely affected the environment, leading to detrimental consequences such as environmental degradation and the release of harmful gases [2–4]. Diesel fuel is the most widely used petroleum-based fuel. Its global demand is skyrocketing because it is more energy-intensive and cost-effective in terms of fuel consumption than other petroleum-based fuels [5].

Biodiesel is considered the optimal substitute for fossil fuels, offering a greenhouse gas emission reduction of approximately 45% compared to diesel [6]. In addition, the use of biodiesel contributes to an enhancement in urban air quality by lowering emissions of hydrocarbons, particulates, and carbon monoxide [7]. This alternative fuel not only diminishes dependence on oil imports but also possesses nontoxic and biodegradable properties [8]. The prevalent technique for biodiesel production involves transesterification, wherein glycerides sourced from animal fats, waste frying oil, algae, or oilseeds are reacted with a low molecular weight alkanol and an alkaline catalyst (NaOH or KOH), or in certain instances, sodium or potassium methoxide.

The predominant obstacle hindering the widespread adoption of biodiesel as a substitute for traditional fossil diesel is the consistently high cost of raw materials, constituting approximately 70%–85% of the overall biodiesel production expenses [9, 10]. While the use of biodiesel holds promise in

Feedstocks for Sustainable Biodiesel Production: Characterization, Selection, and Optimization, First Edition. Chinwe P. Okonkwo and Modestus O. Okwu.

mitigating global warming and pollution associated with the combustion of fossil fuels [11], the biodiesel industry faces significant challenges, particularly concerning the availability of sustainable feedstock and the need to reduce production costs [6]. The accessibility and geographical location significantly influence the availability of feedstock [12]. In Africa, many oil seeds remain underutilized due to a lack of awareness regarding their industrial applications. However, the recent upswing in seed oil consumption and the encouragement of the bioenergy sector have elevated the importance of utilizing biomass waste from agricultural operations as a crucial international source of raw materials [13].

Vegetable oils play a vital role as a raw material in various industries, necessitating the exploration of new and underutilized plant oil seeds to meet the global demand for oils. It is essential to avoid relying solely on edible oil sources for fuel sustainability, as this could lead to issues such as food scarcity and price hikes [14]. Addressing the increasing global need for oil underscores the importance of researching underutilized plant oilseeds [13]. *H. crepitans* seeds, an example of an underutilized plant oilseed, can be subjected to solvent extraction to obtain oil. Belonging to the Euphorbiaceae family, these seeds come from a tree with ovate leaves that can reach up to 2 ft in width [15]. An unexplored potential feedstock for biogas production could be the leaves of this tree, which grows to nearly 30 m (100 ft) in height, featuring a trunk with a girth exceeding 1 m (3.3 ft) and covered in small conical prickles [16].

A tall tree with a rounded crown featuring long-stalked dark-green leaves characterizes the Hura plant. The seed capsules, globular in shape, are divided into 15 sections and measure 7.6 cm (3 in.) in diameter. In some countries, these capsules were utilized as sandboxes for blotting ink. Some indigenous communities employ a mixture of poisonous latex and sand to stun fish. Both the heartwood and sapwood exhibit a pale yellowish-brown or pale olive-gray hue. The wood has a fine to medium texture, with a straight to interlocking grain. Hura wood is characterized as low-strength and low-density with air-dried wood of average density. It is abstemiously challenging to air dry due to tension wood, resulting in a fuzzy surface for green material. However, the wood finishes well and is easily worked through gluing and nailing. *H. crepitans* is noted to be prone to wood termites and highly susceptible to blue stains. In certain Nigerian provinces, the *H. crepitans* tree is utilized for shelter. Further visual representations of the seed, leaves, and tree are provided in the accompanying Figures 6.1–6.4.

Figure 6.1 Pictorial view of (a) *Hura crepitans*fruit (*Source:* Magnefl/Wikimedia Commons/CC BY-SA 4.0); (b) *Hura crepitans*leaf (*Source:* Oldcrookedjaw/Wikimedia Commons/CC BY-SA 3.0); (c) *Hura crepitans*spiny trunk (*Source:* Joel Bradshaw/Wikimedia Commons/Public domain); (d) *Hura crepitans*seed (*Source:* Hans Hillewaert/Wikimedia Commons/CC BY-SA 3.0).

Figure 6.2 Dry Sandbox fruit, broken capsules, and seed (*Source:* Akintunde et al. [17]/with permission of Elsevier).

Figure 6.3 Pictorial view of (a) fresh fruits; (b) Dried fruits; (c) Seeds with coats; (d) Dehulled seeds (*Source:* Omilakan et al. [18]/with permission of Springer Nature).

Figure 6.4 (a) *Hura crepitans* tree, (b) *H. crepitans* leaves and seeds (*Source:* Olugbemide et al. (2022) [43]/with permission of Springer Nature).

6.2 Review of Related Research Conducted Using *Hura crepitans*

The oil concentration of this precious biodegradable feedstock ranges from 37.2% to 70.4% [18, 19]. *H. crepitans* seed oil contains more than 50% linolenic acid, making it an ideal substitute for produced biodiesel. Previous research on its use in biodiesel synthesis has been published [20]. However, there is a lack of information in the modeling of the transesterification process for seed oil to methyl esters. Various nonedible oils have been used to produce biodiesel [21–24]. Because of their high oil yield, they are thought to be useful in industrial applications [19]. Some recent studies conducted using *H. crepitans* in diverse applications is presented in Table 6.1.

Table 6.1 Research conducted using *Hura crepitans* in diverse applications.

Citations	Authors	Title	Applications	Year
[24]	Ugbaja et al.	Biochemical appraisal of the underutilized *Hura crepitans* seed oil: functional and inflammatory responses in albino rats	AA	2022
[17]	Akintunde et al.	Combustion and emission study of sandbox seed oil biodiesel performance in a compression ignition (CI) engine	B	2021
[25]	Velazquez-González et al.	*Hura crepitans*seeds for control of *Eimeria* spp. in lambs as an alternative to conventional therapies	AA	2022
[26]	Owojuyigbe et al.	*Hura crepitans* stem bark extract: a potential remedy to sub-acute liver damage	AA	2022
[12]	Ajala et al.	A nonlinear autoregressive exogenous neural network (NARX–NN) model for the prediction of solvent-based oil extraction from *Hura crepitans* seed	GP	2023
[27]	Temitayo et al.	Investigation on corrosion inhibition of *Hura crepitans* for 2101 duplex stainless steel in an acidic environment	GP	2022
[6]	Ogbu et al.	Performance evaluation of carbon-based heterogeneous acid catalyst derived from *Hura crepitans*seed pod for esterification of high FFA vegetable oil	B	2018
[18]	Omilakin et al.	Process modeling of solvent extraction of oil from *Hura crepitans* seeds: adaptive neuro-fuzzy inference system versus response surface methodology	GP	2020

(*Continued*)

Table 6.1 (Continued)

Citations	Authors	Title	Applications	Year
[43]	Olugbemide et al.	Reaction kinetics modeling for dry anaerobic digestion of lignocellulosic *Hura crepitans* leaves	GP	2022
[1]	Oraegbunam et al.	Sandbox oil biodiesel production modeling and optimization with neural networks and genetic algorithm	B	2023

AA, antimicrobial activity; B, biodiesel; FP, food products; GP, general purpose; Ph, pharmaceuticals; WT, water treatment.

6.3 Materials and Method

H. crepitans Biodiesel Production Stages – The process of producing biodiesel from *H. crepitans* involves several steps, which include: Seed collection and preparation: *H. crepitans* seeds are collected and cleaned of any debris, such as leaves or twigs. The seeds are then left to dry in the sun until the moisture content is very low; Oil extraction: The oil is extracted from the seeds using a mechanical press. The oil is filtered to remove any impurities; such as seed particles or debris; Transesterification: The next step involves the application of a transesterification procedure to the extracted oil. This process involves the interaction of the oil with an alcohol, such as methanol or ethanol, in the presence of a catalyst, like sodium hydroxide or potassium hydroxide. Through the transesterification process, the oil undergoes conversion into biodiesel and glycerol. Subsequently, the biodiesel and glycerol are isolated. The ensuing steps encompass washing and drying and the residual impurities or catalyst in the biodiesel are eliminated through a water-washing process. The biodiesel is then dried to ensure that no water is present; Quality control: The biodiesel is tested for quality to ensure it meets the required standards. The tests may include measurements of viscosity, density, flash point, and other parameters; Storage and distribution: The biodiesel is then stored in a suitable container and distributed for use as a fuel. Further details on the two important aspects of the production stages – Mechanical Pressing and Transesterification are presented thus.

6.3.1 Various Techniques of Biodiesel Production

Various techniques, including mechanical pressing, pressurized solvent extraction, transesterification, ultrasound-assisted extraction, microwave-assisted extraction, percolation, pyrolysis, supercritical fluid extraction, subcritical fluid extraction, and aqueous enzymatic oil extraction, are employed for extracting oil from *H. crepitans* seeds [28]. Among these, solvent extraction and mechanical pressing are the predominant methods widely utilized for commercial oil extraction [29]. Mechanical pressing is the most common method, and it appears to be the easiest to use, especially for small-scale production. The procedure calls for the seed to be sourced and then dried. The dried seed is then crushed into a fine powder in the following step. The powdered *H. crepitans* is then mixed with water and pressed to extract the oil. The extracted oil is known to be of high quality [30]. The choice of solvent for extracting oil from plant seeds is crucial for its effectiveness, as it directly impacts the quality and quantity of the extracted oil [31–35]. The pods will be cut open and the seeds extracted, while the creamy white cotyledons inside the endocarp will be removed gently. It will then be sun-dried for a week before the seeds are ground into a fine powder with an electronic blender. It will then be placed in a polythene bag for extraction and further research. Extraction is carried out using a suitable solvent for the desired feedstock while maintaining the required time, temperature, catalyst concentration, and agitation speed. They are typically performed with a soxhlet apparatus fitted to a dry cellulose thimble. At the end of the extraction, the solvent will be distilled. The yield will be evaluated using Eq. (6.1).

$$\% \text{ Oil yield} = \frac{\text{mass of oil produced}}{\text{mass of seed used}} \times \frac{100}{1} \tag{6.1}$$

6.3.2 Transesterification

Transesterification is the most important process in the use of oils as a replacement for fossil diesel. It produces biodiesel with a high viscosity and cetane index [36]. The heterogeneous catalysis' selectivity and eco-friendliness make it an appealing approach to biodiesel production [37]. Transesterification of various oils can be used to produce esters and glycerin by adding alcohol, most commonly methanol or ethanol [38]. The addition of a catalyst aids in the breakdown of triglyceride into diglyceride and then monoglyceride. Glycerin is made from the produced monoglyceride. Alkali, acid, and enzymatic catalysts are the most commonly used in the production of biodiesel [37].

6.4 Benefits of Using Biodiesel Obtained from *Hura crepitans*

6.4.1 Information on Environmental Benefits

It lacks sulfur or aromatics, and employing biodiesel in a standard diesel engine leads to a significant decrease in unburned hydrocarbons, carbon monoxide, and particulate matter emissions. The biodiesel sector is working on an additive to mitigate NOx emissions. In blends of B5 or lower, the rise in NOx is minimal.

6.4.1.1 Energy Security

These fuels exhibit a favorable energy balance. The concept of a "positive energy balance" indicates that the energy expended in the fuel production process is lower than the energy content of the produced fuel, along with many of the associated by-products utilized in the process.

6.4.1.2 Cheap Source of Biodiesel

Utilizing low-cost and nonedible seed oils like *H. crepitans*, animal fats, and waste oils as raw materials for biodiesel production proves to be an efficient strategy for cutting down biodiesel expenses. Consequently, the quest for underutilized nonedible seed oils as biodiesel feedstock concludes. The benefits of *H. crepitans* biodiesel, in comparison to mineral-derived diesel fuel, include reduced exhaust emissions, making it a feasible alternative to diesel sourced from minerals, enhanced biodegradability, diminished toxicity, improved lubricity, a higher flash point, and lower vapor pressure [26].

6.4.2 Economic Benefits

Biofuels obtained from *H. crepitans* have the potential to deliver substantial microeconomic advantages to both urban and rural areas, impacting the trade balance positively. According to the 2001 USDA study, a surge in biodiesel demand by 200 million gallons annually would elevate total cash crop proceeds by approximately 5 billion United States dollars by 2010, leading to a yearly increase in net farm income on average of 300 millionUnited States dollars. Over the 10-year span, the cost of a bushel of *H. crepitans* would witness an average yearly increase of 17 cents [39].

6.4.2.1 Energy Policy Act Benefits

The United States Department of Agriculture (USDA) office responsible for the Congressional Budgethas affirmed that one of the most economical alternative fuel choices to fulfill the EPAct compliance mandates of the federal government is biodiesel. Given its compatibility with current diesel engines, biodiesel offers a swift and uncomplicated method to convert existing diesel vehicles into a more environmentally friendly and cleaner-burning fleet [40].

6.4.3 Common Challenges Associated with Using *Hura crepitans* as a Source of Biodiesel

H. crepitans, commonly known as the sandbox tree or dynamite tree, is a tropical tree that produces seeds containing oil that can be used as a source of biodiesel. However, there are several challenges associated with using *H. crepitans* as a source of biodiesel, they include:

Toxicity: The oil from *H. crepitans* contains toxic compounds, such as phorbol esters, that can be harmful to humans and the environment. Therefore, careful handling and processing are required to avoid exposure to these toxins.

Yield: The oil yield of from novel *H. crepitans* seeds is comparatively low compared to other biodiesel sources, which makes it less economically viable.

Sustainability: *H. crepitans* is a wild plant species, and its cultivation for biodiesel production may lead to environmental concerns such as deforestation, soil erosion, and loss of biodiversity.

Accessibility: *H. crepitans* is mainly found in tropical regions, and its distribution is limited, which makes it less accessible and difficult to scale up for commercial production.

Processing: The oil from *H. crepitans* seeds requires specialized processing techniques, such as solvent extraction and transesterification, which can be costly and require specialized equipment.

Generally, while *H. crepitans* has the potential to be a source of biodiesel, but the challenges associated with its use make it less favorable compared to other biodiesel sources that are more sustainable, accessible, and economically viable.

6.4.4 Factors Affecting Production of *H. crepitans*

6.4.4.1 Effect of Molar Ratio of Alcohol

The molar ratio of alcohol plays a crucial role in determining the yield of *H. crepitans*. A surplus of alcohol expedites the transformation of fats into esters, leading to an augmented yield up to a certain threshold. However, beyond this point, further increases in alcohol concentration do not contribute to higher yields but instead escalate the expenses associated with alcohol recovery. It is noteworthy that the alcohol content ratio varies depending on the catalyst employed. For instance, when utilizing an alkali catalyst, the transesterification of oils or fats necessitates a 6:1 alcohol ratio to catalyze the reaction. In cases where oil samples exhibit a high level of free fatty acid (FFA), alkali catalysts may not be as effective. This is attributed to the fact that acid catalysts can tolerate the presence of FFA and water content in the oil samples [41, 42]. The careful consideration of both alcohol concentration and the choice of catalyst is pivotal in optimizing the transesterification process for *H. crepitans*, ensuring efficient and cost-effective biodiesel production.

6.4.4.2 Effect of Water and FFA Contents

The detrimental effects of water surpass those of FFAs due to its tendency to induce soap formation and frothing, thereby elevating viscosity. In addition, the foams and gel formation during this process hampers the effective separation of biodiesel from glycerol. In the course of transesterification, the presence of both water and FFAs consistently yields unfavorable outcomes, leading to catalyst consumption, soap formation, and a decline in catalyst efficiency. Furthermore, the impact of FFA and water extends to the methyl ester reduction. To tackle these challenges, a proposed solution involves employing the supercritical methanol method (with a molar ratio for methanol-to-oil of 42:1, four-minute treatment time, and at 623 K, 43 MPa), which was compared to acid-catalyzed and alkali methods. Notably, the methanol method which is supercritical mitigates the adverse impact of water to a greater extent [40]. This underscores the importance of considering water content in biodiesel production methods and highlights the potential efficacy of the supercritical methanol approach in addressing these challenges.

6.4.4.3 Reaction Time

The reaction time is a significant parameter influencing the production of biodiesel using *H. crepitans* as feedstock. It refers to the duration the feedstock undergoes transesterification, the process of converting fats or oils into biodiesel. The reaction time is a critical factor that affects the completeness of the reaction and, consequently, the yield and quality of biodiesel. An optimal reaction time is essential for achieving maximum biodiesel production efficiency. If the reaction time is too short, the transesterification may not be complete, resulting in lower biodiesel yield. Conversely, an excessively long reaction time may lead to over processing, potentially causing the formation of unwanted by-products or affecting the overall quality of the biodiesel.

Researchers and biodiesel producers conduct experiments to determine the ideal reaction time when using *H. crepitans* as feedstock. These studies consider various factors, including the type of catalyst, alcohol-to-feedstock ratio, and reaction temperature, to find the optimum conditions for transesterification. The goal is to strike a balance that ensures thorough conversion of feedstock into biodiesel while avoiding unnecessary prolongation of the reaction time.

The initial stage of the reaction exhibits a sluggish pace, which is attributed to the initial challenges in effectively mixing and dispersing alcohol and oil. Subsequently, the reaction accelerates significantly. However, optimal ester conversion is attained in less than 90 minutes. Prolonging the reaction time beyond this point does not enhance product yield, specifically biodiesel/mono alkyl ester. Moreover, an extended reaction duration contributes to a decline in the end product (biodiesel) due to the reversible nature of the transesterification reaction, leading to ester loss and soap formation [40–42]. In the context of biodiesel production from *H. crepitans*, adjusting the reaction time is part of the optimization process to achieve the desired biodiesel characteristics and maximize overall process efficiency.

6.4.4.4 Reaction Temperature

The reaction temperature is a crucial factor influencing the production of biodiesel using *H. crepitans* as feedstock. Optimal temperatures are essential for the transesterification process, where the conversion of fats or oils into biodiesel occurs. The reaction temperature affects the reaction yield, kinetics, and the overall efficiency of biodiesel production. In general, higher temperatures can accelerate the transesterification reaction, potentially leading to increased biodiesel yield.

However, excessively high temperatures may also result in side reactions or thermal degradation, affecting the quality of the biodiesel produced. Therefore, finding the right balance and optimizing the reaction temperature is essential for maximizing biodiesel production efficiency while maintaining product quality.

Researchers and biodiesel producers often conduct experiments to determine the optimal temperature range for transesterification using *H. crepitans* as feedstock. These studies consider factors such as the type of catalyst used, the alcohol-to-feedstock ratio, and the reaction time in conjunction with temperature to achieve the desired biodiesel characteristics and yield. Adjusting the reaction temperature within a suitable range is a key aspect of process optimization in biodiesel production from *H. crepitans*.

6.4.4.5 Catalyst Concentration

The concentration of the catalyst is a critical factor that influences the biodiesel production process using *H. crepitans* as feedstock. The catalyst plays a pivotal role in facilitating the transesterification process, wherein fats or oils undergo conversion into biodiesel. The concentration of the catalyst in the reaction mixture significantly affects the efficiency and overall effectiveness of biodiesel production. Achieving an optimal catalyst concentration is vital for obtaining a high biodiesel yield. Insufficient catalyst may decelerate the reaction or lead to incomplete conversion of the feedstock, resulting in reduced yields. Conversely, an excess of catalyst does not necessarily enhance the reaction and might escalate costs without proportional benefits. Researchers and biodiesel producers frequently conduct experiments to pinpoint the ideal catalyst concentration when utilizing *H. crepitans* as feedstock. These studies take into account various factors, including reaction temperature, alcohol-to-feedstock ratio, and reaction time, to identify the optimal conditions for transesterification. Determining the appropriate catalyst concentration is critical for ensuring the efficiency of the reaction and the production of biodiesel with desired characteristics.

In the field of biodiesel production from *H. crepitans*, adjusting the catalyst concentration stands out as a pivotal aspect of process optimization. This adjustment contributes significantly to enhancing the overall efficiency and economic viability of the biodiesel production process.

6.4.4.6 Agitation Speed

The speed of agitation represents a crucial factor that influences the production of biodiesel using *H. crepitans* as feedstock. This parameter pertains to the velocity at which the reaction mixture undergoes stirring or agitation during the transesterification process, wherein fats or oils are converted into biodiesel. The role of agitation speed is pivotal in fostering thorough mixing of reactants, ensuring effective contact between the catalyst and feedstock. Achieving an optimal agitation speed is indispensable for attaining a high biodiesel yield by improving mass transfer, facilitating the reaction, and enhancing overall reaction kinetics. Insufficient agitation speed may lead to uneven mixing and slower reaction rates, potentially resulting in incomplete conversion and diminished biodiesel yields. Conversely, excessive agitation speed could give rise to issues such as foaming or emulsification, adversely affecting the overall efficiency of the biodiesel production process.

Researchers and biodiesel producers routinely conduct experiments to determine the ideal agitation speed when employing *H. crepitans* as feedstock. These studies take into account various factors, including the catalyst type, reaction temperature, alcohol-to-feedstock ratio, and reaction time, with the aim of optimizing the transesterification process. Identifying the appropriate agitation speed is critical for ensuring a uniform and efficient reaction, thereby contributing to the overall success of biodiesel production from *H. crepitans*.

The speed of agitation holds significance in the production of the final product, whether it be monoalkyl ester or biodiesel, as it facilitates the reaction by stirring the mixture of oil and catalyst. Conversely, elevated stirring speeds promote the formation of soap due to the opposing behavior of the transesterification reaction [40–42].

6.4.5 Potential Cost Challenges of Producing Biodiesel from *H. crepitans*

Harvesting and processing, transportation costs, and infrastructure costs are some of the potential cost challenges associated with producing biodiesel from *H. crepitans*.

Harvesting and processing: Harvesting and processing *H. crepitans* seeds can be time-consuming and costly. To extract the oil, the seeds must be collected, cleaned, dried, and pressed, which requires specialized equipment and expertise.

Transportation of *H. crepitans* seeds and biodiesel can also increase production costs, particularly if the production site is far from the processing plant or distribution center.

Infrastructure: Building and maintaining the infrastructure required for biodiesel production, such as processing plants and distribution networks, can be costly. This can raise the overall cost of *H. crepitans* biodiesel production.

6.5 Conclusion

The detailed exploration of the potential of *H. crepitans* as a source of biodiesel. This chapter presents an exposition on the subject, covering the diverse areas of application of the feedstock and the oil conversion process into biodiesel. It can be concluded from the study that *H. crepitans* has the potential to be an important source of biodiesel. The plant is abundant in many parts of the world, and the oil extracted from it has a relatively high percentage of fatty acids that can be used to produce biodiesel. Furthermore, the process of extracting oil from the plant is quite simple and can be considered using a variety of methods, including mechanical pressing and solvent extraction. However, the chapter also highlights some of the challenges associated with using *H. crepitans* as a source of biodiesel. One of the main challenges is that the oil extracted from the plant has a relatively high percentage of FFAs, which can make it difficult to produce biodiesel using conventional methods. In addition, the high viscosity of the oil can also pose challenges during the production process. Despite these challenges, the study suggests that *H. crepitans* has the potential to be a valuable source of biodiesel, particularly in regions where the plant is abundant. This study also provides valuable contributions to the collected works on the subject and underscores the importance of exploring new sources of renewable energy in the quest for a more sustainable future.

References

1 Oraegbunam, J.C., Ishola, N.B., Sotunde, B.A. et al. (2023). Sandbox oil biodiesel production modeling and optimization with neural networks and genetic algorithm. *Green Technol. Sustainability* 1 (1): 100007.

2 Mandegari, M., Farzad, S., and Görgens, J.F. (2018). A new insight into sugarcane biorefineries with fossil fuel co-combustion: techno-economic analysis and life cycle assessment. *Energy Convers. Manag.* 165: 76–91.

3 Farzad, S., Mandegari, M., and Görgens, J.F. (2020). Life cycle assessment of lignocellulosic biorefineries. In: *Recent Advances in Bioconversion of Lignocellulose to Biofuels and Value Added Chemicals Within the Biorefinery Concept*, 259.

4 Tabatabaei, M., Aghbashlo, M., Dehhaghi, M. et al. (2019). Reactor technologies for biodiesel production andprocessing: a review. *Prog. Energy Combust. Sci.* 74: 239–303.

5 Samuel, O.D., Okwu, M.O., Tartibu, L.K. et al. (2021). Modelling of *Nicotiana tabacum* L, oil biodiesel production: comparison of ANN and ANFIS. *Front. Energy Res.* 8: 612165.

6 Ogbu, I.M., Ajiwe, V.I.E., and Okoli, C.P. (2018). Performance evaluation of carbon-based heterogeneous acid catalyst derived from Hura crepitans seed pod for esterification of high FFA vegetable oil. *BioEnergy Res.* 11: 772–783.

7 Oladipo, B. and Betiku, E. (2020). Optimization and kinetic studies on conversion of rubber seed (*Hevea brasiliensis*) oil to methyl esters over a green biowaste catalyst. *J. Environ. Manag.* 268: 110705.

8 Falowo, O.A., Apanisile, O.E., Aladelusi, A.O. et al. (2021). Influence of nature of catalyst on biodiesel synthesis via irradiation-aided transesterification of waste cooking oil-honne seedoil blend: modeling and optimization by Taguchi design method. *Energy Convers. Manag.* X: 100119.

9 Gui, M.M., Lee, K., and Bhatia, S. (2008). Feasibility of edible oil vs, non-edible oil vs. waste edible oil as biodiesel feedstock. *Energy* 33 (11): 1646–1653.

10 M.J.Haas, T.A.Foglia, Alternate feedstocks and technologies for biodiesel production, in: *The Biodiesel Handbook*, Vol.2005, 2005, pp. 42–61.

11 Giakoumis, E.G., Rakopoulos, C.D., Dimaratos, A.M., and Rakopoulos, D.C. (2012). Exhaust emissions of diesel engines operating under transient conditions with biodiesel fuel blends. *Prog. Energy Combust. Sci.* 38 (5): 691–715.

12 Ajala, O.O., Oyelade, J.O., Oke, E.O. et al. (2023). A nonlinear autoregressive exogenous neural network (NARX-NN) model for the prediction of solvent-based oil extraction from Hura crepitans seeds. *Chemical Product and Process Modeling* 18 (4): 647–655.

13 Adepoju, T. and Eyibio, U. (2016). Comparative study of response surface methodology (RSM) and artificial neural network (ANN) on oil extraction from *Citrus sinensis* oilseed and its quality characterization. *Chem. Res. J.* 37–50.

14 Naylor, R.L. and Higgins, M.M. (2018). The rise in global biodiesel production: implications for food security. *Glob. Food Sec.* 16: 75–84. https://doi.org/10.1016/j.gfs.2017.10.004.

15 Abdulkadir, M.N., Amoo, I.A., and Adesina, A.O. (2013). Chemical composition of *Hura crepitans* seeds and antimicrobial activities of its oil. *Int. J. Sci. Res.* 2: 440–445.

16 Okolie, P.N., Uaboi-Egbenni, P.O., and Ajekwene, A.E. (2012). Extraction and quality evaluation of sandbox tree seed (*Hura crepitans*) oil. *World J. Agric. Sci.* 8: 359–365.

17 Akintunde, S.B., Obayopo, S.O., Adekunle, A.S. et al. (2021). Combustion and emission study of sandbox seed oil biodiesel performance in a compression ignition (CI) engine. *Energy Rep.* 7: 3869–3876.

18 Omilakin, R.O., Ibrahim, A.P., Sotunde, B., and Betiku, E. (2020). Process modeling of solvent extraction of oil from Hura crepitans seeds: adaptive neuro-fuzzy inference system versus response surface methodology. *Biomass Convers. Biorefinery* 1–14.

19 Ibrahim, A.P., Omilakin, R.O., and Betiku, E. (2019). Optimization of microwave-assisted solvent extraction of non-edible sandbox (*Hura crepitans*) seed oil: a potential biodiesel feedstock. *Renew. Energy* 141: 349–358.

20 Oraegbunam, J.C., Oladipo, B., Falowo, O.A., and Betiku, E. (2020). Clean sandbox (*Hura crepitans*) oil methyl esters synthesis: a kinetic and thermodynamic study through pH monitoring approach. *Renew. Energy* 160: 526–537.

21 Betiku, E., Okeleye, A.A., Ishola, N.B. et al. (2019). Development of a novel mesoporous biocatalyst derived from kola nut pod husk for conversion of Kariya seed oil to methyl esters: a case of synthesis. *Model. Optim. Stud. Catal. Lett.* 149 (7): 1772–1787.

22 Betiku, E. and Ishola, N.B. (2020). Optimization of sorrel oil biodiesel production by base heterogeneous catalyst from kola nut pod husk: neural intelligence-genetic algorithm versus neuro-fuzzy-genetic algorithm. *Environ. Prog. Sustain. Energy* 39 (4): e13393.

23 Ideris, F., Nomanbhay, S., Kusumo, F. et al. (2021). Optimization of ultrasound-assisted oil extraction from *Canarium odontophyllum* kernel as a novel biodiesel feedstock. *J. Clean. Prod.* 288: 125563.

24 Ugbaja, R.N., Simeon, A.O., Ugwor, E.I. et al. (2022). Biochemical appraisal of the underutilized Hura crepitans seed oil: functional and inflammatory responses in albino rats. *Grasas Aceites* 73 (3): e476.

25 Velazquez-González, M.Y., Loya-Olguin, J.L., Valdes-Garcia, Y.S. et al. (2022). *Hura crepitans*seeds for control of *Eimeria* spp. in lambs as an alternative to conventional therapies. *Vet. Sci.* 9 (9): 488.

26 Owojuyigbe, O.S., Larbie, C., Firempong, C.K. et al. (2022). *Hura crepitans* stem bark extract: a potential remedy to sub-acute liver damage. *J. Ethnopharmacol.* 284: 114768.

27 Temitayo, O.O., Vawe, H., Johnson, J. et al. (2022). Investigation on corrosion inhibition of *Hura crepitans* for 2101 duplex stainless steel in an acidic environment. *Adv. Sci. Technol. Res. J.* 16 (4): 56–63.

28 Oladipo, B. and Betiku, E. (2019). Process optimization of solvent extraction of seed oil from *Moringa oleifera*: an appraisal of quantitative and qualitative process variables on oil quality using D-optimal design. *Biocatal. Agric. Biotechnol.* 20: 101187.

29 Jisieike, C.F. and Betiku, E. (2020). Rubber seed oil extraction: effects of solvent polarity, extraction time and solid-solvent ratio on its yield and quality. *Biocatal. Agric. Biotechnol.* 24: 101522.

30 Okeleye, A.A. and Betiku, E. (2019). Kariya (*Hildegardia barteri*) seed oil extraction: comparative evaluation of solvents, modeling, and optimization techniques. *Chem. Eng. Commun.* 206: 1181–1198.

31 Mueanmas, C., Nikhom, R., Petchkaew, A. et al. (2019). Extraction and esterification of waste coffee grounds oil as non-edible feedstock for biodiesel production. *Renew. Energy* 133: 1414–1425.

32 Devi, V. and Khanam, S. (2019). Comparative study of different extraction processes for hemp (*Cannabis sativa*) seed oil considering physical, chemical and industrial-scale economic aspects. *J. Clean. Prod.* 207: 645–657.

33 Fernandes, S.S., Tonato, D., Mazutti, M.A. et al. (2019). Yield and quality of chia oil extracted via different methods. *J. Food Eng.* 262: 200–208.

34 Al Juhaimi, F., Uslu, N., Babiker, E.E. et al. (2019). The effect of different solvent types and extraction methods on oil yields and fatty acid composition of safflower seed. *J. Oleo Sci.* ess19131. https://doi.org/10.5650/jos.ess19131.

35 Mohammadpour, H., Sadrameli, S.M., Eslami, F., and Asoodeh, A. (2019). Optimization of ultrasound-assisted extraction of *Moringa peregrina* oil with response surface methodology and comparison with Soxhlet method. *Ind. Crop. Prod.* 131: 106–116.

36 Giakoumis, E.G. and Sarakatsanis, C. (2019). A comparative assessment of biodiesel cetane number predictive correlations based on fatty acid composition. *Energies* 12: 422.

37 Fonseca, J.M., Teleken, J.G., de Cinque Almeida, V., and da Silva, C. (2019). Biodiesel from waste frying oils: methods of production and purification. *Energy Convers. Manag.* 184: 205–218.

38 Quah, R.V., Tan, Y.H., Mubarak, N.M. et al. (2019). An overview of biodiesel production using recyclable biomass and non-biomass derived magnetic catalysts. *J. Environ. Chem. Eng.* 4: 103219.

39 Zhao, X., Qi, F., Yuan, C. et al. (2015). Lipase-catalysed process for biodiesel production: enzyme immobilization, process simulation and optimization. *Renew. Sust. Energ. Rev.* 44: 182–197.

40 National Biodiesel Board, U. S. A. (2015). Introduction to biodiesel and glossary of terms. In: *Bioenergy*, 37–40. Academic Press.

41 Mathiyazhagam, M. and Ganapathic, A. (2011). Factors affecting biodiesel production. *Res. Plant Biol.* 1 (2): 1–5.

42 Zhang, X.L., Yan, S., Tyagi, R.S., and Surampalli, R.Y. (2013). Biodiesel production from heterotrophic microalgae through transesterification and nanotechnology application in the production. *Renew. Sust. Energ. Rev.* 26: 216–223.

43 Olugbemide, A. D., Likozar, B., Oberlintner, A., Novak, U., and Ekebafe, L. (2022). Reaction kinetics modeling for dry anaerobic digestion of lignocellulosic Hura crepitans leaves. *Chemical Papers*, 76 (10): 6263–6269.

7

Sustainable Biodiesel Production from *Cyperus esculentus* (Tiger Nut)

7.1 Background of the Study

The excessive reliance on fossil fuels has precipitated both an energy crisis and environmental challenges. In response to these issues, biomass is emerging as a highly promising contender for new alternative energy sources, primarily due to its renewability, carbon neutrality, and substantial productivity. The global urgency to address these challenges has fueled concerted efforts to identify and harness biomass resources. Diverse conversion technologies are available for transforming various biomass sources into biofuel [1–3]. Biodiesel, composed of fatty acid methyl esters (FAME), stands out as a widely acknowledged and potential low-carbon substitute for diesel derived from fossil fuels. This recognition stems from its attributes such as low toxicity, eco-friendliness, and sourcing from nonedible plant and algal oils as well as animal fats [4]. As the world seeks sustainable energy alternatives, the exploration of biomass and its conversion into biodiesel represents a crucial step in mitigating environmental impact and addressing the pressing challenges posed by fossil fuel consumption. The multifaceted nature of biomass as a renewable resource contributes to its growing significance in the global pursuit of cleaner and more sustainable energy solutions.

C. esculentus, commonly known as underground chestnut, tiger nut, or walnut, and several other classical names, and belongs to the Cyperaceae family. They are known as a perennial herb [5]. Being a versatile crop, *C. esculentus*, presents a multitude of benefits, including high yield, superior quality, economic viability, and positive environmental attributes. Its underground components serve dual purposes, functioning as valuable feed or green manure, while its aerial parts are both fatty and edible [6]. Notably, the tubers of the common *C. esculentus* are notably rich in oil, rendering the plant highly valuable in the vegetable oil

Feedstocks for Sustainable Biodiesel Production: Characterization, Selection, and Optimization, First Edition. Chinwe P. Okonkwo and Modestus O. Okwu.

industry [7, 8]. Presently, based on statistics, the leading country where *C. esculentus* is cultivated the most is the United States, Canada is considered the second leading country, followed by Spain. Beyond these primary regions, it is extensively grown in countries such as Egypt, Ghana, China, Russia, and several others [9]. This widespread cultivation underscores the global recognition of *C. esculentus* for its diverse applications and economic significance. As a crop with a rich oil content, it holds substantial promise not only for its nutritional value but also as a valuable resource in the vegetable oil industry. The multifaceted utility of *C. esculentus* positions it as a noteworthy perennial herb with global importance.

C. esculentus, identified as an invasive and troublesome neophyte, has found extensive applications in the edible oil, sugar, grain, and feed industries [10–13]. Despite its potential for various industries, meeting the demand for the development of *C. esculentus*-related sectors remains a challenge. This is primarily due to the reliance on labor-intensive or semi-mechanized harvesting methods characterized by high labor intensity, elevated harvesting costs, and notably low efficiency. Complicating matters is the high sugar and moisture content of the tubers, making them prone to rot and mildew, thus posing challenges in storage and transportation [14]. Nonetheless, *C. esculentus* stands out as a significant and multifunctional crop with diverse economic contributions. Its stems and leaves offer utility in fabric weaving and serve as nutritious green feed. Extracting essential oils from the tuber system yields fragrant and lipid-rich substances, comprising dietary fiber, protein, starch, fat, including other valuable constituents. These extracted oils find application in the production of various products such as cooking oil, soymilk, snack foods, and more [15, 16]. The starch derived from *C. esculentus*, appearing as an off-white powder, bright white and odorless powder which further enhances its economic value. The yield of the starch varies within the range value of 14–37%, depending on the size and variety of *C. esculentus* tubers [17]. The granules of *C. esculentus* starch exhibit an architecture that is characterized by irregular regions of crystalline and noncrystalline, similar to starches from other sources [18]. This diversity in applications and valuable components underscores the economic potential and versatility of *C. esculentus* in various industries (Table 7.1).

To enhance the biofuel yield, there is a proposition emphasizing the significance of the lipid content in biomass. The lipid components of biomass are readily convertible when biomass is processed into biofuel. Consequently, the preference in biofuel production is often for biomass with a higher lipid content [19]. The oil extracted from *C. esculentus* contains 12 distinct types of fatty acids (FAs), with unsaturated FAs constituting approximately 84% of the total oil content [20].

Table 7.1 Proximate analysis (wt. %).

Proximate analysis	Unit
Ash	16.5
Moisture	1.5
Carbohydrate	45.2
Lipid	30.8
Volatile	76.5
Protein	3.7
Fixed carbon	2.3

Figure 7.1 Pictorial view of *Cyperus esculentus*.

Understanding the lipid metabolism in *C. esculentus* becomes pivotal in meeting the escalating demand for vegetable oil as it plays a crucial role in addressing the ever-increasing need for this essential resource. This knowledge becomes instrumental in optimizing processes for obtaining biofuel from biomass, aligning with the broader goal of sustainable energy production (Figure 7.1).

7.2 Review of Literature

Table 7.2 shows a summarized version of the literature

Table 7.2 Review of related research conducted using Cyperus esculentus in diverse applications.

Citations	Authors	Title	Applications	Year
[14]	Qu et al.	Design and test of a crawler-type tiger-nut combine harvester	FP	2023
[10]	Xia et al.	Optimization and experiment of hot air drying process of *Cyperus esculentus* seeds	FP	2023
[21]	Liu et al.	Preparation and physicochemical properties of *Cyperus esculentus* starch from its tubers using ultrasound-assistedalkali method	FP	2023
[22]	De Ryck et al.	Genetic and morphological variation of Belgian *Cyperus esculentus* L. clonal populations and their significance for integrated management	GP	2023
[23]	Zou et al.	Analysis of Lhc family genes reveals development regulation and diurnal fluctuation expression patterns in *Cyperus esculentus*, a Cyperaceae plant	GP	2023
[19]	Tian et al.	Lewis acid-catalyzed in situ transesterification/esterification of tiger nut in sub/supercritical ethanol: an optimization study	B	2019
[7]	Zhu et al.	Analysis of oil synthesis pathway in *Cyperus esculentus* tubers and identification of oleosin and caleosin genes	GP	2023
[24]	Alade et al.	Assessing the safety of tiger nut drinks produced from *Cyperus esculentus* L. seeds sold in ota	FP	2022
[25]	Feys et al.	Impact of electrocution on shoot and tuber vitality of yellownutsedge (*Cyperus esculentus*)	GP	2023
[26]	Raji	Optimization of roasting-assistedoil extraction from tiger nut using response surface methodology	FP	2023

AA, antimicrobial activity; B, biodiesel; FP, food products; GP, general purpose; Ph, pharmaceuticals; WT, water treatment.

7.3 Biodiesel Production Process

Previous research has demonstrated the extraction of oil from tiger nuts with the use of organic solvents like petroleum benzene, *n*-hexane, and petroleum ether. The subsequent generation of biofuel usually includes the transesterification based on catalyzed alkali of the obtained tigernut oil. The conventional biofuel production process from tiger nuts follows a two-stage approach, encompassing the initial stage of oil extraction and the subsequent transesterification process. The production process is further divided into:Harvesting Tiger Nuts; Cultivation and harvesting mature tiger nuts from the *C. esculentus* plant; Cleaning and Drying: Clean the harvested tiger nuts to remove impurities. Allowing cleaned tiger nuts to dry thoroughly; Oil Extraction: Extract oil from the dried tiger nuts using organic solvents such as petroleum benzene, *n*-hexane, or petroleum ether; Transesterification: Conduct transesterification (alkali-catalyzed) of the tiger nut oil extracted. This involves reacting the oil with an alcohol (such as methanol or ethanol) in the presence of a catalyst (alkali) to produce biodiesel (FAME) and glycerol; Separation and Purification: Separate the biodiesel from the glycerol and other by-products.Purify the biodiesel through processes such as washing to remove impurities; Drying Biodiesel: Dry the biodiesel to eliminate any remaining water content; Quality Testing: Conduct quality tests on the biodiesel to ensure it meets the required standards and specifications; Storage: Store the produced biodiesel in appropriate containers under suitable conditions to maintain its quality; By-Product Utilization: Explore potential uses for the by-products generated during the process, such as glycerol; Scaling Up: Consider scaling up production based on the desired quantity of biodiesel.This step-by-step process provides a general outline for the production of biodiesel from tiger nuts. It is important to note that specific details and conditions may vary based on the scale of production, equipment used, and other factors. In addition, adherence to safety and environmental regulations is crucial throughout the entire process. A comprehensive representation of this production process is illustrated in Figure 7.2, detailing the sequential steps involved in the transformation of tiger nut oil into biofuel. This procedure is pivotal in elucidating the intricate process of biofuel production and serves as a foundational reference for further exploration and optimization of tiger nut-based biofuel technologies.

Esterification and sub/supercritical transesterification have emerged as highly promising processes in biodiesel production. Alcohols operating under sub- or supercritical conditions possess distinctive characteristics, including low viscosity, high diffusion coefficient, and a low dielectric constant. These unique properties equip them with the capability to effectively surmount mass transfer constraints, allowing for the rapid completion of transesterification or esterification reactions in a significantly condensed timeframe [27]. The utility of alcohols under sub/supercritical conditions is particularly advantageous in overcoming challenges related to mass transfer, making them valuable choices for transesterification or esterification processes.

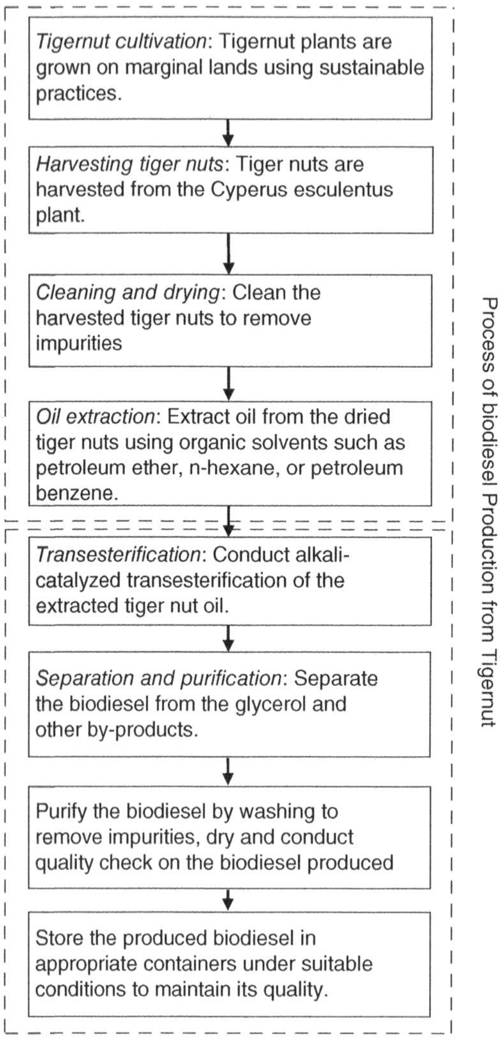

Figure 7.2 Flowchart showing the step-by-step process of producing biodiesel from tigernut.

The prevalent use of methanol and ethanol in the production of biofuel has been extensively documented, especially in the context of processing edible oils, used cooking oil, and lipid-rich algae biomass [28]. The versatility of these alcohols in various feedstock scenarios underscores their applicability and effectiveness in biodiesel production. The reported successes in utilizing methanol and ethanol highlight their importance as key components in the synthesis of biofuels from diverse sources, further contributing to the exploration of sustainable and efficient alternatives to traditional fossil fuels.

7.4 Factors Affecting Biodiesel Production with TigerNut as Feedstock

The production of biodiesel from tiger nuts as feedstock is influenced by several factors. These factors can impact the yield, quality, and efficiency of the biodiesel production process. Here are some key factors: the presence of water, the molar ratio of alcohol, reaction temperature, catalyst, reaction time, agitation rate, and amount of free FAs in the oil samples. All the factors mentioned have a significant impact on the reaction.

7.4.1 Effect of Alcohol Molar Ratio

The impact of the alcohol molar ratio on the production of biodiesel using tiger nuts as feedstock is a pivotal consideration. The concentration of alcohol plays a critical role in influencing the conversion of fats into esters during the biodiesel production process. An excess amount of alcohol accelerates this conversion, leading to a more rapid transesterification reaction. Consequently, the biodiesel yield tends to increase up to a certain concentration, indicating a positive correlation between alcohol concentration and biodiesel production efficiency.

The specific molar ratio of alcohol required can vary depending on the catalyst utilized in the process. Notably, the choice of catalyst becomes crucial, especially in scenarios where oil samples exhibit high levels of free fatty acids (FFA). In such instances, transesterification reactions may not be responsive to alkali catalysts. Instead, acid catalysts prove effective in catalyzing the reaction, necessitating a higher alcohol content than alkali catalysts. The key advantage lies in the ability of acid catalysts to tolerate water and FFA content present in oil samples, ensuring a robust reaction even under challenging conditions [29].

Understanding the nuanced influence of the molar ratio of alcohol is fundamental in optimizing biodiesel production processes from tiger nuts. The choice of catalyst and its compatibility with the composition of oil samples, particularly concerning FFA levels, underscores the need for precision in alcohol concentration for achieving efficient and high-yield biodiesel production.

7.4.2 Effect of Water and FFA Contents

The presence of FFA exerts a notable influence on the production of biodiesel utilizing tiger nuts as feedstock. FFA are a crucial factor that can significantly impact the efficiency and outcome of the biodiesel production process. When the levels of FFA are elevated in the oil samples derived from tiger nuts, the conventional alkali catalysts employed in transesterification reactions may prove less effective. In cases where oil samples exhibit higher FFA content, it becomes imperative to consider alternative catalysts, such as acid catalysts, to catalyze the reaction effectively. Unlike alkali catalysts, acid catalysts demonstrate a higher tolerance for water and FFA, ensuring a more robust reaction under challenging conditions. The choice of catalyst and its compatibility with the specific composition of tiger nut oil samples, particularly with regard to FFA concentrations, becomes a critical aspect in determining the success of the biodiesel production process.

Furthermore, addressing the impact of FFA necessitates careful consideration of the transesterification reaction conditions, including reaction temperature, molar ratio of alcohol to oil, and reaction time. By understanding and optimizing these parameters, biodiesel producers can enhance the overall efficiency and yield of the production process, particularly when dealing with tiger nuts characterized by varying levels of FFA. The presence of FFA and water typically exerts an adverse influence on transesterification, leading to the formation of soap and the consumption of the catalyst, thereby diminishing its efficacy. In addition, water and FFA contribute to the reduction of methyl ester during the transesterification process [29].

7.4.3 Reaction Time

The duration of the reaction period significantly influences the production of biodiesel from tiger nuts. As the reaction time increases, there is a notable enhancement in the conversion rate of fatty acid esters. Initially, when the oil and alcohol are mixed, the reaction commences at a gradual pace due to the need for thorough dispersion. Subsequently, the reaction gains momentum and proceeds at an accelerated rate. However, it is crucial to note that prolonged reaction times can lead to a reduction in the overall yield of the end product, namely, biodiesel. This is attributed to the reversible nature of transesterification, wherein esters can be lost, and soap formation may occur over extended reaction periods. The intricate balance between reaction time and biodiesel yield underscores the need for optimization in order to achieve an efficient and economically viable production process [29].

7.4.4 Reaction Temperature

The production of biodiesel from tiger nuts is notably affected by the reaction temperature, a critical factor in determining overall yield. Elevated reaction temperatures exhibit a noteworthy impact, illustrated by the reduced viscosity of oils. This phenomenon at higher temperatures contributes to an accelerated reaction rate, consequently leading to a reduction in reaction time. The interplay between temperature and viscosity plays a pivotal role in optimizing the efficiency of the transesterification process. However, it is essential to navigate the delicate balance of reaction temperatures. Beyond the optimal level, higher reaction temperatures were found to induce an unfavorable outcome. Specifically, the accelerated saponification of triglycerides at elevated temperatures results in a counterproductive decrease in biodiesel yield. This nuanced relationship underscores the importance of precise temperature control to achieve an optimal balance between reaction kinetics and biodiesel production efficiency [29]. In summary, understanding the intricate influence of reaction temperature on the biodiesel production process from tiger nuts allows for the implementation of controlled conditions, ensuring both efficiency and maximum yield.

7.4.5 Catalyst Concentration

The concentration of the catalyst plays a pivotal role in shaping the biodiesel production process from tiger nuts. This influence is one of the reasons for adopting the practice of premixing the catalyst with methanol before its introduction to the oil or fats. This preparatory step is crucial in achieving controlled catalytic conditions and optimizing the ensuing transesterification reaction.

The impact of catalyst concentration on the conversion of triglycerides into biodiesel is noteworthy. A direct correlation is observed, wherein an increase in catalyst concentration with oil samples results in an enhanced conversion of triglycerides into fatty acid esters. This positive relationship underscores the catalytic role in facilitating the transformation of raw materials into the desired biodiesel product.

Conversely, insufficient catalyst concentration leads to a scenario where triglycerides undergo only partial transformation into fatty acid esters, thereby affecting the overall yield. Striking the right balance is essential, and empirical studies have revealed that an optimal catalyst concentration, particularly with NaOH at 1.5 wt. %, yields the highest biodiesel product. However, exceeding this concentration has been demonstrated to have adverse effects on the final product yield, emphasizing the need for precise control over catalyst concentrations during the biodiesel production process [29].

In essence, understanding the nuanced relationship between catalyst concentration and biodiesel yield provides valuable insights for achieving optimal conditions and ensuring the efficiency of the production process from tiger nuts.

7.4.6 Agitation Speed

The speed of agitation stands as a critical factor influencing the production of biodiesel from tiger nuts. The agitation speed plays a pivotal role in expediting the transformation of the raw materials, oil, and catalyst mixture into the final product, either mono alkyl ester or biodiesel. The significance of agitation lies in its ability to promote thorough mixing and enhance the interaction between the oil and catalyst, facilitating a more efficient reaction.

However, it is crucial to navigate the delicate balance of agitation speed, as excessive stirring can lead to unintended consequences. Specifically, higher stirring speeds have been observed to encourage the formation of soap. This phenomenon is a result of the transesterification reaction exhibiting opposite behavior under heightened agitation conditions [29].

Understanding the intricate relationship between agitation speed and the biodiesel production process is vital for process optimization. Striking the right balance ensures a uniform reaction and helps avoid undesired outcomes such as soap formation, contributing to the overall success of biodiesel production from tiger nuts.

7.5 Benefits, Challenges, and Future Prospects of *Cyperus esculentus*

7.5.1 Benefits of *Cyperus esculentus*

Tiger nuts, although classified as root tubers, exhibit characteristics of both tubers and nuts, sharing a chemical relation. The multifaceted benefits of *Cyperus esculentus*, a significant agricultural product, encompass various domains:

Health benefits and medicinal properties: Tiger nuts have been established as a reliable source of bioactive compounds with diverse health benefits. Enriched with flavones, minerals, essential FA, and vitamins C, D, and E, among other nutrients, they contribute to enhancing memory retention and maintaining proper nerve coordination. These properties, in turn, positively impact the immune system [18].

Food applications: C. esculentus finds versatile applications in the culinary domain. The incorporation of tiger nuts with wheat flour results in the production of fresh composite pasta. The addition of *C. esculentus* powder significantly elevates the fiber, fat, and mineral content of the pasta product. Introducing 30% tiger nut powder as a replacement ensures a higher fiber content compared to products with a mere 3% incorporation. Strengthening the gluten protein structure becomes essential for reducing cooking loss and improving pasta firmness. Furthermore, roasted *C. esculentus* flour occasionally serves as an alternative ingredient in the creation of biscuits and various other bakery goods [30]. The dual nature of tiger nuts, combining the characteristics of tubers and nuts, coupled with their nutritional richness, positions *C. esculentus* as a valuable resource with applications spanning health and culinary settings.

Starch replacement with enhanced properties: Cyperus esculentus starch serves as an excellent substitute for traditional starch due to its comparable viscosity properties. Noteworthy characteristics include cohesion, clarity, softness, and adhesiveness. The starch from *C.esculentus* demonstrates remarkable compatibility and binding capabilities, coupled with advantageous features like robust swelling power, high-density water absorption, and effective binding. When utilized as an oxidant and binder in biofuel production, it showcases outstanding performance in terms of binder activity, mechanical strength, fluidity, and release properties [30–34].

High oil yield for biodiesel production: With an oil content of approximately 36%, *Cyperus esculentus* presents itself as a promising crop for biodiesel production [35]. This high oil yield contributes to the feasibility and efficiency of biofuel manufacturing processes.

Environmental benefits for a cleaner and healthier environment: Embracing *Cyperus esculentus* offers substantial environmental advantages that translate into a healthier population. The cultivation and utilization of this crop result in reduced emissions of nontoxic materials, thereby lowering the risk associated with respiratory problems and cancer. Notably, there are zero emissions of hazardous gases such as sulfur oxide (SO) and carbon monoxide (CO), leading to a diminished risk of global warming. The inherent qualities of *Cyperus esculentus*, being nontoxic and biodegradable while lacking sulfur and aromatics, contribute to efficient combustion. In addition, the high oxygen content further enhances the overall environmental friendliness of this crop [36–38]. The implementation of *Cyperus esculentus* is possible in various applications, especially in biofuel production, and thus emerges as a sustainable choice with multifaceted benefits.

Reducing greenhouse gas emissions through Cyperus esculentus biofuels: The incorporation of biofuels extracted from residual crops, particularly *Cyperus esculentus*, is in line with government incentives designed to lower fuel prices. This strategic initiative not only addresses the economic concerns of diverse sectors, spanning factories, automobiles, and agriculture but also plays a pivotal role in curbing greenhouse gas emissions [39]. The promotion of biofuel usage by authorities signifies a commitment to fostering a sustainable energy system that resonates with overarching environmental conservation objectives.

Addressing environmental challenges amidst urban development: The contemporary challenges presented by rapid urban development and burgeoning industrial enterprises, coupled with the exponential growth in the human population, necessitate urgent and effective environmental conservation measures. *Cyperus esculentus* emerges as a significant asset in this context, poised to tackle two critical challenges – the extensive generation of waste and the degradation of ecosystems. The by-products derived from cultivating *Cyperus esculentus*, when strategically employed for bioenergy production, offer a viable and sustainable resolution to these pressing challenges.

Moreover, as the global demand for energy continues its upward trajectory, bioenergy sourced from crops like *Cyperus esculentus* stands out as an ultimate and versatile solution. While the implementation of bioenergy necessitates meticulous consideration of energy transformation processes, the manifold benefits it offers in terms of environmental conservation make it an avenue well worth exploring [40–42]. The integration of *Cyperus esculentus* into bioenergy strategies thus represents a proactive contribution toward a more sustainable and environmentally conscious approach to meeting the escalating energy demands of our society.

7.5.2 Limitations and Challenges

Elevated labor costs: Presently, the production of biodiesel derived from tiger nuts faces a notable hurdle due to higher production expenses compared to petroleum-based diesel. This cost disparity is primarily attributed to the substantial expenses associated with processing and refining the tiger nut oil [29]. Addressing this limitation requires exploring more cost-effective methods for oil extraction and refining processes to enhance the economic viability of tiger nut-based biodiesel.

Extensive storage requirements: The production, storage, and distribution of biomass, including tiger nut-based biodiesel, necessitate a distinct infrastructure compared to conventional petroleum-based diesel. The unique storage requirements pose a significant challenge, as it demands substantial investments to establish the infrastructure essential for the effective support of tiger nut-based biodiesel utilization. This includes the construction of specialized storage facilities and distribution networks, adding to the overall operational costs.

Environmental and social impacts: The escalating demand for agricultural land, particularly in regions with dry climates coupled with high water requirements, poses environmental and social challenges. The increased cultivation of crops for biodiesel production can lead to deforestation, soil degradation, and disruptions to the natural life cycle. These consequences may result in adverse environmental impacts and community displacement, necessitating careful consideration of sustainable practices and land use planning to mitigate potential issues.

Vegetable oil conversion process: The conversion of vegetable oils, including tiger nut oil, into methyl esters is essential to reduce the viscosity of biofuel, preventing interference with the efficient operation of engines. This process involves treating vegetable oil with methyl alcohol in the presence of a catalyst. While this method is crucial for optimizing the functionality of biodiesel, continuous research, and development efforts are necessary to refine and streamline the conversion process for enhanced efficiency and reduced costs.

In overcoming these challenges, the tiger nut-based biodiesel industry can progress toward a more economically viable, environmentally sustainable, and socially responsible future. This necessitates a comprehensive approach involving technological innovation, infrastructure development, and the implementation of responsible agricultural and production practices.

7.5.3 Opportunities and Prospects of *Cyperus esculentus* for Biodiesel Production

High oil content: *Cyperus esculentus*, commonly known as tigernuts, boasts a substantial oil content of approximately 36%, making it a promising feedstock for biodiesel production. This high oil yield presents a significant opportunity for the development of a sustainable and efficient source of biodiesel.

Reduced greenhouse gas emissions: The use of tigernut derived biodiesel can contribute to a reduction in greenhouse gas emissions. By replacing traditional diesel with a biofuel alternative, there is a potential to mitigate the environmental impact of transportation and industrial activities, aligning with global efforts to combat climate change.

Agricultural diversification: Cultivating tigernuts for biodiesel production provides an opportunity for agricultural diversification. Farmers can benefit from growing a versatile crop that serves dual purposes – producing biodiesel and providing tigernuts for various culinary applications. This diversification can enhance agricultural resilience and contribute to food and energy security.

Sustainable land use practices: Tigernuts are well-suited to grow in diverse climates and require relatively low inputs, contributing to sustainable land use practices. The crop's adaptability to various environmental conditions allows for cultivation on marginal lands, minimizing the competition for prime agricultural areas and promoting responsible land use.

Job creation and economic development: The establishment of tigernut biodiesel production facilities can lead to job creation and economic development in regions where the crop is cultivated. The entire biodiesel production value chain, from farming and harvesting to processing and distribution, has the potential to stimulate local economies and create employment opportunities.

Technology and process optimization: Ongoing research and development in the field of tigernut biodiesel production offer opportunities for technological advancements and process optimization. Innovations in extraction and refining techniques can enhance efficiency, reduce production costs, and make tigernut biodiesel a more competitive and economically viable alternative.

Government incentives and policies: Governments may introduce incentives and policies to support the development and adoption of biodiesel derived from tigernuts. This could include subsidies, tax benefits, or regulatory frameworks that encourage the use of sustainable and renewable biofuels, creating a favorable environment for tiger nut biodiesel in the energy market.

International market potential: As global awareness of sustainable energy solutions grows, tiger nut biodiesel has the potential to tap into the international market. Countries and industries seeking environmentally friendly alternatives may create opportunities for the export of tiger nut biodiesel, contributing to economic growth on a broader scale.

In summary, the opportunities and prospects of *Cyperus esculentus*, or tigernuts, for biodiesel production are multifaceted. With the right investments, research, and supportive policies, tigernut biodiesel could play a pivotal role in meeting the world's energy needs while promoting environmental sustainability and economic development.

7.6 Conclusion

In conclusion, the exploration of *C. esculentus*, particularly tiger nuts, as a biodiesel feedstock holds substantial promise for ushering in a sustainable and environmentally friendly alternative to conventional fossil fuels. The multifaceted nature of this plant, encompassing its agricultural implications, high lipid content, and adaptability to diverse climates and soils, underscores its potential as a valuable resource in the pursuit of greener energy solutions. The current discourse has shed light on the benefits and detriments of employing tiger nuts in biodiesel production, emphasizing their capacity to yield high-quality oil for sustainable fuel. Ongoing research and development initiatives focused on refining production processes signify a commitment to enhancing the efficiency and overall quality of biodiesel derived from *C. esculentus*. As the world grapples with the dual challenges of an energy crisis and environmental degradation, the significance of renewable and eco-friendly energy sources cannot be overstated. The versatility of tiger nuts, not only as a biodiesel feedstock but also as a source of food and animal husbandry, further amplifies its potential impact on sustainable agriculture and energy. In moving forward, continued efforts to optimize the role of *C. esculentus* in biodiesel production will be crucial. This entails further research into hybrid feedstock, cultivation practices, extraction methods, and technological advancements that contribute to the economic viability and scalability of tiger nut-derived biodiesel. Ultimately, embracing tiger nuts as a biodiesel feedstock aligns with the global push toward sustainable practices and offers a promising avenue for reducing our reliance on finite fossil fuels. By leveraging the inherent benefits of this resilient and versatile crop, effective contributions can be made toward a greener and more sustainable future.

References

1 Chen, W., Lin, B., Huang, M., and Chang, J. (2015). Thermochemical conversion of microalgal biomass into biofuels: a review. *Bioresour. Technol.* 184: 314–327.

2 Williams, C.L., Westover, T.L., Emerson, R.M. et al. Sources of biomass feedstock variability and the potential impact on biofuels production. *Bioenergy Res.* 9, 1–14.

3 Yuan, T., Tahmasebi, A., and Yu, J. (2015). Comparative study on pyrolysis of lignocellulosic and algal biomass using a thermogravimetric and a fixed-bed reactor. *Bioresour. Technol.* 175: 333–341.

4 Gardy, J., Nourafkan, E., Osatiashtiani, A. et al. (2019). A core-shell SO4/ mg-Al-Fe3O4 catalyst for biodiesel production. *Appl. Catal., B* 259: 118093.

5 Djikeng, F.T., Djikeng, C.F.T., Womeni, H.M. et al. (2022). Effect of different processing methods on the chemical composition, antioxidant activity and lipid quality of tiger nuts (*Cyperus esculentus*). *Appl. Food Res.* 2 (2): 100124. https://doi.org/10.1016/j.afres.2022.100124.

6 Adewuyi, A., Otuechere, C.A., Oteglolade, Z.O. et al. (2015). Evaluation of the safety profile and antioxidant activity of fatty hydroxamic acid from underutilized seed oil of Cyperus esculentus. *J. Acute Disease* 4 (3): 230–235. https://doi.org/10.1016/j.joad.2015.04.010.

7 Zhu, Y., Wang, Y., Wei, Z. et al. (2023). Analysis of oil synthesis pathway in *Cyperus esculentus* tubers and identification of oleosin and caleosin genes. *J. Plant Physiol.* 284: 153961.

8 Li, T., Sun, Y., Chen, Y. et al. (2022). Characterisation of two novel genes encoding Δ9 fatty acid desaturases (CeSADs) for oleic acid accumulation in the oil-rich tuber of *Cyperus esculentus*. *Plant Sci.* 319: 111243. https://doi.org/10.1016/j.plantsci.2022.111243.

9 Zhang, A., Wang, L., Song, T. et al. (2022). Effects of high pressure homogenization on the structural and emulsifying properties of a vegetable protein: *Cyperus esculentus* L. *LWT153* 112542. https://doi.org/10.1016/j.lwt.2021.112542.

10 Xia, X., Jin, Y., Zhao, H. et al. (2023). Optimization and experiment of hot air drying process of *Cyperus esculentus*seeds. *Agriculture* 13 (3): 617.

11 Liu, H.Y., Wang, L.H., and Liu, Y.Y. (2021). Development status of olea bean products. *Farm Prod. Process.* 14: 97–99.

12 Guo, T.T., Wan, C.T., Huang, F.H. et al. (2021). Research progress on the main nutritional components and physiological functions of tiger nut (*Cyperus esculentus* L.). *Chin. J. Oil Crop Sci.* 43: 1174–1180.

13 Chen, P., Guo, P., Yang, F. et al. (2022). Drying characteristics and quality of wet peeling *Cyperus esculentus* under different drying processes. *Chin. J. Oil Crop Sci.* 44: 1117–1122.

14 Qu, Z., Han, M., Lv, Y. et al. (2023). Design and test of a crawler-type tiger-nut combine harvester. *Agriculture* 13 (2): 277.

15 Adelakun, S.A., Akintunde, O.W., and Ogunlade, B. (2021). Fluoride-induced testicular degeneration and sperm quality deteriorations: salutary role of *Cyperus esculentus*tubers (tiger nut) extract inanimal model. *Rev. Int. Androl.* 19 (3): 201–212. https://doi.org/10.1016/j.androl.2020.01.003.

16 Cui, Q., Wang, L., Wang, G. et al. (2021). Ultrasonication effects on physicochemical and emulsifying properties of Cyperus esculentusseed (tiger nut) proteins. *LWT142* 110979. https://doi.org/10.1016/j.lwt.2021.110979.

17 Liu, H., Yan, Y., Liu, X. et al. (2020). Effects of various oil extraction methods on the gelatinization and retrogradation properties of starches isolated from tigernut (*Cyperus esculentus*) tuber meals. *Int. J. Biol. Macromol.* 156: 144–152.

18 Yu, Y., Lu, X., Zhang, T. et al. (2022). Tiger nut (*Cyperus esculentus* L.): nutrition, processing, function and applications. *Foods* 11 (4): 601. https://doi.org/10.3390/foods11040601.

19 Tian, Y., Wang, F., Xie, L.F. et al. (2019). Lewis acid-catalyzed in situ transesterification/esterification of tigernut in sub/supercritical ethanol: an optimization study. *Fuel* 245: 96–104.

20 Glew, R.H., Glew, R.S., Chuang, L.T. et al. (2006). Amino acid, mineral and fatty acid content of pumpkin seeds (*Cucurbita* spp) and *Cyperus esculentus* nuts in the Republic of Niger. *Plant Foods Hum. Nutr.* 61: 51–56.

21 Liu, Z., Tian, S., Lv, C., and Chen, Z. (2023). Preparation and physicochemical properties of *Cyperus esculentus*starch from its tubers using ultrasound assisted alkali method. *BioResources* 18 (1).

22 De Ryck, S., Reheul, D., De Riek, J. et al. (2023). Genetic and morphological variation of Belgian *Cyperus esculentus* L. clonal populations and their significance for integrated management. *Agronomy* 13 (2): 572.

23 Zou, Z., Xiao, Y., Zhang, L., and Zhao, Y. (2023). Analysis of Lhc family genes reveals development regulation and diurnal fluctuation expression patterns in *Cyperus esculentus*, a Cyperaceae plant. *Planta* 257 (3): 1–16.

24 Alade, M.B., Ahuekwe, E.F., Adeyemi, A.O., and Nwinyi, O.C. (2022). Assessing the safety of tiger nut drinks produced from *Cyperus esculentus* L. seeds sold in ota. In: *Bioenergy and Biochemical Processing Technologies: Recent Advances and Future Demands*, 163–174. Cham: Springer International Publishing.

25 Feys, J., De Cauwer, B., Reheul, D. et al. (2023). Impact of electrocution on shoot and tuber vitality of yellow nuts edge esculentus. *Agriculture* 13 (3): 696.

26 Raji, A.O. (2023). Optimization of roasting assisted oil extraction from tiger nut using response surface methodology. *Chem. Afr.* 1–10.

27 Jin, B., Duan, P., Xu, Y. et al. (2014). Lewis acid-catalyzed in situ transesterifification/esterifification of microalgae in supercritical ethanol. *Bioresour. Technol.* 162: 341–349.

28 Palacios-Nereo, F.J., Olivares-Carrillo, P., Rios, A.P., and Quesada-Medina, J. (2016). High-yield noncatalytic supercritical transesterifification of soybean oil to biodiesel induced by gradual heating in a batch reactor. *J. Supercrit. Fluids* 111: 135–142.

29 Said, M.S.M., Ghani, W.A.W.A.K., Tan, H.B., and Ng, D.K.S. (2020). Prediction and optimisation of syngas production from air gasification of Napier grass via stoichiometric equilibrium model. *Energy Convers. Manage.: X* 100057. https://doi.org/10.1016/j.ecmx.2020.100057.

30 Albors, A., Raigon, M., García-Martinez, M., and Martín-Esparza, M. (2016). Assessment of techno- functional and sensory attributes of tiger nut fresh egg tagliatelle. *LWT* 74: 183–190. https://doi.org/10.1016/j.lwt.2016.07.047.

31 Akonor, P.T., Tortoe, C., Oduro-Yeboah, C. et al. (2019). Physicochemical, microstructural, and rheological characterization of tigernut (*Cyperus esculentus*) starch. *Int. J. Food Sci.* 2019: 3830651. https://doi.org/10.1155/2019/3830651.

32 Fathoni, A., Hartati, N., Wahyuni, H.F. et al. (2020). Characterization of cassava starch and its potential for fermentable sugar production. *IOP Conf. Ser.: Earth Environ. Sci* 439: 012024. https://doi.org/10.1088/1755-1315/439/1/012024.

33 Liu, X.-X., Liu, H.-M., Li, J. et al. (2019). Effects of various oil extraction methods on the structural and functional properties of starches isolated from tigernut (*Cyperus esculentus*) tuber meals. *Food Hydrocolloids* 95: 262–272. https://doi.org/10.1016/j.foodhyd.2019.04.044.

34 Olayemi, O.J., Apeji, Y.E., and Isimi, C.Y. (2020). Formulation and evaluation of *Cyperus esculentus* (tiger nut) starch-alginate microbeads in the oral delivery of ibuprofen. *J. Pharm. Innovation* 1: 1–10. https://doi.org/10.1007/s12247-020-09509-2.

35 Gambo, A. and Da'u, A. (2014). Tiger nut (*Cyperus esculentus*): composition, products, uses and health benefits – a review. *Bayero J. Pure Appl. Sci.* 7 (1): 56–61.

36 Malode, S.J., Prabhu, K.K., Mascarenhas, R.J. et al. (2021). Recent advances and viability in biofuel production. *Energy Convers. Manage.: X* 10: 100070.

37 Calvey, C.H., Su, Y.-K., Willis, L.B. et al. (2016). Nitrogen limitation, oxygen limitation, and lipid accumulation in Lipomycesstarkeyi. *Bioresour. Technol.* 200: 780–788. https://doi.org/10.1016/j.biortech.2015.10.104.

38 Ahmad Dar, R., Ahmad Dar, E., Kaur, A., and Gupta, P.U. (2018). Sweet sorghum-a promising alternative feedstock for biofuel production. *Renewable Sustainable Energy Rev.* 82: 4070–4090. https://doi.org/10.1016/j.rser.2017.10.066.

39 Lucia, U. and Grisolia, G. (2019). Time: a constructal viewpoint & its consequences. *Sci. Rep.* 9: 1–7.

40 Lucia, U. and Grisolia, G. (2018). Cyanobacteria and microalgae: thermoeconomic considerations in biofuel production. *Energies* 11: 156. https://doi.org/10.3390/en11010156.

41 Ni, M., Xu, Y., Wang, C. et al. (2020). A novel thermo-controlled acetaminophen electrochemical sensor based on carboxylated multi-walled carbon nanotubes and thermosensitive polymer. *Diamond Relat. Mater.* 107: 107877. https://doi.org/10.1016/j.diamond.2020.107877.

42 Tumminia, G., Guarino, F., Longo, S. et al. (2020). Grid interaction and environmental impact of a net zero energy building. *Energy Convers. Manage.* 203: 112228. https://doi.org/10.1016/j.enconman.2019.112228.

8

Sustainable Biodiesel Production Using *Cucurbita pepo* as a Feedstock

8.1 Background of the Study

The imperative need for alternative energy sources to mitigate the adverse effects of petroleum depletion, global warming, and environmental degradation resulting from oil exploration underscores the significance of developing sustainable options. In response to these challenges, biodiesel has emerged as a renewable fuel with notable advantages – it is nontoxic, safer to handle, biodegradable, requires no engine modifications, and diminishes dependence on foreign oil imports [1]. Representing a renewable and sustainable alternative to fossil fuels, biodiesel can be derived from various sources, including vegetable oils and animal fats. The exploration of diverse feedstocks such as seeds, nuts, leaves, wood, and even tree bark has been pivotal in identifying suitable oil sources for biodiesel production [2].

Within this context, the Cucurbitaceae family, recognized for its significance in providing humans with edible products and essential plant foods, has garnered attention for biodiesel production. This family, particularly the *Cucurbita pepo* (*C. pepo*) species, stands out due to its high oil yield [3]. According to findings by Bikash et al. [3], pumpkin seeds, a part of the Cucurbitaceae family, exhibit an impressive oil content of approximately 48%, surpassing other seed oils commonly considered for biodiesel production. Notably, Hernández-Santos et al. [4] reported a substantial oil yield of about 62.5% from pumpkin seeds, employing an ultrasonic extraction method. Pumpkin seeds, constituting 10–25% of the total fruit mass [5], are abundant, with 20 kg of pumpkins yielding approximately 1.4 kg of seeds [6]. The global production of pumpkins, as reported by the Food and Agriculture Organization in 2018, reached 27 milliontons [7].

Feedstocks for Sustainable Biodiesel Production: Characterization, Selection, and Optimization, First Edition. Chinwe P. Okonkwo and Modestus O. Okwu.

The cultivation of *C. pepo*, including pumpkin varieties, is well-suited for temperate and subtropical climates, further enhancing its potential as a biodiesel feedstock (Figure 8.1). As the world grapples with the need for sustainable energy solutions, the investigation into *C. pepo* seed oil as a biodiesel source holds promise, contributing to the ongoing quest for environmentally friendly and renewable fuel alternatives (Figure 8.2).

Figure 8.1 (a) *C. pepo* shoots; (b) *C. pepo* immature fruit; (c) *C. pepo* mature fruit; (d) *C. pepo* seeds.

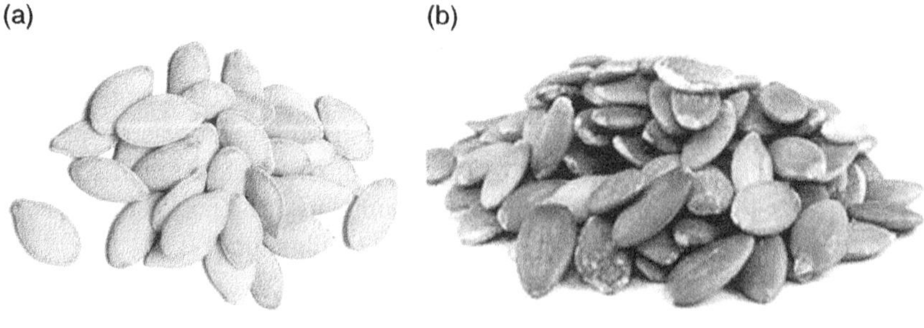

Figure 8.2 (a) Deshelled seeds of *C. pepo*; (b) Shelled seeds of *C. pepo* (*Source:* [8]/with permission of Springer Nature).

The exploration for alternative fuels for diesel engines has predominantly centered on vegetable oil, leveraging its status as a renewable resource in agriculture and its widespread global availability [9]. However, earlier attempts to directly utilize vegetable oils as diesel engine fuel encountered setbacks attributed to challenges such as high viscosities, distinct combustion chemistry, and lubrication oil contamination arising from incomplete combustion [10]. In the context of direct injection engines, these disparities in combustion chemistry manifested as a spectrum of issues, including poor cold flow properties, inadequate atomization, coking tendencies, carbon deposits, cold starting challenges, and increased wear [10]. Consequently, there arises a crucial need for a conversion method that transforms vegetable oil into esters, commonly known as biodiesel [11]. Addressing the drawbacks associated with vegetable oil, one of the most promising strategies involves transesterifying it into esters. The transesterification process yields mono-alkyl esters, referred to as biodiesel, along with glycerol as a by-product [12, 13]. This chemical transformation serves to reduce the molecular weight and viscosity of the fuel while concurrently enhancing its volatility. Importantly, the cetane number and heating value are maintained through out this process [11]. Beyond these advantages, transesterification offers additional benefits, including diminished harmful exhaust emissions, improved biodegradability, inherent lubricity, a higher flash point, and the utilization of a domestic resource. In reality, the pursuit of a viable alternative to conventional diesel fuels involves not only the selection of renewable resources like vegetable oil but also the implementation of transformative processes such as transesterification. By overcoming the inherent challenges associated with vegetable oil, this approach enhances the overall performance and environmental characteristics of biodiesel, contributing to a more sustainable and domestically sourced solution for diesel engines.

Recently, environmentalists have engaged in discussions regarding the adverse implications of producing biodiesel from edible oils on a large scale. This heightened concern stems from the potential to disrupt the global food supply and market equilibrium [11]. This chapter focuses on the exploration of biodiesel production from *C. pepo* oil, aiming to harness its potential as a viable feedstock in biodiesel production. The focus is on critical aspects such as oil content, the methyl esters profile of the resulting biodiesel, and the fuel properties. Despite the manifold benefits of pumpkins, including contributions to health, the economy, and serving as a substitute raw material for industries, they remain underutilized. The objective is to shed light on potential solutions to these challenges and offer sustainable production methods [8].

8.2 Review of Literature

Table 8.1 shows the uses of Cucurbita pepo in connected research works.

Table 8.1 Review of related research using *Cucurbita pepo* in a variety of applications.

Citations	Authors	Title	Applications	Year
[1]	James et al.	Studies on *Cucurbita maxima* Seed Oil for its Potentials as Feedstock for Biodiesel Production in Nigeria	B	2021
[14]	Schinas et al.	Pumpkin (*C. pepo* L.) seed oil as an alternative feedstock for the production of biodiesel in Greece	B	2009
[15]	Goodhead et al.	Production of Biodiesel from Fluted Pumpkin Oil	B	2021
[16]	Choudhury et al.	Production and evaluation of physicochemical, rheological,and tribological properties of *C. pepo L.* seed oil	GP	2021
[8]	Oloyede et al.	Exploring the Industrial Potentialof the Nigerian Pumpkins(*C. pepo L.*)	IP	2021
[2]	Ibeto et al.	Comparative study of the physicochemical characterization of some oils as potential feedstock for biodiesel production	B	2012
[17]	Ayyildiz et al.	Pumpkin (*C. pepo L.*) Seed Oil	GP	2019
[18]	Vinothkannan et al.	Performance and Emission Characteristic Analysis of *Cucurbita Pepo L.* and *Tectona Grandis* Seed Oil Biodiesel Blends In Ci Engine With Additive	IP	2020
[3]	Bikash et al.	Physicochemical assessment of Pumpkin (*C. pepo L.*) seed oil as a viable feedstock for biodiesel production	B	2018
[19]	Hundie et al.	Optimization of biodiesel production parameters from *Cucurbita maxima* waste oil using microwave-assisted via Box–Behnken design approach	B	2023

B, Biodiesel; GP, General purpose; IP, Industrial Potential.

8.3 Materials and Method

8.3.1 Plant Material and Chemical Preparation

For this investigation, waste from *C. pepo* was sourced from a well-known market in the southeastern region of Nigeria. The collected samples were methodically organized in bags and transported to the University Chemistry Laboratory for subsequent chemical preparations and experimental procedures. To ensure uniformity in weight, the *C. pepo* waste underwent a thorough process involving washing and sun drying over a period of seven days. The dried waste was then finely ground into a powdered form using a mortar and pestle, followed by sieving through a 50-mesh carbon steel sieve to attain consistent particle sizes and maximize the surface area for efficient oil extraction.

The resulting powdered samples were carefully sealed in glass bottles and stored at room temperature in preparation for further analysis. It is noteworthy that all chemicals and reagents utilized in this study were of analytical grade, ensuring the accuracy and reliability of the experimental outcomes. This systematic approach to sample collection, preparation, and analysis lays the foundation for a comprehensive examination of the potential of *C. pepo* waste as a source for biodiesel production.

Soxhlet extractor, round-bottom flask, conical flask, funnel, condenser, weighing balance, hot plate, cotton bags, viscometer, stopwatch, capillary tube, thermometer, pycnometer, digital weighing balance, air cooler (cooling bath), test tube (test jar), strips of pH indicator paper, pH color chart, Beaker and Pensky Martens closed tester were all used in the sample preparation, extraction, and characterization.

8.3.2 Extraction of Oil

The extraction of *C. pepo* seed followed the methodology proposed by Nayak and Vyas [20], albeit with certain modifications. Initially, *C. pepo* seed powder (15 g) was combined with 45 mL of *n*-hexane in conical flasks. The mixture was then placed in a Soxhlet apparatus, initially immersed in *n*-hexane to ensure the complete diffusion of the solvent into the sample. Subsequently, the thimble was positioned in a system containing *n*-hexane at 40 °C for a duration of four hours [21].

Upon the completion of the extraction process, the obtained oil was subjected to filtration using a vacuum filter. To facilitate the settling of any suspended particles, the filtered oil was left undisturbed for 48 hours. The oils dissolved in *n*-hexane were then recovered through a rotary evaporator at 40 °C. The resultant solvent-free oils were transferred to an airtight container for further characterization using standard testing methods to evaluate their suitability for biodiesel production. The quantity of oil obtained was determined based on the weight of the *C. pepo* waste powder, expressing the recovery as a percentage of the oil yield.

Following the extraction, the obtained oil was allowed to cool and subsequently placed in a round-bottom flask. This marked the initial step in the characterization process, wherein the solvent-free oils were carefully stored in an airtight container. The subsequent characterization involved subjecting the oils to standard testing methods to assess their viability for biodiesel production. The overall methodology ensures a systematic and thorough evaluation of the extracted *C. pepo* seed oil for its potential application in biodiesel production.

8.3.3 Pretreatment of the Oil (Acid Degumming)

The procedure began by measuring the quantity of *C. pepo* oil, which was subsequently heated to a temperature of 65 °C using a heating mantle. A solution containing approximately 25% citric acid was prepared by dissolving 250 g of citric acid in 1000 cm^3 of distilled water. Subsequently, a ratio of 5:1 citric acid to oil was achieved by combining the prepared citric acid solution with the oil. The resulting mixture was thoroughly mixed for a duration of 10 minutes. To facilitate the separation of gums and other impurities that settled at the bottom, the mixture underwent centrifugation for 20 minutes.

Following centrifugation, the oil present at the top of the centrifuge test tubes was carefully decanted, separating it from the undesirable impurities. The decanted oil was then subjected to a reheating process [22]. This methodological approach ensures the removal of impurities from the *C. pepo* oil, enhancing its purity and suitability for further processes, thereby contributing to the overall quality of the biodiesel production process.

8.3.4 Transesterification Reaction

The transesterification process yields two distinct products: the primary output is biodiesel, while glycerol is generated as a by-product. After an hour of continuous stirring, the resulting mixture is subjected to cooling to facilitate gravity separation. After a 24-hour duration within the container, two discernible phases, comprising glycerol and biodiesel, become evident. Glycerol is observed in the lower phase, while biodiesel, owing to its lower density, floats atop the glycerol layer. To effectuate the separation of these two components, a separating funnel is employed. Post-separation, the biodiesel undergoes a washing step using 60% of its volume in warm distilled water. This mixture is gently agitated for a period of three minutes, and subsequently, allowing the mixture to settle yields a two-layered composition. The biodiesel is then meticulously separated from this two-layered mixture. This washing process is iterated twice more, enhancing the clarity of the ester produced. The primary objective of this washing procedure is twofold: to eliminate impurities and residual methanol from the biodiesel. By meticulously executing these separations and washing steps, the transesterification process is optimized, ensuring the production of a biodiesel product of heightened purity and quality.

8.3.5 Evaluation of Process Parameters

The impact of various process parameters, including catalyst quantity, time, temperature, methanol/oil ratio, and agitation speed, was systematically explored through established methodologies.

8.3.5.1 Engine Performance Assessment

To assess the effectiveness of the biodiesel samples, an evaluation of engine performance was conducted, comparing key metrics with those of a TECHNO R175A diesel engine. The parameters examined included fuel consumption rate, emission levels, powering capabilities, and overall performance. The fuel consumption rate was determined by passing each biodiesel sample (50 mL) through the engine at predetermined intervals. Simultaneously, the emitted fumes from each sample during the engine performance test were directed through a conical flask containing 25 mL of 1 M sodium hydroxide solution to quantify the emission levels. Subsequently, the resulting solution underwent titration against a 1 M solution of hydrochloric acid.

This comprehensive engine performance test aimed to provide insights into the comparative efficiency and environmental impact of the biodiesel samples in contrast to a conventional TECHNO R175A diesel engine. The analysis encompassed critical aspects such as fuel consumption, emissions, and overall engine capabilities, contributing valuable data to the assessment of biodiesel viability for practical applications. The vapor's acidity was computed with the formula presented in the system of equations below:

$$\text{Fume acidity} = (a - b) \times (\text{molarity of NaOH}) \times (\text{molar mass of NaOH}) \quad (8.1)$$

To calculate the rate of fuel consumption, the formula is presented in Eq. (8.2):

$$\text{Rate of fuel consumption} = \frac{\text{volume of fuel consumed}}{\text{time taken}} \quad (8.2)$$

8.4 Results and Discussions

8.4.1 Characterization of the Oil Produced from *Cucurbita pepo* as Feedstock

In the investigation of biodiesel production utilizing *C. pepo* oil, an in-depth characterization of the extracted oil was conducted. The study encompassed an assessment of various physicochemical properties, including but not limited to density, ash content, moisture content, kinematic viscosity, acid value, saponification value, peroxide value, iodine value, and others. The determination of these properties followed the established standards outlined by the American Society for Testing and Materials (ASTM). The analytical process involved a careful examination of the density to gauge the oil's mass per unit volume. Moisture content and ash content were scrutinized to evaluate the water and mineral content within the oil, respectively. Kinematic viscosity, a critical parameter in biodiesel production, was assessed to understand the oil's flow characteristics.

Table 8.2 Physicochemical properties of the extracted oils.

Properties	*C. pepo* seed oil
Color	Greenish brown
Moisture (%)	3.500 (%)
Pour Point	−10 (°C)
Yield	48.870 (%)
Refractive index	1.476 at 29 °C
Kinematic viscosity	12.96 at 40 °C (mm^2/s)
Acid value	6.710 (mgKOH/g)
Energy value (kJ/kg)	44 147
Peroxide value	2.5 (meq/kg)
Saponification	209.38 (mgKOH/kg)
Flash Point	174 (°C)
Iodine value	182.74 (g/100 g of oil)
Density	914 (kg/m^3)

Saponification value, indicative of the fatty acid average molecular weight, and acid value, representing the acidity of the oil, were accurately measured. The iodine value, reflecting the oil's unsaturation level, and peroxide value, indicative of oxidative stability, were also scrutinized to comprehensively understand the oil's chemical composition and potential as a biodiesel feedstock. The adherence to ASTM standards ensures the reliability and comparability of the obtained data, facilitating a robust understanding of the physicochemical properties inherent in *C. pepo* oil. This comprehensive characterization serves as a foundational step in evaluating the suitability of *C. pepo* oil for biodiesel production, providing valuable insights for further optimization and development in the field of sustainable bioenergy. See Table 8.2 for the Physicochemical Properties of the extracted oils.

8.4.2 Further Discussion on Results

The oil's density, denoting its mass per unit volume, was determined to be 914 kg/m^3 at a temperature of 28 °C. Kinematic viscosity, a property reflecting a fluid's resistance to flow over a surface, was established at 12.96 mm^2/s for *C. pepo*. Higher viscosity implies greater resistance to flow. The calorific value, or energy value, of a fuel signifies the amount of energy released during its complete combustion, expressed in kJ/kg^{-1}. For *C. pepo*, this value was not specified. The moisture content of *C. pepo* was measured at 3.5%. The temperature at which a liquid transitions to a semi-solid state is termed the pour point. At this temperature, the liquid loses its flow ability and becomes stationary. In the case of *C. pepo*, its pour point was identified as −10 °C. The iodine value, a measure of unsaturation in fats and oils, signifies their degree of saturation. A higher iodine value indicates greater unsaturation. This parameter for *C. pepo* was not explicitly stated [23, 24].

Oils can be categorized based on their iodine values, with drying oils exhibiting values above 125, semi-drying oils falling between 110 and 140, and nondrying oils having values below 110. The propensity of oils to dry is associated with a vulnerability to rancidity, rendering them unsuitable for prolonged storage. In the context of direct usage as straight vegetable oil in engines, oils prone to drying are more likely to polymerize under the engine's heat, making them less practical for extended application. An additional quality parameter, the acid value of the oil, surpassed the ASTM standard of 0.8 mgKOH/g. The acid value serves as an indicator of the presence of corrosive FFA and oxidation products in the oil. This metric is crucial for assessing oil quality, as a lower free fatty acid content generally correlates with higher oil quality. Vegetable oils characterized by elevated free fatty acid levels can significantly impede the transesterification process with methanol, particularly when employing an alkaline catalyst. Moreover, it poses challenges to the separation of fatty acid esters and glycerols. Given these considerations, it is implied that the conversion of such oils to biodiesel would be more effective through a two-stage process involving esterification and transesterification. This approach proves beneficial in addressing the challenges associated with high FFA content, ensuring a more efficient and reliable conversion of oils into biodiesel.

The saponification value of the oil, exceeding 100 mgKOH/g, serves as an indicator for detecting potential adulteration. Specifically, a saponification value of 210.38 mgKOH/kg suggests a notable concentration of fatty acids in the oil [25]. This high value implies the likelihood of soap formation and potential challenges in product separation if the oil is utilized for biodiesel production. Furthermore, it suggests that the use of such oils for biodiesel production might yield very low methyl ester yields. The peroxide value of the oil, recorded at 2.5 meq/kg, is indicative of its instability, with high peroxide levels making oils susceptible to rancidity [26]. Consequently, if these oils are intended for biodiesel production, they should be utilized promptly after production or extraction to mitigate the risk of rancidity. The flash point, denoting the temperature at which fuel particles ignite, was determined for *C. pepo* oil and found to be 174 °C. This exceeds the minimum ASTM recommended range of 130 °C, indicating that there is no risk of fire outbreaks in the event of an accident. This flash point value also aligns with the recommended flash point of >100 °C by SNI 04 [27]. These findings underscore the safety and compliance of *C. pepo* oil for potential applications, reinforcing its suitability for biodiesel production.

Table 8.3 Physicochemical properties of *C. pepo* biofuels and its blend.

Fuels	Density (kg/m^3)	Kinematic viscosity (poise)	Calorific value (kJ/kg)	Flashpoint (°C)	Fire point (°C)	Cetane number
Diesel	825	2.87	42 000	62	76	44.00
C. pepo methyl ester	790	4.21	39 128	135	145	53.00
Biodiesel blend B_{20}	824.6	3.16	41 316	84	90.4	54

The recorded viscosity of 12.96 mm^2/s was notably low, a characteristic that is highly recommended. This low viscosity is advantageous, as viscosity tends to increase with molecular weight, influencing factors such as injection lubrication, engine deposits, and fuel atomization [28, 29]. A low viscosity is desirable for optimal engine performance and efficiency. The determined refractive index of 1.476 closely aligns with values reported for other seed oils, specifically 1.48 for *Telfairia occidentalis* seed oil, 1.47 for soybean oil, and 1.47 for maize oil [30, 31]. The refractive index of oils is influenced by various factors, including their molecular weight, fatty acid chain length, unsaturation level, and degree of conjugation [6]. This similarity in refractive index indicates a comparable composition and molecular structure among these seed oils, highlighting the consistency of *C. pepo* oil within this context (Table 8.3).

The findings affirm that the transesterification process involving vegetable oils and methanol yields noteworthy improvements in their fuel properties. Notably, the viscosity of the oil decreases as it undergoes conversion into methyl esters, aligning with the standard diesel ASTM range of 1.9–6.0 mm^2/s. This reduction in viscosity is attributed to the replacement of heavy alcohol (glycerol) in the oil with a simpler alcohol (methanol). The elevated viscosity of vegetable oil, when considered as an alternative fuel in diesel engines, poses a notable drawback [32–34]. Higher viscosity levels increase the likelihood of encountering operational issues [35]. Consequently, the primary objective of transesterification in the context of fuel production from vegetable oils and animal fats is to mitigate this viscosity concern. Interestingly, when the resulting methyl esters were blended with petroleum diesel, there was a significant increase in their viscosity. The viscosity values of the feedstocks' methyl esters/diesel blend (20:80) fell within the 2D grade diesel category (2.0–4.3 mm^2/s), indicating their suitability for fueling mobile equipment. On the contrary, higher blends were categorized within the 4D grade diesel range (5.0–34.0 mm^2/s), making them suitable for powering stationary equipment [36]. This viscosity adjustment through transesterification and blending is instrumental in tailoring the fuel properties to specific applications, enhancing the versatility and applicability of biodiesel derived from vegetable oils and animal fats.

8.5 Benefits, Challenges, and Future Prospects of Using *Cucurbita pepo* for Biodiesel Production

8.5.1 Benefits of Using *Cucurbita pepo* for Biodiesel Production

Utilizing *C. pepo* (squash or pumpkin) for biodiesel production offers various benefits, making it an intriguing feedstock for sustainable energy. Here are detailed insights into the advantages of using *C. pepo* for biodiesel production:

Abundant and renewable resource: *C. pepo* is a widely cultivated plant, and its fruits are produced abundantly. The availability of this crop makes it a sustainable and renewable resource for biodiesel production, contributing to long-term energy sustainability.

High oil content: The seeds of *C. pepo* are known to contain a significant amount of oil. This high oil content is a crucial factor in biodiesel production, as it directly influences the yield and efficiency of the biofuel extraction process.

Biodegradability and low environmental impact: Biodiesel derived from *C. pepo* is inherently biodegradable, and its combustion produces lower levels of green house gas emissions compared to traditional fossil fuels. This characteristic aligns with environmental conservation goals, reducing the overall impact on ecosystems.

Cultivation flexibility: *C. pepo*is a versatile plant that can be cultivated in various climates and soil types. Its adaptability to different agricultural conditions provides flexibility for farmers and contributes to its potential as a widespread and accessible biodiesel feedstock.

Economic viability and income generation: The cultivation of *C. pepo* for biodiesel production can create economic opportunities for farmers. The dual purpose of growing squash or pumpkins for both food consumption and biodiesel extraction can enhance the economic viability of cultivating this crop.

Reduced dependency on fossil fuels: The use of *C. pepo* for biodiesel helps reduce dependence on finite fossil fuels. By promoting a shift toward bio-based energy sources, it contributes to energy security and mitigates the environmental impact associated with fossil fuel extraction and combustion.

Waste utilization: Biodiesel production from *C. pepo* can involve utilizing not only the seeds but also other parts of the plant, such as pulp or waste materials. This approach supports a more comprehensive utilization of the crop and reduces waste in the agricultural production cycle.

Potential for localized production: The widespread cultivation of *C. pepo* globally opens up opportunities for localized biodiesel production. This decentralization can contribute to regional energy independence and reduce the need for long-distance transportation of fuel.

In summary, the use of *C. pepo* for biodiesel production presents a range of benefits encompassing environmental sustainability, economic advantages for farmers, and a reduced reliance on traditional fossil fuels. Exploring the full potential of this versatile feedstock can contribute to the development of more sustainable and eco-friendly energy systems.

8.5.2 Challenges of Using *Cucurbita pepo* for Biodiesel Production

While using *C. pepo* (squash or pumpkin) as a feedstock for biodiesel production offers various benefits, there are also challenges associated with this process. Here is a detailed examination of the challenges faced when utilizing *C. pepo* for biodiesel production:

Low oil content in seeds: One significant challenge is the relatively low oil content in the seeds of *C. pepo* compared to some other oilseed crops. This can pose a challenge in achieving high biodiesel yields, as the oil content directly influences the quantity of biodiesel that can be extracted.

Seasonal variability: *C. pepo* cultivation is often seasonal, and the availability of seeds for biodiesel production may be limited to specific times of the year. This seasonality can result in challenges related to consistent feedstock supply for continuous biodiesel production.

Complex oil extraction process: The extraction of oil from *C. pepo* seeds can be a complex process. Specialized equipment and methods are often required, and the efficiency of oil extraction may vary, impacting the overall feasibility of large-scale biodiesel production.

Competing uses for C. pepo: *C. pepo* is commonly cultivated for various purposes, including food consumption. The competition for resources between biodiesel production and other uses, such as culinary or agricultural applications, can present challenges in balancing the overall utilization of the crop.

Limited research and development: Compared to more traditional biodiesel feedstocks, there might be limited research and development dedicated to optimizing the biodiesel production process using *C. pepo*. This lack of extensive research can hinder the development of efficient and cost-effective production techniques.

Land use considerations: The cultivation of *C. pepo* for biodiesel production requires agricultural land, and considerations regarding land use and potential competition with food crops must be addressed. Striking a balance between biodiesel production and maintaining food security can be a complex challenge.

Biodiesel quality and stability: The composition of the oil from *C. pepo* may affect the quality and stability of the biodiesel produced. Addressing issues related to the properties of the biodiesel, such as oxidative stability and cold flow properties, may require additional processing steps.

Economic viability: Assessing the economic viability of biodiesel production from *C. pepo* is crucial. Factors such as seed yield, oil content, extraction costs, and market demand can impact the overall economic feasibility of utilizing this feedstock on a larger scale.

Challenges associated with scaling-up of system: Transitioning from small-scale experiments to large-scale production can present challenges in terms of scalability. The technical, logistical, and economic challenges associated with scaling up biodiesel production using *C. pepo* need to be carefully addressed.

In summary, while *C. pepo* holds potential as a biodiesel feedstock, overcoming these challenges requires concerted efforts in research, technology development, and sustainable agricultural practices. Addressing these challenges is crucial for unlocking the full potential of *C. pepo* in contributing to sustainable and diversified biodiesel production systems.

8.5.3 The Future Prospects of Using *Cucurbita pepo* (Squash or Pumpkin) for Biodiesel Production

The future prospect of utilizing this interesting feedstock is promising, with potential benefits and opportunities. The detailed exploration of the future prospects associated with utilizing *C. pepo* for biodiesel production include:

Biodiversity and crop rotation: Incorporating *C. pepo* into biodiesel production can contribute to biodiversity and sustainable agricultural practices. The cultivation of diverse crops, including those with dual purposes like *C. pepo* for both food and biodiesel, supports crop rotation strategies, enhancing soil health and minimizing agricultural monoculture.

Genetic improvement: Future research and development efforts may focus on genetic improvement of *C. pepo* varieties specifically tailored for biodiesel production. This could involve breeding programs aimed at increasing seed oil content, enhancing resistance to pests and diseases, and improving overall crop yields.

Technology advancements: Continued advancements in extraction technologies and biodiesel production methods may further enhance the efficiency of utilizing *C. pepo* as a biodiesel feedstock. Innovations in oil extraction processes, including novel solvent systems or mechanical techniques, can contribute to higher yields and lower production costs.

Sustainable agriculture practices: The integration of *C. pepo* into sustainable agriculture practices aligns with the growing emphasis on environmentally friendly and regenerative farming. Biodiesel production from this crop supports the principles of agroecology, emphasizing the importance of sustainable and holistic agricultural systems.

Climate resilience: *C. pepo*'s adaptability to diverse climates positions it as a climate-resilient biodiesel feedstock. As climate change continues to impact traditional crops, the cultivation of versatile crops like *C. pepo* can offer an alternative that is more resilient to changing environmental conditions.

Waste utilization and circular economy: Future prospects may involve exploring the utilization of waste materials from *C. pepo*, such as peels or biomass, in the production of additional bioenergy or value-added products. This aligns with the principles of a circular economy, maximizing resource use and minimizing waste.

Community and rural development: The cultivation of *C. pepo* for biodiesel production has the potential to contribute to community and rural development. It can create local economic opportunities, generate income for farmers, and support sustainable livelihoods, especially in regions where the crop is well-suited for cultivation.

Global energy transition: As the world undergoes an energy transition toward renewable sources, the adoption of alternative biodiesel feedstocks, such as *C. epo*, can play a role in diversifying the bioenergy sector. This diversification contributes to reducing dependency on traditional fossil fuels.

Policy support and incentives: Future prospects for *C. pepo* in biodiesel production are influenced by policy support and incentives. Governments and regulatory bodies may implement policies that encourage the cultivation of biodiesel feedstocks, providing financial incentives and fostering a conducive environment for sustainable bioenergy practices.

In summary, the future prospects of using *C. pepo* for biodiesel production are optimistic, driven by advancements in technology, sustainable agricultural practices, and the global shift toward renewable energy sources. Continued research and development efforts, coupled with supportive policies, can further enhance the role of *C. pepo* in contributing to a sustainable and diversified bioenergy systems.

8.6 Conclusion

In conclusion, this chapter underscores the compelling potential of *C. pepo* seed oil as a promising feedstock for biodiesel production, positioning it as a noteworthy alternative to conventional fossil fuels. By considering the transesterification reaction of mono-alkyl alcohols derived from both vegetable oil and animal fats, the study systematically explores the biodiesel potential of *C. pepo*. The distinctive attributes of *C. pepo*, including its high oil content, rapid growth rate, and adaptability to diverse climates, further emphasize its viability as a sustainable feedstock. The comprehensive investigation involving systematic seed collection, processing, and oil extraction using a mechanical method, followed by the meticulous examination of key physical and chemical properties, provides robust evidence supporting the feasibility of *C. pepo* seeds for commercial biodiesel production. The quality assessment, adhering to ASTM standards, includes the synthesis of fatty acid methyl ester (FAME) and an exhaustive examination of critical physicochemical properties such as density, acid value, viscosity, pour point, and flash point. These properties are essential benchmarks for evaluating the suitability of biodiesel for practical applications. The experimental results convincingly demonstrate that *C. pepo* seeds possess the necessary characteristics for serving as a viable feedstock, offering substantial promise for the development of commercial biodiesel. In light of the imperative need to explore sustainable and environmentally friendly fuel alternatives, the findings from this study contribute valuable insights to the ongoing quest for greener energy systems.

References

1 Yusuf, J., Kantoma, D., and Mosunmola, O.R. (2021). Studies on Cucurbita maxima seed oil for its potentials as feedstock for biodiesel production in Nigeria. *Nigerian Res. J. Chem. Sci.* 9: 215–233.

2 Ibeto, C.N., Okoye, C.O.B., and Ofoefule, A.U. (2012). Comparative study of the physicochemical characterization of some oils as potential feedstock for biodiesel production. *Int. Sch. Res. Notices* 2012: 621518.

3 Bikash,B., Choudhury,N.D., Bora,D.K., and Kalita,K. (2018).Physicochemical assessment of Pumpkin (Cucurbita pepo L.) seed oil as a viable feedstock for biodiesel production. In: *Conference Proceedings of the Second International Conference on Recent Advances in Bioenergy Research* (ed. S. Kumar, R. Sani, and Y. Yadav). Springer Proceedings in Energy. Singapore: Springer. https://doi.org/10.1007/978-981-10-6107-3_2 27.

4 Hernández-Santos, B., Rodríguez-Miranda, J., Herman-Lara, E. et al. (2016). Effect of oil extraction assisted by ultrasound on the physicochemical properties and fatty acid profile of pumpkin seed oil (Cucurbita pepo). *Ultrason. Sonochem.* 31: 429–436. https://doi.org/10.1016/j.ultsonch.2016.01.029.

5 Wang, S., Karthickeyan, V., Sivakumar, E., and Lakshmikandan, M. (2020). Experimental investigation on pumpkin seed oil methyl ester blend in diesel engine with various injection pressure, injection timing and compression ratio. *Fuel* 264: 116868. https://doi.org/10.1016/j.fuel.2019.116868.

6 Khandelwal, S. and Chauhan, Y.R. (2012). Biodiesel production from non edible-oils: a review. *J. Chem. Pharm. Res.* 4 (9): 4219–4230.

7 FAO data, 2019 retrieve from FAO data set. http://www.fao.org/faostat/en/#data/QC/metadata (accessed 25 June 2020).

8 Oloyede, F.M. (2021). Exploring the industrial potential of the Nigerian pumpkins (*Cucurbita pepo*). In: *Food Security and Safety: African Perspectives*, 169–178. https://doi.org/10.1007/978-3-030-50672-8_9.

9 Hanna, L. and Raimo, A. (2012). Pyrolysis tall oil- derived fatty and resin acid mixtures. *Int. Sch. Res. Notices* 8. https://doi.org/10.5402/2012/409157.

10 Bello, E.I., Anjorin, S.A., and Agge, M. (2013). Production of biodiesel from fluted pumpkin (*Telfairia occidentalis* Hook F.) seeds oil. *Int. J. Mech. Eng.* 2: 22–31.

11 Leung, D.Y.C. and Guo, Y. (2006). Transesterification of wet and used frying oil: optimization for biodiesel production. *Fuel Process. Technol.* 89 (2): 152–159.

12 Knothe, G.L., Van Gerpen, J., and Krahl, J. (2005). *The Biodiesel Handbook*. Illinois, USA: AOCS Press Champaign.

13 Mittelbach, M.P. and Remschmidt, C. (2004). *Biodiesel Comprehensive Handbook*. Graz, Austria: M. Mittelbach, Karl-Franzens-Universitst Graz.

14 Schinas, P., Karavalakis, G., Davaris, C. et al. (2009). Pumpkin (*Cucurbita pepo* L.) seed oil as an alternative feedstock for the production of biodiesel in Greece. *Biomass Bioenergy* 33 (1): 44–49.

15 Goodhead, T.O., Wami, E.N., and Ukpaka, C.P. (2021). Production of biodiesel from fluted pumpkin oil. *J. Pet. Eng. Technol.* 11 (2): 50–58.

16 Choudhury, N.D., Saha, N., Bhaumik, S., and Kataki, R. (2021). Production and evaluation of physicochemical, rheological, and tribological properties of *Cucurbita pepo* L. seed oil. *Biomass Convers. Biorefin.* 13: 1104–1114.

17 Ayyildiz, H.F., Topkafa, M., and Kara, H. (2019). Pumpkin (*Cucurbita pepo* L.) seed oil. In: *Fruit Oils: Chemistry and Functionality*, 765–788. https://doi.org/10.1007/978-3-030-12473-1_41.

18 Vinothkannan, V. and Pushparaj, T. (2020). Performance and emission characteristic analysis of *Cucurbita pepo* L. and *Tectona grandis* seed oil biodiesel blends in CI engine with additive. *Energy Sources Part A* 1–15. https://doi.org/10.1080/15567036.2020.1849453.

19 Hundie, K.B. (2022). Optimization of biodiesel production parameters from cucurbita maxima waste oil using microwave assisted via Box-Behnken design approach. *J. Chem.* 2022. https://doi.org/10.1155/2022/8516163.

20 Nayak, M.G. and Vyas, A.P. (2019). Optimization of microwaveassisted biodiesel production from papaya oil using response surface methodology. *Renewable Energy* 138: 18–28.

21 Ravi, H.K., Vian, M.A., Tao, Y. et al. (2019). Alternative solvents for lipid extraction and their effect on protein quality in black soldier fly (*Hermetia illucens*) larvae. *J. Cleaner Prod.* 238: 117861.

22 Nick, F. and Greg, A. (2008). Degumming and bleaching: effect on selected constituents of palm oil. *J. Oil Palm Res.* 16: 57–63.

23 Aremu, M.O., Ibrahim, H., and Bamidele, T.O. (2015). Physicochemical characteristics of the oils extracted from some Nigerian plant foods – a review. *Chem. Eng. Res. Des.* 32: 36–35.

24 Knothe, G. (2002). Structure indices in FA chemistry. How relevant is the iodine value? *J. Am. Oil Chem. Soc.* 79 (9): 847–854.

25 Omolara, O.O. and Dosumu, O.O. (2009). Preliminary studies on the effect of processing methods on the quality of three commonly consumed marine fishes in Nigeria. *Biokemistri* 21 (1): 1–7.

26 Nzikou, J.M., Mvoula-Tsieri, M., Matos, L. et al. (2007). *Solanum nigrum* L. seeds as an alternative source of edible lipids and nutriment in Congo Brazzaville. *J. Appl. Sci.* 7 (8): 1107–1115.

27 Sulistyo, H., Suardjaja, I.M., and Rahayu, S.S. (2006). Transesterification of candlenut oil with ethanol to biodiesel. *Proceeding on Regional Symposium on Chemical Engineering (RSCE'06).* Singapore, December 2006.

28 Aremu, M.O. and Akinwumi, O.D. (2014). Extraction, compositional and physicochemical characteristics of cashew (*Anarcadium occidentale*) nuts reject oil. *Asian J. Appl. Sci. Eng.* 3: 33–40.

29 Fakhri, N.A. and Qadir, H.K. (2011). Effect of temperature change on the physico- chemical properties. *J. Environ. Sci. Eng.* 5: 844–849.

30 Abayeh, O.M., Garba, I.H., Adamu, H.M., and Abayeh, O.J. (2013). Quality characteristics of *Luffaa egyptiaca*seed oil. *Int. J. Sci. Eng. Res.* 4: 11–15.

31 Atadashi, I.M., Aroua, M.K., Abdul Aziz, A.R., and Suleiman, N.M.N. (2012). The effect of water on biodiesel production and refining technology. *SciVerse Sci. Direct* 3456–3470.

32 Okonkwo, C.P., Ajiwe, V.I.E., Obiadi, M.C. et al. (2023). Production of biodiesel from the novel non-edible seed of *Chrysobalanus icaco* using natural heterogeneous catalyst: modeling and prediction using artificial neural network. *J. Cleaner Prod.* 385: 135631.

33 Ajiwe, V.I.E., Mbonu, S.O. and Enukorah, E.A.O. (2006). Biofuel from *Irvinga gabonesis* seed oil and its methyl ester, *Proceedings of the NASEF'06 International Conference*, Awka, Solar Energy Society of Nigeria, 247–252.

34 Jaichandar, S. and Annamalai, K. (2011). The status of biodiesel as an alternative fuel for diesel engine – an overview. *J. Sustain. Energy Environ.* 2: 71–75.

35 Knothe, G. and Steidley, K.R. (2018). The effect of metals and metal oxides on biodiesel oxidative stability from promotion to inhibition. *Fuel Process. Technol.* 177: 75–80.

36 Okonkwo, C.P., Ajiwe, V.I., Ikeuba, A.I. et al. (2023). Production and performance evaluation of biodiesel from *Elaeis guineensis* using natural snail shell-based heterogeneous catalyst: kinetics, modeling and optimisation by artificial neural network. *RSC Adv.* 13 (28): 19495–19507.

9

Sustainable Biodiesel Production with Sweet Sorghum as a Renewable Feedstock

9.1 Introduction

The demand for energy is growing exponentially due to both economic development and population growth worldwide. Approximately 75% of greenhouse gas emissions come from the energy sector, which is mostly dependent on fossil fuels. It has also resulted in a 1.1 °C increase in the average global temperature, along with numerous detrimental effects on the environment [1]. As fossil fuel supplies run out, worries about the security of energy and the environment are heightened. In light of this, the international community has chosen an energy transition strategy aimed at achieving carbon neutrality and lowering greenhouse gas emissions in order to limit global warming to 1.5 °C by 2050.[2]. Utilizing renewable energy sources is now absolutely required to meet the current challenge. Enhancing biomass energy recovery efficiency is a critical component in maintaining the sustainability of our environment [3, 4]. The most cost-effective crops for producing biodiesel are sweet sorghum, maize, and sugarcane. Cereals grown in various parts of the world came in second by a significant margin. Cellulose appears to be a promising raw material source for the production of biofuels, as it can be used to produce liquid biofuels of the second generation [5].

Sweet sorghum is a crop with great potential for producing bioethanol and biodiesel because of its easy juice extraction, sugar-rich stalks, high biomass yield, and adaptability to various production systems [6]. Sweet sorghum is a sugar-rich plant that can be used as an alternate source of renewable bioenergy because it has no effect on global warming or food security [7]. Due to its qualities, it is the perfect raw material for the long-term, sustainable production of biofuels. They are more likely to produce value-added chemicals because they are rich in components such as fermentable carbohydrates, insoluble lignocellulosic parts,

Feedstocks for Sustainable Biodiesel Production: Characterization, Selection, and Optimization, First Edition. Chinwe P. Okonkwo and Modestus O. Okwu.

and bioactive compounds [8]. Sweet sorghum is specifically used for biofuel and biochemical purposes, while all sorghum species offer important health benefits [6, 9–12]. Agro-industrial feedstocks, which are abundant and inexpensive, comprise energy crops, industrial wastes, and agricultural wastes. These feedstocks have the potential to be renewable nonfood sources for the synthesis of value-added biochemicals [13]. Utilizingsorghum as a nonfood feedstock platform shows promise in producing value-added biochemicals when compared to other agro-industrial feedstocks. Grain, fodder, fiber, and sweet sorghum are the four groups of sorghum that have been developed and used thus far based on genotype traits and usage [9, 14, 15].

Sweet sorghum can be used to create a variety of desired chemicals because it is rich in soluble and insoluble carbohydrates as well as bioactive compounds [16]. Thereby producing additional feedstock for later industrial uses. Second, it uses less water and nitrogen fertilizer than maize (*Zea mays* L.) [17], demonstrating a reduction in input costs. Third, due to its broad tolerance to drought and resistance to salinity and alkalinity, it can be grown in a variety of environments. Crucially, compared to results in nonsaline land, only less than 25% of the biomass accumulation was decreased when cultivated in margin and sanitized lands [18]. More specifically, China has about 49.65 million hectares of marginal land available for the large-scale production of sweet sorghum, which yields 13.57 milliontons of grain [19]. Hence, offering sustainable methods for effectively using the resources of semi-arid agricultural land.

A number of insightful conversations about the utilization of sweet sorghum biomass have taken place [20]. Presently, the primary obstacle to the industrialization of sweet sorghum juice use could be linked to the short-term fermentation sugar preservation period due to microbial contamination. Many approaches have been tried in an attempt to prevent the issue [19, 20]. Interestingly, fresh sweet sorghum juice can be preserved for a long time (about 90 days at room temperature) using a combination of chemical and evaporation preservation, resulting in only 5.0% fermentable sugar loss. At last, using the 90-day-preserved sweet sorghum juice as a feedstock, *Saccharomyces cerevisiae* M 3013 can produce 121.1 ± 4.6 g/L ethanol with a yield of 0.45 ± 0.02 g/g.

It has also been attempted recently to bioconvert insulin or beverage waste to high fructose syrup [21]. Sweet sorghum juice contains a variety of contaminants, necessitating multiple processing steps to produce high fructose syrup later on. The use of nonfood sweet sorghum juice is the best feedstock for producing high fructose syrup, according to a multi-technology integration strategy that includes the hydrolytic process, clarification process, decolorization process, desalinization process, and evaporation process. Although sweet sorghum biomass has been used to produce a variety of biochemical products and effective technologies have been developed to increase the yield and quality of these desired products, there is still a need to address the low economic competitiveness of these products. The most economical approach, in our opinion, is to continue enhancing the biomass of sweet sorghum's economic competitiveness by extending new industrial application chains, engineering sweet sorghum varieties for specific uses, and building a reliable microbial cell factory.

Although sweet sorghum biomass has been used to produce a variety of biochemical products and effective technologies have been developed to increase the yield and quality of these desired products, there is still a need to address the low economic competitiveness of these products. The most economical approach, in our opinion, is to continue enhancing the biomass of sweet sorghum's economic competitiveness by extending new industrial application chains, engineering sweet sorghum varieties for specific uses, and building a reliable microbial cell factory (Figures 9.1 and 9.2).

Figure 9.1 Sweet sorghum farm.
Source: ICRISAT/PS Rao.

Figure 9.2 Sweet sorghum seeds.

9.2 Review of Literature

A concise review of the existing literature on sweet sorghum reveals its versatile applications and potential contributions across various domains. Sweet sorghum is a multipurpose crop known for its resilience and adaptability to diverse agro-climatic conditions. Table 9.1 presents related research focused on sweet sorghum in diverse areas of application.

Table 9.1 Review of related research focused on Sweet sorghum in diverse areas.

Citations	Authors	Title	Applications	Year
[8]	Hu et al.	Conversion sweet sorghum biomass to produce value-added products.	GP	2022
[4]	Hamadou et al.	Optimization of energy recovery efficiency from sweet sorghum stems by ethanol and methane fermentation processes coupling	GP	2023
[21]	Sebayang et al.	Optimization of bioethanol production from sorghum grains using artificial neural networks integrated with ant colony	GP	2017
[22]	Fasahati et al.	Process design and economics for production of advanced biofuels from genetically modified lipid-producing sorghum	B	2019
[7]	Kasegn et al.	Production of bioethanol from sweet sorghum [*Sorghum bicolor* L.] juice using yeast isolated from fermented sweet sorghum juice	GP	2023
[23]	Ray et al.	Sweet sorghum for bioethanol production: scope, technology, and economics. In Bioethanol Production from Food Crops	FP	2019
[24]	Xiao et al.	Sweet sorghum for phytoremediation and bioethanol production	GP	2021
[25]	Alphy et al.	Sweet sorghum juice as an alternative carbon source and adaptive evolution of *Lactobacillus brevis* NIE9.3.3 in sweet sorghum juice and biodiesel derived crude glycerol to improve 1, 3 propanediol production	B	2021
[26]	Ammar et al.	Propionic acid production by *Propionibacterium freudenreichii* using sweet sorghum bagasse hydrolysate	AA	2020

Table 9.1 (Continued)

Citations	Authors	Title	Applications	Year
[27]	Stamenković et al.	Production of biofuels from sorghum	B	2020
[6]	López-Sandin et al.	Evaluation of bioethanol production from sweet sorghum varietyroger under different tillage and fertilizer treatments	GP	2021

AA, antimicrobial activity; B, biodiesel; FP, food products; GP, general purpose; Ph, pharmaceuticals; WT, water treatment.

9.3 Sweet Sorghum Cultivation

Sweet sorghum as a versatile and multipurpose crop, can be cultivated for various purposes, including food, fodder, and bioenergy production. It grows well in the semi-arid tropics, where other crops cannot flourish. It is also a great choice for cultivation in more difficult dry land growing conditions. All continents, tropical, subtropical, temperate, and semi-arid regions, as well as low-quality soils and little water requirements, are suitable for their easy growth. It is a four-month-long plant that can be grown two or three times a year. It can withstand both biotic and abiotic stresses, including salinity, temperature, and drought. It can be cultivated as a rain-fed crop. Summary of the cultivation practices associated with sweet sorghum include:

9.3.1 Climate, Rainfall, and Soil Requirements

Considering the climate, sweet sorghum is well-suited to warm and tropical climates. It thrives in temperatures ranging from 25 to 35 °C. Concerning rainfall, adequate rainfall is essential during the growing season, with a preference for well-distributed rainfall. However, sweet sorghum is known for its drought tolerance and can adapt to semi-arid conditions. In terms of soil, well-drained soils with good fertility are ideal for sweet sorghum cultivation. It can grow in a variety of soil types, including loam, sandy loam, and clay loam.

9.3.2 Variety Selection

Different sweet sorghum varieties are available, each with unique characteristics such as sugar content, biomass yield, and maturity period.

Selecting a variety depends on the specific goals of cultivation, whether it is for syrup production, bioethanol, or fodder.

9.3.3 Land Preparation and Planting

It is important to prepare the land by plowing and harrowing to create a fine seedbed. The sweet sorghum seeds is then planted directly in the field at a depth of about 2.5–5 cm. Planting can be done in rows with a spacing of 60–75 cm between rows, depending on the variety.

9.3.4 Nutrient Management

Sweet sorghum has moderate nutrient requirements. Conduct soil tests to determine the nutrient levels and apply fertilizers accordingly. Nitrogen, phosphorus, and potassium are essential for optimal growth. Proper nutrient management contributes to higher biomass and sugar yield.

9.3.5 Water Management

Adequate water is crucial, especially during critical growth stages like germination, flowering, and grain filling. Employ irrigation methods such as drip, sprinkler, or furrow irrigation, depending on water availability and field conditions.

9.3.6 Weed Control

Implement effective weed control measures to prevent competition for nutrients and water. Mulching and timely weeding are common practices to suppress weed growth.

9.3.7 Disease and Pest Management

Sweet sorghum is generally resistant to many pests and diseases. However, occasional monitoring is recommended. Common pests include aphids, stem borers, and armyworms. Integrated pest management strategies can be employed.

9.3.8 Harvesting

Harvest time varies depending on the intended use. For syrup or bioethanol production, harvest when the stalks are in the dough stage. For fodder, harvesting can occur at the soft dough or hard dough stage.

Post-harvest processing: Extract juice from the harvested stalks for syrup or bioethanol production. For grain production, allow the sorghum heads to dry on the plant before harvesting.

9.3.9 Crop Rotation

Practice crop rotation to maintain soil health and prevent the buildup of pests and diseases associated with continuous sorghum cultivation. Successful sweet sorghum cultivation requires a combination of suitable varieties, good agricultural practices, and proper management of water, nutrients, and pests. As with any crop, staying informed about the latest research and agricultural techniques is crucial for optimizing yields and sustainability.

9.3.10 Traditional Use by Farmers and Prospects

Sorghum is typically grown for fodder under both irrigation and rain-feeding systems. It is estimated that sorghum provides between 60% and 70% of the forage needed during the rainy season. In various places, sweet sorghum is grown in small plots. A number of nations are being forced to search for renewable energy sources due to the unpredictability of fossil fuel supplies, the steadily rising price of crude oil, and the need to protect the environment. It is the perfect crop to grow in areas where sugarcane is grown in order to use the existing sugarcane machinery during the off-season and to supplement molasses for the production of ethanol. The lack of sufficient amounts of feedstock suitable for off-season crushing in the sugar industries, photoperiod sensitivity, and the restricted availability of genotypes adapted to various agro-climatic conditions with all built-in resistances against biotic and abiotic stresses are the main obstacles to its large-scale cultivation. It is necessary to develop sweet sorghum cultivators with all the desirable traits, such as high stalk yield per unit time, input, energy, and land area in different agro-climatic areas of the country, in order to meet the raw material demand of the industry, especially after sugarcane crushing. Research on the parental line of sweet sorghum is desperately needed, particularly to improve the genetic potential of females for high sugar content. Sorghum population improvement offers a long-term approach to breeding in order to produce superior varieties and hybrid parents. Grain crop quality and productivity have improved significantly as a result of sorghum genotypes being converted to longer days, increasing genetic diversity [28].

9.4 Processing Technology for Sweet Sorghum

Processing sweet sorghum involves converting its various components into valuable products, such as syrup, bioethanol, and other by-products. The processing technology for sweet sorghum generally includes the following steps:

Harvesting: Sweet sorghum is typically harvested when the stalks are in the dough stage for syrup production or bioethanol extraction. If the goal is grain production, the heads are left to dry on the plant.

Juice extraction: The harvested stalks are crushed to extract the sweet juice. This can be done using a mechanical press, roller mill, or a specialized sweet sorghum juicer.

The extracted juice contains sugars, mainly sucrose, which can be further processed for syrup production or fermented for bioethanol.

Clarification and filtration: The raw juice is often cloudy and contains impurities. Clarification involves the addition of agents like lime to improve juice quality and facilitate impurity settling.

Filtration processes, such as sedimentation or using filter presses, help remove solid particles and further clarify the juice.

Evaporation: The clarified juice is concentrated through evaporation to increase sugar content. This step is crucial for syrup production, where the concentrated juice is boiled to achieve the desired consistency.

Syrup production: Concentrated juice is processed into syrup by heating and evaporating water, resulting in a sweet, thick liquid. This syrup can be used as a sweetener in various food products.

Fermentation (bioethanol production): For bioethanol production, the juice is subjected to fermentation. Yeast is added to convert sugars into ethanol through a fermentation process. After fermentation, the mixture undergoes distillation to separate ethanol from water, producing bioethanol.

By-product utilization: Residual materials, such as bagasse (the crushed stalks), can be used as feedstock for energy generation or as a source of fiber for various industrial applications. The extracted grain can be further processed for food or animal feed.

Quality control: Throughout the processing stages, quality control measures are implemented to ensure the final products meet safety and regulatory standards.

Monitoring parameters like sugar content, pH, and ethanol concentration is essential for consistent product quality.

Waste management: Proper disposal or utilization of by-products and waste materials is crucial for minimizing environmental impact. Residual materials can be composted or used for soil amendment.

Scaling up and commercialization: To make sweet sorghum processing economically viable, efforts are made to scale up production and commercialize the technology. This involves optimizing processes for efficiency and cost-effectiveness.

The processing technology of sweet sorghum is adaptable to various end products, making it a versatile and sustainable crop for bioenergy and food production. Advances in technology and research continue to enhance the efficiency and viability of sweet sorghum processing.

9.5 Sweet Sorghum Supply Chain

Managing the supply chain and building capacity in the context of sweet sorghum cultivation and processing involves a tailored approach to ensure efficiency, sustainability, and overall success. The concept applies specifically to sweet sorghum in the following ways:

Supply chain management (SCM) for sweet sorghum: This involves considering supplier relationship management which entails cultivating strong relationships with seed suppliers, equipment providers, and other key stakeholders in the sweet sorghum supply chain. It is equally important to ensure a steady and quality supply of seeds, fertilizers, and necessary equipment for cultivation.

Harvest and transport logistics (HTL): Considering HTL, it is important to develop efficient logistics for harvesting sweet sorghum at the optimal stage. The way forward is to implement transportation systems that minimize delays between harvest and processing, preserving the quality of the sorghum.

Processing facility optimization: At this stage, it is important to ensure that processing facilities are strategically located to minimize transportation costs and facilitate timely processing. Then, technology is integrated for real-time monitoring of processing stages, from juice extraction to bioethanol production.

Quality control measures: To control and measure quality, it is important to implement rigorous checks at various stages of processing to maintain high standards for syrup, bioethanol, and other derived products. Then, traceability systems can be utilized to identify and address any deviations in product quality.

Waste management and by-product utilization: This aspect is very important as waste to wealth has become very important when considering biodegradable feedstocks. To actualize this process, strategies can be developed for the efficient utilization of by-products, such as using bagasse for energy generation or animal feed. Then, eco-friendly waste management practices is implemented to minimize environmental impact.

Market integration: It is important to establish strong connections with end-users or industries relying on sweet sorghum-derived products. This can be actualized by staying informed about market trends and demand so as to adjust cultivation and processing volumes accordingly.

Capacity building for sweet sorghum

Training programs for farmers: Training initiatives can be developed for farmers to enhance their skills in sweet sorghum cultivation, covering best practices in planting, irrigation, and pest management. Farmers can also be educated on sustainable agriculture techniques to optimize yield and quality.

Technology adoption in agriculture: In this era of Industry 4.0, it is important to introduce farmers to advanced agricultural technologies, such as precision farming and IoT-enabled devices, to improve efficiency in cultivation. With this, guidance can be provided on the selection and use of modern farming equipment.

Processing facility workforce training: It is equally important to offer specialized training for workers involved in sweet sorghum processing. This training can cover aspects such as juice extraction, fermentation, and quality control. It is equally important to foster a culture of safety and compliance within processing facilities.

Research and development collaboration: Research and development is very important to advance in this noble area, it is important to collaborate with research institutions to stay abreast of technological advancements in sweet sorghum cultivation and processing. The development of improved sweet sorghum varieties can be encouraged through research and breeding programs.

Environmental and sustainable practices: Stakeholders can be educated on the importance of sustainable and environmentally friendly practices in sweet sorghum cultivation and processing. Practices that reduce water usage can be implemented by, minimizing chemical inputs, and promoting overall environmental responsibility.

Government and community engagement: It is also important to collaborate with government agencies to access support programs and incentives for sweet sorghum cultivation. Local communities can be engaged to address any concerns and promote the positive impacts of sweet sorghum cultivation on the regional economy. In summary, an integrated approach to SCM and capacity building is crucial for the success of sweet sorghum cultivation and processing. By optimizing these aspects, stakeholders can contribute to the sustainability and profitability of the entire sweet sorghum value chain.

9.6 Government Policy Support

Government policy support would play a crucial role in promoting the cultivation, processing, and utilization of sweet sorghum. These policies can create an enabling environment for farmers, processors, and other stakeholders in the sweet sorghum value chain. Several key areas where government policy support can be instrumental include.

Agricultural Policies

- *Research and development funding*: Funds can be allocated for research and development programs focused on sweet sorghum varieties, cultivation practices, and processing technologies; Partnerships between research institutions and the private sector can be supported to drive innovation in sweet sorghum agriculture.

- *Subsidies and incentives*: Subsidies can be provided for sweet sorghum seeds, fertilizers, and other inputs to encourage farmers to adopt sweet sorghum cultivation; Financial incentives can be introduced for farmers involved in sweet sorghum production.
- *Extension services*: Extension services can be strengthened to educate farmers about best practices in sweet sorghum cultivation, pest management, and sustainable agriculture techniques; Demonstration farms can be established to showcase successful sweet sorghum cultivation methods.

Environmental and Sustainability Policies

- *Water management initiatives*: Water management policies can be implemented that support the cultivation of sweet sorghum, especially in regions prone to water scarcity; Water-efficient irrigation techniques can equally be promoted for sweet sorghum cultivation.
- *Renewable energy support*: Policies that incentivize the production of bioenergy from sweet sorghum, such as bioethanol can be introduced, by providing subsidies or ensuring favorable market conditions; Renewable energy targets can be set that include contributions from bioenergy sources, encouraging investment in sweet sorghum for biocthanol production.

Trade and Market Policies

- *Market access and price support*: Market access for sweet sorghum products can be facilitated by streamlining regulations and reducing trade barriers; Price stabilization mechanisms can be implemented to ensure fair and consistent returns for sweet sorghum farmers.
- *Value addition and processing incentives*: Incentives can be offered for the establishment of sweet sorghum processing facilities, encouraging investment in technologies for juice extraction, bioethanol production, and other value-added products; It is equally important to strategize on initiatives that promote the development of a diversified product range from sweet sorghum, such as syrup, animal feed, and bioplastics.

Education and Training Policies

- *Training programs*: Training programs can be put in place for farmers, processors, and other stakeholders involved in sweet sorghum cultivation and processing; Possibility of collaboration with educational institutions by integrating sweet sorghum-related courses into agricultural curricula.

Regulatory Framework

- *Standards and certification*: Quality standards can be established for sweet sorghum products to ensure consistency and consumer safety; Certification process can equally be facilitated for sweet sorghum-derived products, making them eligible for export markets.

- *Environmental regulations*: Environmental regulations that encourage sustainable farming practices, waste management, and the responsible use of inputs in sweet sorghum cultivation can be implemented and actioned.

By implementing supportive policies in these areas, governments can create a conducive environment for the growth of the sweet sorghum industry, contributing to food security, sustainable agriculture, and renewable energy production.

9.7 Key Findings in Utilizing Sweet Sorghum for Biodiesel Formation

Sweet sorghum's key characteristics, including its ability to propagate seeds, produce crops mechanically, remain noninvasive, and inhibit biological nitrification, have drawn a lot of interest as a possible nonedible feedstock for the production of biodiesel. At current prices, the cost of producing bioethanol from sweet sorghum is less than that of sugarcane molasses [29]. Sorghum is highly prized as an alternative fuel. A significant amount of research has been done to find out how to store and supply sweet sorghum for the biorefinery continuously. As a result, energy and environmental impacts, such as intensive land use, soil, water, and air pollution from fertilizer use, and plant protection, must be considered in environmental sustainability assessments [30]. This study shows that the sweet sorghum can be used in the production of biodiesel an alternative to crude oil and these biodiesels produced when used can help reduce the greenhouse gas been introduced into the atmosphere that cause climate change [31].

9.8 Conclusion

In conclusion, the study underscores the critical need for sustainable and renewable energy sources, particularly in the context of biodiesel production from nonedible feedstocks. The environmental crisis and concerns about energy security have heightened the urgency for research on alternative fuels to replace fossil fuels. Sweet sorghum, identified as a versatile and high-yielding crop, emerges as a promising feedstock for biodiesel production, offering a viable solution to the challenges posed by the competition between fuel and food. The study emphasizes that sweet sorghum's adaptability to diverse agro-climatic conditions makes it a robust candidate for biodiesel production. Its unique characteristics, such as requiring minimal fertilizers and irrigation while exhibiting high photosynthetic activity, position sweet sorghum as an ideal substrate for microbial oil production. The investigation into farming practices, future prospects, and the biodiesel production process from sweet sorghum provides a comprehensive understanding of the agronomic, biochemical, and engineering aspects involved. By presenting valuable insights into sweet sorghum as a novel feedstock for biodiesel, the research aims to guide policymakers, researchers, and industry stakeholders toward the development of a more sustainable and environmentally friendly energy future. In essence, this study contributes to the growing body of knowledge that supports the feasibility and potential of sweet sorghum as a key player in the transition to sustainable and renewable energy sources, with biodiesel production presenting itself as a viable and environmentally conscious alternative to traditional petroleum-based fuels.

References

1 IEA (2021). *World Energy Outlook 2021*. Paris: IEA.
2 IEA (2022). *World Energy Investment 2022*. Paris: IEA.
3 Kevser, M., Tekbaş, M., Doğan, M. et al. (2022). Nexus among biomass energy consumption, economic growth, and financial development: evidence from selected 15 countries. *Energy Rep.* 8: 8372–8380. https://doi.org/10.1016/j.egyr.2022.06.033.
4 Hamadou, B., Djomdi, D., Zieba Falama, R. et al. (2023). Optimization of energy recovery efficiency from sweet sorghum stems by ethanol and methane fermentation processes coupling. *Bioengineered* 14 (1): 228–244.
5 Kashapov, N.F., Nafikov, M.M., Gilmanshin, I.R., and Nigmatzyanov, A.R. (2017, September). Obtaining alternative fuel from sweet sorghum in the conditions of the Republic of Tatarstan. In: *IOP Conference Series: Materials Science and Engineering*, vol. 240, No. 1, 012034. IOP Publishing.
6 López-Sandin, I., Zavala-García, F., Levin, L. et al. (2021). Evaluation of bioethanol production from sweet sorghum variety roger under different tillage and fertilizer treatments. *Bioenergy Res.* 14 (4): 1058–1069.
7 Kasegn, M.M., Simachew, A., Redda, Y.T., and Gebremedhn, H.M. (2023). Production of bioethanol from sweet sorghum [*Sorghum bicolor* L.] juice using yeast isolated from fermented sweet sorghum juice. *Int. Microbiol.* 27: 491–504.
8 Hu, W., Zhou, L., and Chen, J.H. (2022). Conversion sweet sorghum biomass to produce value-added products. *Biotechnol. Biofuels Bioprod.* 15 (1): 1–11.
9 Zhang, W., Song, M., Yang, Q. et al. (2018). Current advance in bioconversion of methanol to chemicals. *Biotechnol. Biofuels* 11 (1): 260. https://doi.org/10.1186/s13068-018-1265-y.
10 Mathur, S., Umakanth, A.V., Tonapi, V.A. et al. (2017). Sweet sorghum as biofuel feedstock: recent advances and available resources. *Biotechnol. Biofuels* 10: 146. https://doi.org/10.1186/s13068-017-0834-9.

11 Xie, Q. and Xu, Z.H. (2019). Sustainable agriculture: from sweet sorghum planting and ensiling to ruminant feeding. *MolPlant* 12 (5): 603–606. https://doi.org/10.1016/j.molp.2019.04.001.

12 Mirfakhar, M., Asadollahi, M.A., Amiri, H., and Karimi, K. (2017). Phenolic compounds removal from sweet sorghum grain for efficient biobutanol production without nutrient supplementation. *IndCropProd* 108: 225231. https://doi.org/10.1016/j.indcrop.2017.06.028.

13 Chen, W., Wu, F.W., and Zhang, J.H. (2016). Potential production of non-food biofuels in China. *Renew. Energy* 85: 939–944. https://doi.org/10.1016/j.renene.2015.07.024.

14 Antar, M., Lyu, D., Nazari, M. et al. (2021). Biomass for a sustainable bioeconomy: an overview of world biomass production and utilization. *Renew. Sustain. Energy Rev.* 139: 110691. https://doi.org/10.1016/j.rser.2020.110691.

15 Ahmad Dar, R., Ahmad Dar, E., Kaur, A., and Gupta, P.U. (2018). Sweet sorghum – a promising alternative feedstock for biofuel production. *Renew. Sustain. Energy Rev.* 82: 4070–4090. https://doi.org/10.1016/j.rser.2017.10.066.

16 Vanamala, J.K.P., Massey, A.R., Pinnamaneni, S.R. et al. (2018). Grain and sweet sorghum (*Sorghum bicolor* L. Moench) serves as a novel source of bioactive compounds for human health. *Crit. Rev. Food Sci.* 58 (17): 2867–2881. https://doi.org/10.1080/10408398.2017.1344186.

17 Maw, M.J.W., Houx, J.H. III, and Fritschi, F.B. (2019). Nitrogen content and use efficiency of sweet sorghum grown in the lower Midwest. *Agron. J.* 111 (6): 2920–2928. https://doi.org/10.2134/agronj2018.08.0489.

18 Tang, S., Wang, Z., Chen, C. et al. (2018). The prospect of sweet sorghum as the source for high biomass crop. *J. Agric. Sci. Bot.* 2 (3): 5–11. https://doi.org/10.35841/2591-7897.2.3.5-11.

19 Jiang, D., Hao, M.M., Fu, J.Y. et al. (2019). Potential bioethanol production from sweet sorghum on marginal land in China. *J. Cleaner Prod.* 220: 225–234. https://doi.org/10.1016/j.jclepro.2019.01.294.

20 Appiah-Nkansah, N.B., Li, J., Rooney, W., and Wang, D.H. (2019). A review of sweet sorghum as a viable renewable bioenergy crop and its techno-economic analysis. *Renew. Energy* 143: 1121–1132. https://doi.org/10.1016/j.renene.2019.05.066.

21 Sebayang, A.H., Masjuki, H.H., Ong, H.C. et al. (2017). Optimization of bioethanol production from sorghum grains using artificial neural networks integrated with ant colony. *Ind. Crops Prod.* 97: 146–155.

22 Fasahati, P., Liu, J.J., Ohlrogge, J.B., and Saffron, C.M. (2019). Process design and economics for production of advanced biofuels from genetically modified lipid-producing sorghum. *Appl. Energy* 239: 1459–1470.

23 Ray, R.C., Uppuluri, K.B., Trilokesh, C., and Lareo, C. (2019). Sweet sorghum for bioethanol production: scope, technology, and economics. In: *Bioethanol Production from Food Crops*, 81–100. Academic Press.

24 Xiao, M.Z., Sun, Q., Hong, S. et al. (2021). Sweet sorghum for phytoremediation and bioethanol production. *J. Leather Sci. Eng.* 3 (1): 1–23.

25 Alphy, M.P., Anjali, K.B., Vivek, N. et al. (2021). Sweet sorghum juice as an alternative carbon source and adaptive evolution of *Lactobacillus brevis* NIE9. 3.3 in sweet sorghum juice and biodiesel derived crude glycerol to improve 1, 3 propanediol production. *J. Environ. Chem. Eng.* 9 (5): 106086.

26 Ammar, E.M., Martin, J., Brabo-Catala, L., and Philippidis, G.P. (2020). Propionic acid production by *Propionibacterium freudenreichii* using sweet sorghum bagasse hydrolysate. *Appl. Microbiol. Biotechnol.* 104: 9619–9629.

27 Stamenković, O.S., Siliveru, K., Veljković, V.B. et al. (2020). Production of biofuels from sorghum. *Renew. Sustain. Energy Rev.* 124: 109769.

28 Vinutha, K.S., Rayaprolu, L., Yadagiri, K. et al. (2014). Sweet sorghum research and development in India: status and prospects. *Sugar Tech* 16: 133–143.

29 Ibeto, C.N., Ofoefule, A.U., and Agbo, K.E. (2011). A global overview of biomass potentials for bioethanol production: a renewable alternative fuel. *Appl. Sci. Res.* 6: 410–425.

30 Gupta, A. and Verma, J.P. (2015). Sustainable bio-ethanol production from agro- residues: a review. *Renew. Sustain. Energy Rev.* 41: 550–567.

31 Guo, Y., Hu, S.Y., Li, Y.R. et al. (2010). Optimization and analysis of a bioethanol agro-industrial system from sweet sorghum. *Renew. Energy* 35: 2902–2909.

10

Sustainable Biodiesel Production with Jatropha as a Renewable Feedstock

10.1 Introduction to Jatropha for Biodiesel Production

The plant *Jatropha curcas* (*J. curcas*) has drawn interest as a potential feedstock for the production of biodiesel as a result of its capacity to grow in poor soils and its high oil content. Nonedible oils typically extracted from oilseeds, nuts, or fruits without the use of chemicals are not rich in nutrients. The castor bean seed, rubber seed tree, and jatropha tree are the three principal types of nonedible oilseed crops [1]. Jatropha falls into this category and is well-recognized for its high oil content [2]. It is a resilient plant that may grow in dry and semi-arid climates because of its ability to withstand heat, drought, and unfavorable soil conditions. It requires extremely little energy to flourish in nonarable soil and dry regions, which require very little water and energy [3]. Jatropha has grown for centuries in South Saharan Africa, South America, Asia, and many countries of the world. It is used for a variety of reasons, including as a hedge or barrier and for medicinal purposes. Jatropha seeds have an oil content of 35%–40% and produce 1892 L/ha of oil annually [4]. With a procedure known as transesterification, the oil from Jatropha seeds can be extracted and transformed into biodiesel [5]. When made from Jatropha oil, biodiesel possesses characteristics that are comparable to those of regular diesel and can be used as a separate fuel in diesel engines or blended with regular diesel [5]. *J. curcas*, among the numerous researched feedstocks, has the finest agronomic features, making it the best triglyceride-based biomass [6].

Jatropha oil is a good alternative fuel for the transportation industry because it does not require any significant changes to the current engine technology. Jatropha cultivation has a number of advantages, including increased energy security, a decrease in greenhouse gas (GHG) emissions, and financial gains for farmers. The Jatropha plant's fruit contains viscous oil that is used to make cream

Feedstocks for Sustainable Biodiesel Production: Characterization, Selection, and Optimization, First Edition. Chinwe P. Okonkwo and Modestus O. Okwu.

and soap in industries such as the cosmetic industry, and the interesting feedstock loss serves as a supernumerary for kerosene and diesel [7]. The plant has the ability to flourish in deteriorated soils, easing the burden on fertile grounds used for food production. Also, the growth of jatropha helps reduce poverty by opening up job opportunities for rural populations.

Jatropha cultivation for biodiesel generation, depends on a number of variables, including the cost of cultivation, oil quality, and yield. The management techniques, soil type and climate, and other factors all affect the plant's yield and oil content. Flashpoint, cloud point, cetane number, kinematic viscosity, cold filter, heating value, plugging point, density, oxidation stability, and iodine number are a few other characteristics that make Jatropha a good biofuel [8–10]. Corn, soybeans, and sugarcane are a few of the sustainable sources used to produce biofuels. One of the most in-demand plants is reportedly jatropha [11–14]. Moreover, the transesterification process' effectiveness as well as the fuel characteristics of biodiesel can be considerably impacted by the quality of jatropha oil for biodiesel production. Another important consideration is the expense of cultivation of the plant as it greatly determines the effectiveness of biodiesel produced from *J. curcas* compared to the traditional biodiesel [5] (Figure 10.1).

Figure 10.1 A pictoral view of *Jatropha curcas* (*Source:* Mang Kelin/Adobe Stock).

10.2 Review of Literature

10.2.1 Exposition of *Jatropha curcas*

Several experiments and comparisons have been performed on the effectiveness of jatropha for biodiesel production. The effects of co-processing were investigated by Biswas et al. [15] using Jatropha oil, high-density polyethylene, and vacuum residue. The use of jatropha oil as a principal fuel in the automobile and aviation industries has drawn significant interest in the energy sector [16]. The Jatropha crop is not entirely utilized during the manufacturing process of Jatropha biofuel (JBF). Seventy-five percent (75%) of Jatropha oil was converted to biodiesel fuels in experiments [17–19]. It has been demonstrated that the remnants from jatropha fruit, which make up 75% of the fruit's weight, have a considerable potential for conversion into a range of energy intermediates, including syngas [20]. The problems associated with jatropha for biofuel production include usage as a sustainable biofuel, oil extraction, conversion, production, and utilization [21].

Due to their broad availability of nonedible oils in many parts of the world, particularly in wastelands, where food oil crops cannot be easily farmed, these oils are currently becoming more popular. Moreover, they are considered more affordable than edible oils, boost production effectiveness, reduce food competition, and stop more trees from being cut down [22]. The economics of the feedstocks and the oil content are the key factors in choosing the oils for fatty acid alkyl esters (FAAE) synthesis [23–26]. Based on the FAAE content, the viability of biodiesel production can be effectively planned. Sunflower is not left out as one of the main feedstocks used today around the world, one other edible oil in use is soybean [27]. Yet, Jatropha, a nonedible oil, has become a viable feedstock in poor nations where there are food and crop shortages [28, 29].

10.2.2 Comparison of Jatropha with Other Biodiesel Feedstocks

One of several potential feedstocks for the manufacturing of biodiesel, jatropha, is frequently contrasted with other feedstocks based on a variety of criteria, including yield, oil content, and compatibility for growing. Jatropha has been compared to soybeans, palm oil, rapeseed, and other biodiesel feedstocks in some studies. According to Aransiola et al. [30], soybean is one of the most extensively produced oilseed crops globally and a common source of biodiesel fuel. Jatropha oil concentration is higher than the 18%–22% range for soybean oil. Yet, soybean farming is more effective and productive than Jatropha, producing more oil per hectare.

Moreover, soybean biodiesel is more suitable for usage in cold areas than Jatropha biodiesel since it has better cold flow characteristics. Another popular biodiesel feedstock is palm oil, particularly in Southeast Asia. The oil content of palm oil is between 20% and 30% higher than that of jatropha. However, the production of palm oil has been linked to deforestation and altered land usage, which has had a huge negative impact on the environment and society. On the other hand, jatropha can be cultivated on marginal grounds, which relieves the burden of food production on fertile lands [31]. Another popular biodiesel feedstock is rapeseed, which is widely grown in Europe and Canada. Rapeseed usage has been thoroughly researched in the literature [32]. When compared to Jatropha, the percentage of rapeseed oil is very higher, ranging from 35% to 45% [33].

Rapeseed cultivation, however, involves higher pesticide and fertilizer inputs, raising environmental issues [34]. Jatropha has a lower influence on the environment due to its lower input requirements and ability to grow on subpar soil, as was already mentioned. Each biodiesel feedstock has benefits and drawbacks, and the selection of a feedstock depends on a number of variables, including yield, oil content, cultivation requirements, and sustainability. While having a lower oil content than some of the other feedstock, jatropha may be cultivated on marginal ground, which relieves strain on fertile lands and benefits rural populations economically. Jatropha studies that matter studies pertaining to the jatropha are reviewed in this section. The majority of research, including study conducted by Atabani et al. [35], and Lim et al. [36], study was focused on the techniques of producing biodiesel from the feedstock. There are other studies that focus more on the availability and opportunities of jatropha as compared with the study conducted by Lim et al. Mazumdar et al. [37] exposed the availability and opportunities of Jatropha. Incorporating techniques for improving the efficiency of biodiesel produced from biodegradable feedstocks has been discussed in high-impact journals. A good example is the study conducted by Alherbawi et al. [38]. This has necessitated further discussion focused on sustainable biofuels.

10.2.3 Sustainable Biofuels and Benefits over Fossil Fuels

Given that they are produced using natural resources, biofuels are regarded as environmentally friendly and sustainable energy sources [39]. Biofuels are made from biodegradable crops and plants. They can be used in a variety of energy-related fields, such as transportation, power generation, and electricity generation. Different scenarios are being considered to estimate the creation of biofuels from diverse sources within the future energy system [40]. The biofuel policy promotes the creation and consumption of fuels derived from biomass. The development of biofuels has grown globally despite the fact that they can have both positive and negative effects [41–44].

The most extensively extracted and used biofuel is bioethanol, also known as ethanol. When combined with gasoline, it can replace fuel in motor vehicles. To enhance octane levels and diminish carbon monoxide and other emissions that

Table 10.1 General classification of fuels.

Fuel type	Primary fuels	Secondary fuels
Gaseous fuels	Biogas, natural gas	Methane, propane, hydrogen, water gas, coal gas, blast furnace gas, etc.
Liquid fuels	Biodiesel, bioalcohols, vegetable oil, petroleum	Ethanol, diesel, gasoline, kerosene, LPG, coal tar, naphtha
Solid fuels	Wood, coal, charcoal, etc.	Charcoal, coke

contribute to smog, ethanol, alcohol, is blended with gasoline [45]. Plant-based fuels are the best option because they are derived from renewable sources that can be cultivated globally and emit less carbon compared to fossil fuels [46]; plant-based fuels are gaining prominence. The transition toward biofuels is becoming increasingly prevalent as it enables a reduction in dependence on commodities, particularly given the speculative nature of crude oil prices [47]. Compared to fossil fuels, biofuels offer more advantages [48–52]. Biofuels have several benefits, but they also have certain drawbacks [53–55], including: Establishing a transparent and enduring policy framework is essential to instill confidence in both the industry and financiers. The evolving requirements of the agricultural and forestry sectors, particularly in terms of biomass feedstock from residues and crops, demand a substantial reorientation of the existing business model. The prevailing technologies for harvesting, storage, and transportation are inadequate for the efficient processing and widespread dispersion of biomass on a large scale. Consequently, the utilization of edible oil fuels is associated with heightened risks of triggering a food crisis in developing nations or adversely impacting consumer prices in affluent nations [56–58]. Using nonedible oils is, therefore, typically preferable. Also, it has been demonstrated in numerous studies that biofuels have more benefits than fossil fuels [59]. Biodegradable feedstocks can be used to produce gaseous, liquid, and solid fuels. See Table 10.1.

10.2.4 Benefit of Jatropha over Other Biofuels

Jatropha's potential as a biofuel is still being investigated, and more study is required to fully grasp its advantages. Despite the plentiful supply of feedstocks for biofuels, jatropha is still in great demand for fuel generation worldwide. Moreover, growing Jatropha on previously wooded land or doing so in a way that displaces food crops might have a negative effect on the ecosystem. The following is a deeper discussion of several advantages: Jatropha seeds are an excellent source of biodiesel due to their high oil content. Jatropha seeds have an oil content that ranges from 30% to 50%, which is higher than that of many other biofuel crops, according to Azam et al. [60]. In their investigation, Asif et al. [61] discovered that the Jatropha seeds can produce 30% or more oil when compared with other plants. Jatropha is a resilient plant that is adaptable enough to thrive in a range of soils and climatic situations. It requires little fertilizer and water and is also capable of growing on marginal lands that are unsuitable for food crops. Jatropha was therefore suggested by Azam et al. [60] and Asif et al. [61] as the best plant for making biodiesel.

Jatropha plants can also absorb carbon dioxide from the air, making them a potential aid in the fight against climate change. Jatropha is not a food crop, unlike some other biofuel crops like corn or soybeans. As a result, it does not compete with food crops for resources like land or water, which can lower food prices and improve food security. Jatropha is a resilient plant that can grow in poor soils with little support, making its cultivation relatively inexpensive. This may contribute to lowering the price of producing biofuel and making it more accessible. According to Rezania et al. [5] and Akubude et al. [62], alcoholysis biodiesel extracts from Jatropha or transesterification is possible through the reaction of a fatty acid with an alcohol.

According to Moazeni et al. [63], the aforementioned procedure can produce biodiesel fatty acid methyl esters (FAME) and glycerol with or without the use of a catalyst. The quality of the FAME can be predicted by factors including fatty acid composition, calorific value, saponification number, and iodine value, which is high for Jatropha [64, 65]. Since Jatropha satisfies all of the requirements of the biodiesel standard, it is the ideal option for use as biodiesel [66]. Initial investment requirements for plant cultivation may be rather minimal, as proposed by Van Eijck and Romijn [67]. According to Najaf et al. [68], Jatropha biodiesel provides benefits for lowering SO_2, CO, and CO_2. The benefits of Jatropha is enormous one important relevant of the plant is its ability to grow in water-scarce area and harsh environments [69]. The experimental investigation performed by Dyer et al. [70] showed that the plant does not require so much water to thrive, it is known as a drought resistant plant, this gives *J. curcas* a greater advantage and benefit compared with other plants since cultivation can be done at any given season.

10.2.5 Benefit of Cultivation of Jatropha for Biodiesel Production

Jatropha cultivation has a number of advantages, including a decrease in GHG emissions, increased energy security, and financial gains for farmers. The plant has the ability to flourish in deteriorated soils, easing the burden on fertile grounds used for food production. According to Forson et al. [71], *J. curcas* may thrive in tropical and subtropical regions. The plant can thrive on practically any surface, including rocky, sandy, and salinized soils, they observed. Moreover, it can thrive in wasteland. Furthermore, the cultivation of *J. curcas* can open doors of opportunities in terms of employment in communities, thereby contributing greatly to poverty reduction. Though, there are several factors to be considered when performing a study based on the commercial viability of Jatropha cultivation majorly for biodiesel production. Some of the factors include the cost of cultivation, quality of oil, and yield. The availability of the Jetropha plant is exceptional, and the production process is considered simpler than that of other crops [72]. The soil type, climate, cultivar, and management techniques all affect the plant's yield and oil content. Moreover, the transesterification process' effectiveness as well as the fuel characteristics of biodiesel can be considerably impacted by the quality of jatropha oil. Another important aspect that affects how competitive Jatropha biodiesel is against regular fuel is the cost of growing.

In summary, *J. curcas* has great potential as a sustainable source of biodiesel, but its commercial viability depends on several factors, which have been discussed in the previous chapter. With technological advancements and proper management practices, the cultivation of *J. curcas* can contribute greatly and provide a sustainable solution to the up-and-coming demand for biofuels, particularly in emerging countries where there is limited availability of land. Souza et al. [73] emphasized on the general opportunities of *J. curcas* in terms of producing variety of products such lubricants, as fuels, cosmetics, binders, pesticides, candles, soaps, and others.

10.3 Materials and Method

10.3.1 Production of Biodiesel from *Jatropha curcas*

The production of biodiesel from *J. curcas* is an interesting process. This section is focused on the step-by-step method using a visual aid. The flow chart showing the process of biodiesel production process from Jatropha is presented in Figure 10.2. Information on the studies focused on biodiesel production from Jatropha is available in research documented by forbearer [74–76].

The detailed explanation of the flow chart of Figure 10.2 detailing the process of biodiesel production from Jatropha:

Jatropha cultivation: The first process of the production of biodiesel from Jatropha is the sourcing of the plant. This is easy when it is cultivated on a large portion of land. The plants are grown on marginal lands by means of sustainable practices such as agroforestry and intercropping.

Harvesting and oil extraction: Harvesting is the second stage before extraction of the oil. Ripe Jatropha fruits are harvested, and the oil is extracted. The extraction can be done manually, by using advanced method which involves using mechanical pressing or by solvent extraction.

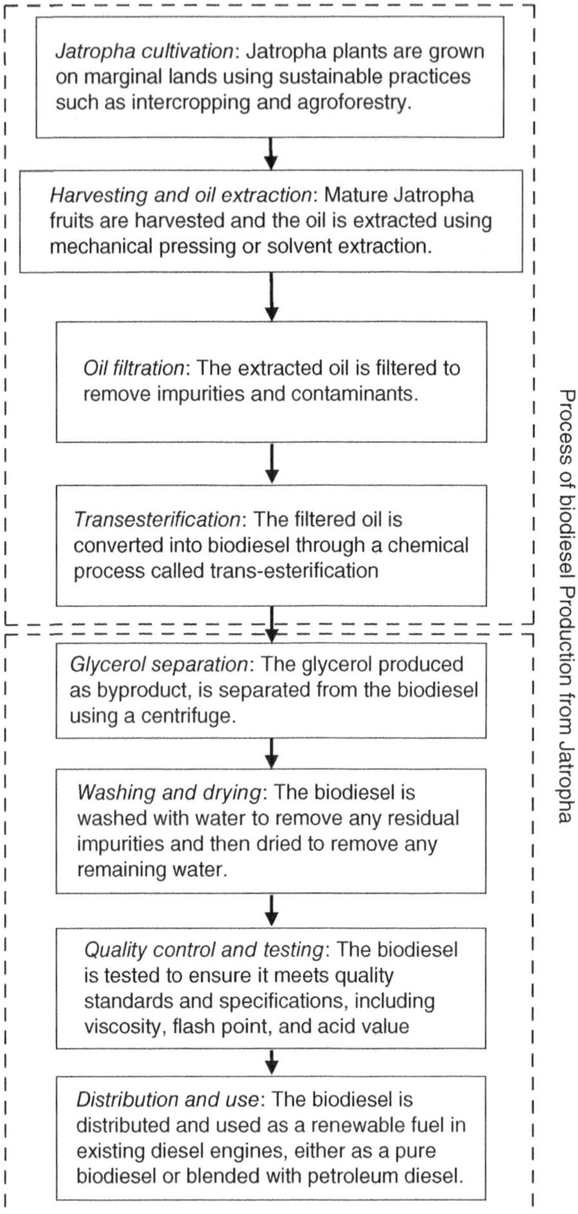

Figure 10.2 Flowchart showing the step-by-stepprocess of producing biodiesel from Jatropha.

Oil filtration: The next stage after extraction of oil is the filtration process. This process has to do with filtering the oil to remove any form of and contaminants and impurities.

Transesterification: Following the filtration process, the purified oil undergoes a chemical transformation into biodiesel through the process of transesterification. This process entails the reaction of the oil with an alcohol, typically methanol, and a catalyst, usually potassium hydroxide or sodium hydroxide.

Glycerol separation: The transesterification process produces glycerol as a by-product, which is separated from the biodiesel using a centrifuge.

Washing and drying: After glycerol separation, the biodiesel produced is then washed with water. This is to remove any additional residual impurities. It is later dried to do away with any remaining water.

Quality control and testing: After washing and drying out water, the biodiesel is tested to ensure that it meets the quality standards and specifications, including acid value. Flashpoint, viscosity, and other quality test.

Distribution and use as a renewable fuel: The biodiesel is distributed and used as a renewable fuel in diesel engines. For easy use, it can be blended with petroleum diesel or used as a pure biodiesel.

Summarily, the biodiesel production process from Jatropha involves several steps, including sourcing of Jotropha seed – the sourcing can be via buying or cultivation, the next step is the extraction of oil, the third process is the transesterification, followed by glycerol separation, then washing and drying process, quality control and testing, and the final process is the distribution and use as a renewable fuel. With proper implementation of sustainable practices and quality control measures, Jatropha biodiesel is able to contribute to a more renewable and sustainable future.

10.3.2 The Transesterification Reaction and Its Mechanism

Jatropha seed containing oil and protein obtained from harvested Jatropha plants will be ground (usually mechanically or via other means) together with processing of the residue via pressing [77]. The traditional extraction technique and transesterification procedure filter the plant oil to create biodiesel, according to Nayab et al. [78]. The process through which triglycerides (fats and oils) are transformed into biodiesel is known as the transesterification reaction. This reaction results in the synthesis of methyl or ethyl ester, which is the chemical name for biodiesel, and a glycerol molecule by exchanging one ester group from the triglyceride molecule with an alcohol molecule.

Transesterification is the best process for converting edible and nonedible oils into biodiesel, according to Ayoob and Fadhil [79]. They said that transesterification, also known as alcoholysis, is the process by which triglycerides interact with alcohol in fat or oil to form glycerol and biodiesel FAME, with or without the use of a catalyst. Eqs. (10.1–10.4) depict the ideal transesterification reaction and its mechanism.

$$\text{Triglycerides} + \text{metanol} \rightarrow \text{glycerol} + \text{fame} \tag{10.1}$$

$$\text{Triglyceride} + \text{R}^1\text{OH} \rightarrow \text{diglyceride} + \text{RCOOR}^1 \tag{10.2}$$

$$\text{Diglyceride} + \text{R}^1\text{OH} \rightarrow \text{monoglyceride} + \text{RCOOR}^1 \tag{10.3}$$

$$\text{Monoglyceride} + \text{R}^1\text{OH} \rightarrow \text{glycerol} + \text{RCOOR}^1 \tag{10.4}$$

A catalyst, usually sodium hydroxide or potassium hydroxide, is used in the process to speed up the reaction and stimulate the separation of the ester and glycerol layers. According to Ayooband Fadhil [79], the transesterification reaction has three key steps: An alkoxide ion and a cation, such as sodium or potassium, are produced when the catalyst, such as sodium hydroxide, dissociates. An intermediary substance known as a diglyceride is created when the alkoxide ion interacts with the triglyceride molecule and breaks the ester link between the glycerol backbone and the fatty acid chains. A single fatty acid chain is released from the triglyceride molecule as a result of this process.

The chemical name for biodiesel is methyl or ethyl ester, which is created when the diglyceride combines with another alcohol molecule (such as methanol or ethanol). The triglyceride molecule's final two fatty acid chains are released as a result of this process. The following succinct summary of the reaction mechanism:

$$\text{Triglyceride} + \text{alcohol} \rightarrow \text{biodiesel} + \text{glycerol} \tag{10.5}$$

As it affects both the effectiveness of the process of conversion and the biodiesel quality produced, the transesterification reaction is a crucial stage in the manufacture of biodiesel. The effectiveness of the transesterification reaction and the quality of the biodiesel produced can be influenced by a number of variables, which include the catalyst type and concentration, the molar ratio of alcohol, reaction temperature, and others. There is a catalyst present during the process.

Due to the reversible nature of this reaction, a substantial amount of alcohol is required to drive the equilibrium reaction forward [80]. The transesterification process leads to the conversion of triglycerides into diglycerides, monoglycerides, and glycerol. Consequently, each glyceride at every stage yields one alkyl ester molecule [81]. Musa highlights that methanol and ethanol are the most prevalent alcohols employed in this process [82]. Ethanol, derived from agricultural feedstocks, is favored over methanol for biodiesel production. However, methanol is commonly chosen due to its cost-effectiveness and advantageous physical and chemical properties.

The cloud and pour points, along with the viscosities of biodiesel produced from methanol, exhibit slightly higher values compared to biodiesel derived from ethanol. Moreover, unlike ethanol, methanol is easily recyclable and does not form an azeotrope [83, 84]. This highlights the versatility of alcohol choices in transesterification, with each alcohol offering distinct advantages and considerations in biodiesel production.

The transesterification process, as outlined by Wang et al. [85], is capable of occurring with or without a catalyst. Catalysts utilized in this process can be enzymatic, alkaline, or acidic. In addition, the solubility of the catalyst in the reactant determines whether the transesterification process is catalyzed homogeneously or heterogeneously [86]. Acidic and alkaline catalysts may exhibit homogeneous or heterogeneous characteristics based on their solubility in the reactants. The reactions can unfold in a single stage with either an acid or a basic catalyst or in two phases with both acidic and basic catalysts, contingent on the quantity of free fatty acids (FFA) present in the oil [87].

As proposed by Tiwari et al. [28], the esterification of FFA with alcohol in acidic catalysts leads to the production of biodiesel and water in the initial stage. The mechanism of the esterification reaction is depicted in Eq. (10.6), and this process is carried out in the presence of a catalyst. The choice of catalyst and its characteristics play a crucial role in determining the efficiency and outcome of the transesterification process in biodiesel production.

$$\text{FFA}\left(\text{carboxylic acid}\right) + \text{alcohol} \rightarrow \text{biodiesel} + \text{water} \tag{10.6}$$

10.4 Economic and Environmental Aspects of Jatropha Biodiesel Production

Jatropha biodiesel production has both economic and environmental aspects that are important to consider.

10.4.1 Economic Aspects of Jatropha Biodiesel Production

In a comprehensive investigation by Tapanes et al. [88], the focus was directed toward biodiesel production using *J. curcas*. The study revealed that the seeds of *J. curcas* boast an impressive oil content, reaching up to 60%, a notably high proportion when compared to other biodegradable crops. Gui et al. [89] further emphasized the economic viability of Jatropha oil production, citing a production cost of approximately US$ 0.39/kilogram (kg). In contrast, an alternative experiment considering soybean oil production reported overall costs of approximately US$ 1.65/kg.

This comparison underscores the economic advantage of Jatropha oil production when compared to other biodegradable crops or feedstock, positioning it as a cost-effective and promising option. The substantial oil content, coupled with favorable production costs, contributes to the plant's myriad prospects in the realm of biodiesel production. The findings from this study not only highlight the economic feasibility of *J. curcas* as a biodiesel feedstock but also emphasize its potential to play a significant role in sustainable and cost-efficient biofuel production. Some of the economic aspects of jatropha production biodiesel include:

Cost-effectiveness: J. curcas is considered to be cost-effective. It can be grown on marginal lands which are considered unsuitable for food production. This makes it a cost-effective option for easy production of biodiesel since it does not compete with food crops in terms of land use.

Income generation: The cultivation of Jatropha has great advantages in terms of income generation. The cultivation is capable of providing an additional source of income for farmers in developing and emerging nations of the world. It can help to alleviate poverty and promote rural development.

Job creation: The cultivation of Jatropha for biodiesel production can help create job opportunities in urban and rural settings, stimulating economic development and growth.

Energy security: This is a great advantage of Jatropha biodiesel. It can serve as alternative energy, thereby reducing a country's reliance on imported fossil fuels. This has a way of improving and reducing the impact of fluctuations in fuel prices on the economy and energy security.

10.4.2 Environmental Aspects of Jatropha Biodiesel Production

Lower GH Gemissions: Jatropha biodiesel possesses the capability to decrease GHG emissions in comparison to fossil fuels known for the emission of carbon dioxide, which is a major GHG and a primary source of pollution. The use of renewable energy based on biodiesel from *J. curcas* will help to mitigate climate change.

Land-use change reduction: Land-use change is reduced by encouraging the use of Jatropha for biodiesel. As Jatropha cultivation requires marginal area, less natural ecosystems like forests or grasslands are converted into agricultural land, lowering the carbon footprint of biodiesel production.

Soil conservation: It has been discovered that jatropha has a beneficial effect on soil fertility and conservation, assisting in lowering soil erosion and maintaining soil health, particularly in degraded regions.

Water conservation: Jatropha is a more sustainable option in areas with scarce water resources because it uses less water than other energy crops like soybean, corn, or sugarcane.

Biodiversity: Jatropha can enhance biodiversity in degraded lands and provide habitat for wildlife. This will greatly support the ecosystem and enhance ecological resilience. In summary, job creation, energy security, cost-effectiveness, revenue generation, reduced land-use change, soil and water conservation, reduced GHG emissions, and biodiversity promotion are just a few of the economic and environmental advantages of jatropha biodiesel production. However, in order to optimize the positive effects and reduce the negative effects of Jatropha biodiesel production, considerable consideration must be given to aspects including land usage, growing techniques, and the employment of appropriate technology.

10.4.3 Cost–Benefit Analysis of Cultivating Jatropha

An evaluation of the economic viability of investing in Jatropha cultivation for biodiesel production and a comparison of the costs and advantages of such an investment constitutes a cost–benefit analysis of Jatropha cultivation in biodiesel. The costs and benefits of Jatropha cultivation for the production of biodiesel can be divided into several classifications:

Investment costs: These are the expenses related to establishing a jatropha farm, including labor, fertilizers, planting, irrigation, land clearing, and others.
Maintenance costs: These are the expenses related to preserving the cultivation of jatropha, including manpower, fertilizers, weed management, etc.
Cost of harvesting: These are expenses involved with the harvesting of Jatropha. It includesthe cost of equipment and labor cost.
Processing costs: These are the expenses related to extracting oil from Jatropha seeds, refining them, and transesterifying the jatropha into biodiesel.
Revenue: The income generated from sales of biodiesel produced from Jatropha seeds and other by-product of Jatropha seeds is referred to as revenue.
Environmental and social benefits: These include other benefits obtained from the cultivation of *J. curcus*, such as energy security, reduced GHG emissions, job creation, and rural development.

Based on the aforementioned factors, the economic feasibility of investing in Jatropha cultivation for the sake of biodiesel production can be determined by performing cost–benefit analysis. The analysis involves comparing the total revenue generated from the sale of biodiesel obtained from jatropha seed oil and other by-products with the total costs of investment. It is equally important to take into consideration the environmental and social benefits of the investment.

The availability of adequate land for Jatropha cultivation, the cost of inputs like fertilizers and labor, the yield of Jatropha seeds, and the market price of biodiesel are just a few of the variables that will affect the outcomes of a cost–benefit analysis. Encouraging internal rate of return and net present value for investors is indicative of an economically viable investment in a well-planned and managed Jatropha farming project. The potential risks and uncertainties related to the cultivation of Jatropha, such as yield unpredictability, fluctuation in price, and environmental issues, should also be considered in the cost–benefit analysis. These hazards can be reduced, and the success of jatropha cultivation for biodiesel production can be ensured with careful planning, implementation, and monitoring.

10.4.4 Environmental Impacts of Jatropha Cultivation and Biodiesel Production

While biodiesel produced from *J. curcas* has several environmental benefits, its cultivation and the process of production can also have negative environmental impacts, which must be carefully considered. Some of the environmental impacts of cultivating Jatropha and the production of biodiesel production from available seed include the following:

Change in land use: The cultivation of Jatropha on a large-scale may require the transformation of natural ecosystems into agricultural land a good example is the grasslands or forests. This can result in fragmentation and habitat loss, this can in every way have a very strong negative impact on ecosystem services and biodiversity.

Soil erosion: Jatropha cultivation on steep slopes or marginal lands can lead to soil erosion and loss of soil fertility, particularly if proper soil conservation practices are not in place.

Water use: Jatropha requires water for growth and production, and large-scale cultivation in regions with limited water resources can lead to water scarcity and competition with other water users, including local communities and ecosystems.

Use of chemical inputs: The use of fertilizers and pesticides in Jatropha cultivation can have negative impacts on soil and water quality and may result in pollution of the surrounding environment.

GHG *emissions:* While Jatropha biodiesel has the potential to reduce GHG emissions, the cultivation and production of Jatropha biodiesel can also result in emissions from the use of fossil fuels in farm machinery, fertilizer production, and transportation.

Biodiversity loss: Large-scale Jatropha cultivation can lead to the loss of biodiversity and ecosystem services, particularly if it replaces natural ecosystems, which provide critical habitat for wildlife.

To minimize these environmental impacts, sustainable practices in Jatropha cultivation and biodiesel production are required. Sustainable practices include:

Use of degraded lands: Jatropha cultivation should be done on degraded lands and lands that are not suitable for food production to avoid competition with food crops.

Conservation of natural ecosystems: Jatropha cultivation should not replace natural ecosystems but rather be done on degraded lands to minimize habitat loss and fragmentation.

Use of sustainable cultivation practices: Sustainable cultivation practices such as conservation tillage, crop rotation, and intercropping should be employed to reduce soil erosion and maintain soil fertility.

Reduced use of chemical inputs: Integrated pest management practices should be employed to reduce the use of chemical inputs and minimize their impact on soil and water quality.

Efficient water use: Sustainable water management practices, such as drip irrigation and rainwater harvesting, should be employed to reduce water use and competition with other water users.

Use of renewable energy: Renewable energy sources such as solar and wind power can be used in farm machinery and biodiesel production to reduce GHG emissions.

It is important to note that Jatropha biodiesel production can have both positive and negative environmental impacts and sustainable practices are required to minimize these impacts. It is important to carefully consider the environmental impacts of Jatropha cultivation and biodiesel production when making decisions regarding its use as a renewable energy source.

10.4.5 Comparison of Jatropha Biodiesel with Other Biodiesel Sources in Terms of Economic and Environmental Sustainability

In terms of economic and environmental sustainability, Jatropha biodiesel has both advantages and disadvantages when compared to other biodiesel feedstocks. Some of the key comparisons are as follows: Yield: Jatropha has a lower yield compared to other feedstocks, such as soybean and palm oil. This means that larger areas of land are required to produce the same amount of biodiesel, which can have implications for land use and environmental impacts; Land Use: Jatropha can be grown on marginal lands that are not suitable for food production, while other biodiesel feedstocks such as soybean and palm oil require fertile agricultural land. This can reduce competition with food crops and help to minimize the environmental impact of biodiesel production; Environmental Impact: Jatropha has the potential to reduce GHG emissions, but its cultivation and production can also have negative environmental impacts, such as habitat loss and soil erosion. Other feedstocks such as algae and waste cooking oil may have lower environmental impacts; Cost: The cost of Jatropha biodiesel production can be higher than other feedstocks such as soybean and palm oil due to lower yields and higher production costs. However, the cost of Jatropha biodiesel production can be competitive with other feedstocks if sustainable cultivation practices and efficient production methods are employed; Feedstock Availability: Jatropha is a nonfood crop and can be grown in many regions of the world, which can reduce the risk of supply chain disruptions compared to other feedstocks such as soybean, which may be subject to supply chain risks; Social Impact: Jatropha cultivation can provide economic benefits to local communities and smallholder farmers in developing countries. However, there are also concerns about the potential displacement of local communities and the impact on food security. Summarily, the economic and environmental sustainability of Jatropha biodiesel compared to other feedstocks depends on several factors, including yield, land use, environmental impact, cost, feedstock availability, and social impact. While Jatropha has the potential to be a sustainable feedstock for biodiesel production, it is important to carefully consider the specific context and local conditions when making decisions about its use.

10.4.6 Current Status and Future Prospects of Jatropha Biodiesel Production

The current status of Jatropha biodiesel production is mixed. While there was a lot of excitement around Jatropha as a potential biodiesel feedstock in the mid-2000s, the crop has faced significant challenges and setbacks. Some of the main issues that have hindered the development of Jatropha biodiesel production include:

Yield variability: Jatropha has proven to be a difficult crop to cultivate, with significant yield variability depending on soil conditions, climate, and other factors.

High production costs: The cost of producing Jatropha biodiesel has been higher than other feedstocks due in part to the challenges of cultivation and the need for specialized equipment.

Limited investment: While there was initial investment in Jatropha biodiesel production, interest has waned in recent years due to the challenges mentioned above, as well as competition from other feedstocks.

Sustainability concerns: There are concerns about the environmental and social sustainability of Jatropha cultivation, particularly if large areas of land are converted from food production to Jatropha cultivation.

Despite these challenges, there is still potential for Jatropha biodiesel production in the future. Advances in technology, such as precision agriculture and genetic engineering, could help to increase yields and reduce production costs. In addition, there is increasing interest in sustainable biofuels, which could create opportunities for Jatropha as a nonfood crop that can be grown on marginal lands. There is also potential for Jatropha to be used in combination with other feedstocks, such as waste cooking oil, to create more sustainable and cost-effective biodiesel blends.

In conclusion, the future prospects of Jatropha biodiesel production depend on several factors, including advances in technology, sustainability considerations, and the development of sustainable biofuel markets. While there are challenges to overcome, Jatropha has the potential to be a valuable feedstock for biodiesel production, particularly in regions where land is not suitable for food crops.

10.5 Challenges, Opportunities, and Prospects for Jatropha Biodiesel Production

10.5.1 Challenges for Jatropha Biodiesel Production Include

Yield variability: Jatropha is a challenging crop to grow, and yields can vary significantly depending on soil quality, climate, and other factors.
High production costs: Jatropha cultivation and biodiesel production can be expensive, particularly compared to other feedstocks like soybean and palm oil.
Sustainability concerns: There are concerns about the environmental and social sustainability of large-scale Jatropha cultivation, particularly if it leads to deforestation or displacement of food crops.
Market competition: Jatropha faces competition from other biodiesel feedstocks, particularly those that are already established and have lower production costs.

10.5.2 Opportunities for Jatropha Biodiesel Production Include

Nonfood crop: Jatropha is not a food crop and can be grown on marginal lands, reducing competition with food crops and reducing pressure on land use.
Diversification: Jatropha biodiesel production can provide a source of income for smallholder farmers and rural communities in developing countries, contributing to rural development and poverty reduction.

Sustainable biofuels: There is growing interest in sustainable biofuels, which could create new opportunities for Jatropha as a nonfood crop that can be grown on marginal lands.

Technological advances: Advances in technology, such as precision agriculture and genetic engineering, could help to increase yields and reduce production costs.

Government support: Governments in some countries have provided support for Jatropha cultivation and biodiesel production, creating incentives and policies that promote the development of the industry.

In summary, while there are challenges to Jatropha biodiesel production, there are also opportunities for the crop to be developed as a sustainable feedstock for biodiesel. Advances in technology and sustainable biofuel markets, along with government support and a focus on sustainable production practices, could help to overcome the challenges and create a more viable future for Jatropha biodiesel production.

10.5.3 Potential for Jatropha in the Context of the Bio-Economy and Sustainable Development

Jatropha has potential in the context of the bio-economy and sustainable development as a nonfood crop that can be grown on marginal lands, providing a source of income for smallholder farmers and rural communities in developing countries. Jatropha cultivation can contribute to sustainable development by creating jobs, reducing poverty, and promoting rural development.

In addition, Jatropha can contribute to the bio-economy by providing a source of renewable energy that can reduce dependence on fossil fuels and mitigate climate change. Jatropha biodiesel has the potential to reduce GHG emissions compared to fossil fuels and can be used in existing diesel engines with little modification. Jatropha cultivation and biodiesel production can also provide valuable co-products, such as glycerol, which can be used in the production of other products.

However, it is important to ensure that Jatropha cultivation and biodiesel production are sustainable and do not cause environmental or social harm. Sustainable practices such as agroforestry, intercropping, and land-use planning can help to minimize environmental impacts, while policies and regulations can ensure that Jatropha cultivation does not displace food crops or contribute to deforestation.

In summary, Jatropha has the potential to contribute to the bio-economy and sustainable development by providing a source of renewable energy, creating jobs and income for rural communities, and reducing GHG emissions. However, it is important to ensure that Jatropha cultivation and biodiesel production are sustainable and do not cause harm to the environment or communities.

10.6 Conclusion

In conclusion, Jatropha has been recognized as a prospective eco-friendly feedstock for the production of biodiesel due to its high oil content and ability to grow on marginal lands. However, the economic and environmental viability of Jatropha as a feedstock is still a topic of debate. The benefits of Jatropha biodiesel have been discussed,including its potential to reduce GHG emissions, provide income for smallholder farmers, and contribute to rural development. Also, the challenges and limitations have been examined. Despite these challenges, there are opportunities for Jatropha biodiesel production, including the development of sustainable cultivation practices, technological advances, and government support. Furthermore, Jatropha has potential in the context of the bio-economy and sustainable development, providing a renewable energy source and contributing to rural development and poverty reduction. In light of the potential benefits of Jatropha biodiesel. It is recommended that further research should focus on the development of sustainable Jatropha cultivation practices, as well as continued technological advancements to improve yield and reduce production costs. Additionally, policies and regulations that promote the sustainable development of Jatropha as a biodiesel feedstock, including land-use planning, environmental safeguards, and social inclusion measures, should be implemented. Conclusively, Jatropha has a significant role to play in the global shift toward renewable energy sources and sustainable development. With continued research and development, the full potential of Jatropha as a biodiesel feedstock can be out locked and *J. curcas* can contribute to a more sustainable future.

References

1 Ntaribi, T. and Paul, D.I. (2019). The economic feasibility of Jatropha cultivation for biodiesel production in Rwanda: a case study of Kirehe district. *Energy Sustain. Dev.* 50: 27–37.

2 Divakara, B.N., Upadhyaya, H.D., Wani, S.P., and Gowda, C.L.L. (2010). Biology and genetic improvement of *Jatropha curcas* L.: a review. *Appl. Energy* 87: 732–742.

3 Ahmad, P. and Wani, M.R. (2014). *Physiological Mechanisms and Adaptation Strategies in Plants under Changing Environment*, vol. 2. Berlin, Germany: Springer Science & Business Media.

4 Athar, M. and Zaidi, S. (2020). A review of the feedstocks, catalysts, and intensification techniques for sustainable biodiesel production. *J. Environ. Chem. Eng.* 8: 104523.

5 Rezania, S., Oryani, B., Park, J. et al. (2019). Review on transesterification of non-edible sources for biodiesel production with a focus on economic aspects, fuel properties and by-product applications. *Energy Convers. Manag.* 201: 112155.

6 Alherbawi, M., AlNouss, A., Mckay, G., and Al-Ansari, T. (2020). Optimum utilization of Jatropha seedcake considering the energy, water and food Nexus. In: *Computer Aided Chemical Engineering*, vol. 48, 229–234. Amsterdam, The Netherlands: Elsevier.

7 Ezzi, A.A., Fayad, M.A., Al Jubori, A.M. et al. (2022). Influence of fuel injection pressure and RME on combustion, NOx emissions and soot nanoparticles characteristics in common-rail HSDI diesel engine. *Int. J.* 15: 100173.

8 Ramos, M.J., Fernández, C.M., Casas, A. et al. (2009). Influence of fatty acid composition of raw materials on biodiesel properties. *Bioresour. Technol.* 100: 261–268.

9 Gopinath, A., Sairam, K., Velraj, R., and Kumaresan, G. (2015). Effects of the properties and the structural configurations of fatty acid methyl esters on the properties of biodiesel fuel: a review. *Proc. Inst. Mech. Eng. Part D J. Automob. Eng.* 229: 357–390.

10 Agarwal, M., Singh, K., and Chaurasia, S.P. (2010). Prediction of biodiesel properties from fatty acid composition using linear regression and ANN techniques. *Ind. Chem. Eng.* 52: 347–361.

11 Koçar, G. and Civas, N. (2013). An overview of biofuels from energy crops: current status and future prospects. *Renew. Sust. Energ. Rev.* 28: 900–916.

12 Oumer, A.N., Hasan, M.M., Baheta, A.T. et al. (2018). Bio-based liquid fuels as a source of renewable energy: a review. *Renew. Sust. Energ. Rev.* 88: 82–98.

13 Saravanan, A.P., Mathimani, T., Deviram, G. et al. (2018). Biofuel policy in India: a review of policy barriers in sustainable marketing of biofuel. *J. Clean. Prod.* 193: 734–747.

14 Adewuyi, A. (2020). Challenges and prospects of renewable energy in Nigeria: a case of bioethanol and biodiesel production. *Energy Rep.* 6: 77–88.

15 Biswas, S. and Sharma, D.K. (2013). Co-cracking of jatropha oil, vacuum residue and HDPE and characterization of liquid, gaseous and char products obtained. *J. Anal. Appl. Pyrolysis* 101: 17–27.

16 Madihi, R., Pourfallah, M., Gholinia, M. et al. (2022). Thermofluids analysis of combustion, emissions, and energy in a biodiesel ($C_{11}H_{22}O_2$)/natural gas heavy-duty engine with RCCI mode (part II: fuel injection time/fuel injection rate). *Int. J. Thermofluids* 16: 100200.

17 Verma, D., Rana, B.S., Kumar, R. et al. (2015). Diesel and aviation kerosene with desired aromatics from hydroprocessing of jatropha oil over hydrogenation catalysts supported on hierarchical mesoporous SAPO-11. *Appl. Catal. A Gen.* 490: 108–116.

18 Chen, H., Wang, Q., Zhang, X., and Wang, L. (2014). Hydroconversion of jatropha oil to alternative fuel over hierarchical ZSM-5. *Ind. Eng. Chem. Res.* 53: 19916–19924.

19 Kham-or, P., Suwannasom, P., and Ruangviriyachai, C. (2016). Effect of agglomerated NiMo HZSM-5 catalyst for the hydrocracking reaction of *Jatropha curcas* oil. *Energy Sources Part A Recover. Util. Environ. Eff.* 38: 3694–3701.

20 Maiti, S., Bapat, P., Das, P., and Ghosh, P.K. (2014). Feasibility study of jatropha shell gasification for captive power generation in biodiesel production process from whole dry fruits. *Fuel* 121: 126–132.

21 Moniruzzaman, M., Yaakob, Z., Shahinuzzaman, M. et al. (2017). Jatropha biofuel industry: the challenges. *Front. Bioenergy Biofuels* 1: 23–256.

22 Mofijur, M., Masjuki, H.H., Kalam, M.A. et al. (2012). Prospects of biodiesel from Jatropha in Malaysia. *Renew. Sust. Energ. Rev.* 16: 5007–5020.

23 Karmakar, B. and Halder, G. (2019). Progress and future of biodiesel synthesis: advancements in oil extraction and conversion technologies. *Energy Convers. Manag.* 182: 307–339.

24 Fonseca, J.M., Teleken, J.G., de Cinque Almeida, V., and da Silva, C. (2019). Biodiesel from waste frying oils: methods of production and purification. *Energy Convers. Manag.* 184: 205–218.

25 Bankovic-Ilic, I.B., Stamenkovic, O.S., and Veljkovic, V.B. (2012). Biodiesel production from non-edible plant oils. *Renew. Sust. Energ. Rev.* 16: 3621–3647.

26 Ahmad, A.L., Yasin, N.H.M., Derek, C.J.C., and Lim, J.K. (2011). Microalgae as a sustainable energy source for biodiesel production: a review. *Renew. Sust. Energ. Rev.* 15: 584–593.

27 Bezergianni, S., Kalogianni, A., and Vasalos, I.A. (2009). Hydrocracking of vacuum gas oil-vegetable oil mixtures for biofuels production. *Bioresour. Technol.* 100: 3036–3042.

28 Tiwari, A.K., Kumar, A., and Raheman, H. (2007). Biodiesel production from jatropha oil (*Jatropha curcas*) with high free fatty acids: an optimized process. *Biomass Bioenergy* 31: 569–575.

29 Berchmans, H.J. and Hirata, S. (2008). Biodiesel production from crude *Jatropha curcas* L. seed oil with a high content of free fatty acids. *Bioresour. Technol.* 99: 1716–1721.

30 Aransiola, E., Betiku, E., Layokun, S., and Solomon, B. (2010). Production of biodiesel by transesterification of refined soybean oil. *Int. J. Biol. Chem. Sci.* 4.

31 Martis, R., Al-Othman, A., Tawalbeh, M., and Alkasrawi, M. (2020). Energy and economic analysis of date palm biomass feedstock for biofuel production in UAE: pyrolysis, gasification and fermentation. *Energies* 13: 5877.

32 Dufour, J., Arsuaga, J.S., Moreno, J. et al. (2013). Comparative life cycle assessment of biodiesel production from cardoon (*Cynara cardunculus*) and rapeseed oil obtained under Spanish conditions. *Energy Fuel* 27 (9): 5280–5286.

33 Firrisa, M.T., van Duren, I., and Voinov, A. (2014). Energy efficiency for rapeseed biodiesel production in different farming systems. *Energy Effic.* 7 (1): 79–95. https://doi.org/10.1007/s12053-013-9201-2.

34 NHA (2016). *Canola or Rapeseed Production Guidelines*. New Holland Agriculture.

35 Atabani, A.E., Silitonga, A.S., Ong, H.C. et al. (2013). Non-edible vegetable oils: a critical evaluation of oil extraction, fatty acid compositions, biodiesel production, characteristics, engine performance and emissions production. *Renew. Sust. Energ. Rev.* 18: 211–245.

36 Lim, B.Y., Shamsudin, R., Baharudin, B.T.H.T., and Yunus, R. (2015). A review of processing and machinery for *Jatropha curcas* L. fruits and seeds in biodiesel production: harvesting, shelling, pretreatment and storage. *Renew. Sust. Energ. Rev.* 52: 991–1002.

37 Mazumdar, P., Singh, P., Babu, S. et al. (2018). An update on biological advancement of *Jatropha curcas* L.: new insight and challenges. *Renew. Sust. Energ. Rev.* 91: 903–917.

38 Alherbawi, M., McKay, G., Mackey, H.R., and Al-Ansari, T. (2021). A novel integrated pathway for jet biofuel production from whole energy crops: a *Jatropha curcas* case study. *Energy Convers. Manag.* 229: 113662.

39 Živkovic, S. and Veljkovic, M. (2018). Environmental impacts the of production and use of biodiesel. *Environ. Sci. Pollut. Res.* 25: 191–199.

40 Ardabili, S., Mosavi, A., and Várkonyi-Kóczy, A.R. (2020). Systematic review of deep learning and machine learning models in biofuels research. *Lect. Notes Netw. Syst.* 101: 19–32.

41 Kaletnik, H., Pryshliak, V., and Pryshliak, N. (2019). Public policy and biofuels: energy, environment and food trilemma. *J. Environ. Manag. Tour.* 10: 479–487.

42 Cheteni, P. (2017). Sustainability development: biofuels in agriculture. *Environ. Ecson.* 8: 83–91.

43 Nguyen, Q., Bowyer, J., Howe, J. et al. (2017). *Global Production of Second Generation Biofuels: Trends and Influences*. Minneapolis, MN: Dovetail Partn. Inc.

44 Stokes, L.C. and Breetz, H.L. (2018). Politics in the US energy transition: case studies of solar, wind, biofuels and electric vehicles policy. *Energy Policy* 113: 76–86.

45 Balwan, W.K. and Kour, S. (2021). A systematic review of biofuels: the cleaner energy for cleaner environment. *Indian J. Sci. Res.* 12: 135–142.

46 Coufalík, P., Matoušek, T., Krumal, K. et al. (2019). Content of metals in emissions from gasoline, diesel, and alternative mixed biofuels. *Environ. Sci. Pollut. Res.* 26: 29012–29019.

47 Navas-Anguita, Z., García-Gusano, D., and Iribarren, D. (2020). Long-term production technology mix of alternative fuels for road transport: a focus on Spain. *Energy Convers. Manag.* 226: 113498.

48 Ruan, R., Zhang, Y., Chen, P. et al. (2019). Biofuels: introduction. In: *Biofuels: Alternative Feedstocks and Conversion Processes for the Production of Liquid and Gaseous Biofuels*, 3–43. Amsterdam, The Netherlands: Elsevier.

49 Mobin, S.M.A. and Alam, F. (2018). A review of microalgal biofuels, challenges and future directions. *Appl. Thermo-Fluid Process. Energy Syst.* 83–108.

50 Rathour, R.K., Ahuja, V., Bhatia, R.K., and Bhatt, A.K. (2018). Biobutanol: new era of biofuels. *Int. J. Energy Res.* 42: 4532–4545.

51 Alizadeh, R., Lund, P.D., and Soltanisehat, L. (2020). Outlook on biofuels in future studies: a systematic literature review. *Renew. Sust. Energ. Rev.* 134: 110326.

52 Lecksiwilai, N. and Gheewala, S.H. (2020). Life cycle assessment of biofuels in Thailand: implications of environmental trade-offs for policy decisions. *Sustain. Prod. Consum.* 22: 177–185.

53 Rasool, U. and Hemalatha, S. (2016). A review on bioenergy and biofuels: sources and their production. *Braz. J. Biol. Sci.* 3: 3–22.

54 Gharabaghi, M., Amrei, H.D., Zenooz, A.M. et al. (2015). Biofuels: bioethanol, biodiesel, biogas, biohydrogen from plants and microalgae. In: *CO2 Sequestration, Biofuels and Depollution*, 233–274. Berlin/Heidelberg, Germany: Springer.

55 Simionescu, M., Albu, L.-L., Raileanu Szeles, M., and Bilan, Y. (2017). The impact of biofuels utilisation in transport on the sustainable development in the European Union. *Technol. Econ. Dev. Econ.* 23: 667–686.

56 Shahare, V.V., Kumar, B., and Singh, P. (2017). Biofuels for sustainable development: a global perspective. In: *Green Technologies and Environmental Sustainability*, 67–89. Berlin/Heidelberg, Germany: Springer.

57 Salian, K. and Strezov, V. (2017). Biofuels from microalgae. In: *Encyclopedia of Sustainable Technologies*, 107–120. Amsterdam, The Netherlands: Elsevier.

58 Guo, M., Song, W., and Buhain, J. (2015). Bioenergy and biofuels: history, status, and perspective. *Renew. Sust. Energ. Rev.* 42: 712–725.

59 Datta, A., Hossain, A., and Roy, S. (2019). An overview on biofuels and their advantages and disadvantages. *Asian J. Chem.* 31 (8).

60 Azam, M.M., Waris, A., and Nahar, N.M. (2005). Prospects and potential of fatty acid methyl esters of some non-traditional seed oils for use as biodiesel in India. *Biomass Bioenergy* 29 (4): 293–302.

61 Asif, S., Ahmad, M., Zafar, M., and Ali, N. (2017). Prospects and potential of fatty acid methyl esters of some non-edible seed oils for use as biodiesel in Pakistan. *Renew. Sust. Energ. Rev.* 74: 687–702.

62 Akubude, V.C., Nwaigwe, K.N., and Dintwa, E. (2019). Production of biodiesel from microalgae via nanocatalyzed transesterification process: a review. *Mater. Sci. Energy Technol.* 2: 216–225.

63 Moazeni, F., Chen, Y.-C., and Zhang, G. (2019). Enzymatic transesterification for biodiesel production from used cooking oil, a review. *J. Clean. Prod.* 216: 117–128.

64 Tan, S.X., Lim, S., Ong, H.C., and Pang, Y.L. (2019). State of the art review on development of ultrasound-assisted catalytic transesterification process for biodiesel production. *Fuel* 235: 886–907.

65 Mofijur, M., Siddiki, S.Y.A., Ahmed, M.B. et al. (2020). Effect of nanocatalysts on the transesterification reaction of first, second and third generation biodiesel sources-A mini-review. *Chemosphere* 270: 128642.

66 Norjannah, B., Ong, H.C., Masjuki, H.H. et al. (2016). Enzymatic transesterification for biodiesel production: a comprehensive review. *RSC Adv.* 6: 60034–60055.

67 Van Eijck, J. and Romijn, H. (2009). Prospects for Jatropha biofuels in Tanzania: an analysis with strategic niche management. In: *Sectoral Systems of Innovation and Production in Developing Countries*. Cheltenham, UK: Edward Elgar Publishing.

68 Najafi, F., Sedaghat, A., Mostafaeipour, A., and Issakhov, A. (2021). Location assessment for producing biodiesel fuel from *Jatropha curcas* in Iran. *Energy* 236: 121446.

69 Gasparatos, A., Mudombi, S., Balde, B.S. et al. (2022). Local food security impacts of biofuel crop production in southern Africa. *Renew. Sust. Energ. Rev.* 154: 111875.

70 Dyer, J.C., Stringer, L.C., and Dougill, A.J. (2012). *Jatropha curcas*: sowing local seeds of success in Malawi?. In response to Achten et al. (2010). *J. Arid Environ.* 79: 107–110.

71 Forson, F.K., Oduro, E.K., and Hammond-Donkoh, E. (2004). Performance of jatropha oil blends in a diesel engine. *Renew. Energy* 29: 1135–1145.

72 Qin, S., Miao, Q., Feng, W.-W. et al. (2015). Biodiversity and plant growth promoting traits of culturable endophytic actinobacteria associated with *Jatropha curcas* L. growing in Panxi dry-hot valley soil. *Appl. Soil Ecol.* 93: 47–55.

73 Souza, S.P., Seabra, J.E.A., and Nogueira, L.A.H. (2018). Feedstocks for biodiesel production: Brazilian and global perspectives. *Biofuels* 9: 455–478.

74 Bahadur, S., Goyal, P., Sudhakar, K., and Bijarniya, J.P. (2015). A comparative study of ultrasonic and conventional methods of biodiesel production from mahua oil. *Biofuels* 6: 107–113.

75 Karmakar, A., Karmakar, S., and Mukherjee, S. (2010). Properties of various plants and animals feedstocks for biodiesel production. *Bioresour. Technol.* 101: 7201–7210.

76 Francis, G., Oliver, J., and Sujatha, M. (2013). Non-toxic jatropha plants as a potential multipurpose multi-use oilseed crop. *Ind. Crop. Prod.* 42: 397–401.

77 Liu, G. and Mai, J. (2022). Habitat shifts of *Jatropha curcas* L. in the Asia-Pacific region under climate change scenarios. *Energy* 251: 123885.

78 Nayab, R., Imran, M., Ramzan, M. et al. (2022). Sustainable biodiesel production via catalytic and non-catalytic transesterification of feedstock materials – a review. *Fuel* 328: 125254.

79 Ayoob, A.K. and Fadhil, A.B. (2020). Valorization of waste tires in the synthesis of an effective carbon based catalyst for biodiesel production from a mixture of non-edible oils. *Fuel* 264: 116754.

80 Ma, F. and Hanna, M.A. (1999). Biodiesel production: a review. *Bioresour. Technol.* 70: 1–15.

81 Fukuda, H., Kondo, A., and Noda, H. (2001). Biodiesel fuel production by transesterification of oils. *J. Biosci. Bioeng.* 92: 405–416.

82 Musa, I.A. (2016). The effects of alcohol to oil molar ratios and the type of alcohol on biodiesel production using transesterification process. *Egypt. J. Pet.* 25: 21–31.

83 Demirbas, A. (2005). Biodiesel production from vegetable oils via catalytic and non-catalytic supercritical methanol transesterification methods. *Prog. Energy Combust. Sci.* 31: 466–487.

84 Yusuf, N., Kamarudin, S.K., and Yaakub, Z. (2011). Overview on the current trends in biodiesel production. *Energy Convers. Manag.* 52: 2741–2751.

85 Wang, X., Liu, X., Zhao, C. et al. (2011). Biodiesel production in packed-bed reactors using lipase–nanoparticle biocomposite. *Bioresour. Technol.* 102: 6352–6355.

86 Ghadge, S.V. and Raheman, H. (2005). Biodiesel production from mahua (*Madhuca indica*) oil having high free fatty acids. *Biomass Bioenergy* 28: 601–605.

87 Jain, S. and Sharma, M.P. (2010). Kinetics of acid base catalyzed transesterification of *Jatropha curcas* oil. *Bioresour. Technol.* 101: 7701–7706.

88 Tapanes, N.C.O., Aranda, D.A.G., de Mesquita Carneiro, J.W., and Antunes, O.A.C. (2008). Transesterification of *Jatropha curcas* oil glycerides: theoretical and experimental studies of biodiesel reaction. *Fuel* 87: 2286–2295.

89 Gui, M.M., Lee, K.T., and Bhatia, S. (2008). Feasibility of edible oil vs. non-edible oil vs. waste edible oil as biodiesel feedstock. *Energy* 33: 1646–1653.

11

Harnessing the Potential of Water Hyacinth for Sustainable Biodiesel Production

11.1 Background of the Study

Water hyacinth (WH) (*Eichhornia crassipes*), commonly referred to as seaweed, is an aquatic plant indigenous to South America that has, over time, proliferated and become widespread in various regions globally. Undesirably, its adaptability to diverse freshwater ecosystems has earned it the reputation of an invasive plant, disrupting human activities and ecosystems [1]. The plant's invasive nature stems from its remarkable ability to thrive in a broad spectrum of freshwater environments, contributing to its interference with natural ecological balances and human engagements. It grows rapidly in nutrient-rich water bodies, such as rivers, lakes, and ponds, and can form dense mats that can cause ecological and economic damage. This interesting specie represent a floating plant that can grow up to 1 m in height, with dark green leaves and showy lavender flowers [1]. As a result of the invasive nature of this seaweed, governments, individuals, and corporate organizations put in so much effort and also invest a significant amount of money to clear the ocean, water, or land where WH grows [2]. Conversely, it has been discovered recently that this seaweed could be managed sustainably and transformed into wealth in their natural ecosystem. WH has several characteristics that make it a potentially valuable feedstock for biodiesel production. First, it has a high oil content, ranging from 20% to 29% by weight, which is comparable to or higher than many traditional biodiesel feedstocks such as soybean, rapeseed, and palm oil [3]. This oil can be extracted and converted into biodiesel using established processes such as transesterification [3]. In addition, WH can grow in a wide range of water bodies, from shallow ponds to deep lakes, and can tolerate various environmental conditions such as temperature, salinity, and nutrient availability. Third, WH can grow in water bodies that are not suitable for conventional agriculture or other land uses, such as areas with poor soil quality or low

Feedstocks for Sustainable Biodiesel Production: Characterization, Selection, and Optimization, First Edition. Chinwe P. Okonkwo and Modestus O. Okwu.

rainfall. This means that its cultivation for biodiesel production would not compete with food crops or other land uses, reducing the risk of negative social and environmental impacts. Finally, the removal of WH from water bodies can have positive environmental and social impacts, such as improving water quality, reducing the risk of flooding, and providing employment opportunities for local communities. Therefore, WH has the potential to be a sustainable and socially responsible feedstock for biodiesel production. However, more research is needed to optimize its production and to assess its full potential and potential negative impacts. WH biomass is good source of energy and can be used for the production of biodiesel, this will open up numerous opportunities for research and development. Also, the end product of WH can generate money for government, companies, and individuals [3]. In the 21st-century energy industry, where there is a paradigm shift in terms of alternative energy source, WH holds a very strong promise as alternatives to fossil fuel. With the rising energy costs, cleaner energy can be obtained as an alternative energy source in this age of industrialization and modernization [4].

In the oil and gas industries, there have been reports of WH completely blocking waterways, making it difficult for floating production storage and offloading (FPSO) to operate freely on the sea. This is also a very serious issue in the agricultural sector as fishing and farming activities are seriously affected due to the rapid growth and spread of the problematic plant, WH [5]. It is a known fact that these species are abundant in Sub-Sahara Africa, managing WH is a very serious issue that requires a proper understanding of the sustainability aspect of the seaweed and the knowledge of transforming the waste into wealth for marketability as an ornamental plant, as a fertilizer to be used on farmland or alternative energy for biofuel and other perspective [6]. The use of WH for biodiesel has caught the attention of researchers globally for over a century [7]. This will help optimize energy costs as many countries have spent millions of hard currencies on fuel importation [8].

Researchers have made attempts to produce biodiesel from other biodegradable feedstocks like Jatropha curucas (*J. curucas*) and vegetables decades ago [9]. The number of feedstocks in the form of *J. curucas* and vegetables available in Sub-Sahara Africa cannot be measured in terms of quantity when compared with the availability of WH with rapid spread. The environmental impacts of WH are complex and depend on the local context. In some cases, it can provide habitat and food for fish, birds, and other wildlife. It can also improve water quality by absorbing nutrients and pollutants from the water. However, in most cases, WH is considered a nuisance and a threat to biodiversity, water resources, and human activities such as fishing, boating, and irrigation. As a fast-growing plant, it is capable of magnifying its biomass in a matter of days under optimal conditions. This means that it can be harvested several times a year, providing a potentially abundant and renewable source of feedstock for biodiesel production [10]. The biomass of WH in the lakes of the Niger Delta in Nigeria extends to approximately half of the lake area [10]. The prevalence of WH infestation is notably high in rivers and seas throughout the Niger Delta region [11]. Its rapid spread is particularly prominent, covering more than 50% of a lake, especially during the rainy season, typically from June to September. The impact of WH on the sea has profound and adverse effects on agricultural practices in Sub-Saharan Africa. For instance, WH detrimentally affects crop yield as obtaining sustainable water for irrigation becomes challenging. Moreover, WH absorbs significant amounts of water, contributing to drying out over time. The consequences of WH's presence have substantial implications for agricultural productivity and water resource availability in the region. This has resulted in serious bottleneck issues in terms of crop production and fish farming [2]. Identifying an area of need for WH especially for the production of biodiesel will go a long way to boost agricultural practices globally. This will help save fish in the ocean and sea, other aquatic animals will be safe too. Figure 11.1 represents WH on the surface of water.

According to Shanab et al. [12], WH has been identified as a promising plant for the production of biofuels such as biogas, biodiesel, and bioethanol. WH will no longer be regarded as a seaweed but a promising plant for the production of fuel as a result of its abundance, high biomass yield and availability [13]. It has also been discovered by the forbearer that WH is the right feedstock for biofuel since it has a low lignin content and a high hemicellulose and cellulose content [6]. According to a study conducted by Lata and Dubey [14] WH was discovered to

Figure 11.1 Water hyacinth on the surface of water.

have a combined hemicellulose and cellulose content of approximately fifty-nine percent (59%) and a very low content of lignin when compared to other biomass [15]. The most beneficial and significant application of WH has been in biofuel production, which is in use in various countries, primarily in Sub-Sahara Africa, Brazil, and India.

In a study conducted by Gunnarsson and Petersen [16], it was revealed that 1 kg of cellulose can yield 1.1 kg of glucose. In addition, another study documented that the yield of glucose from 1 kg of cellulose is 0.56 kg [16]. The versatile applications of WH, including energy production in Eichhornia-based power plants, compost/fertilizer production, and animal feed, highlight its significance in various commercial contexts [14]. Particularly, the engineering features of WH become crucial in its application as power plant energy, contributing to the generation of biogas, biodiesel, bioethanol, and more. This has the potential to reduce dependence on fossil fuels [17].

Considering the characteristics of WH (Eichhornia sp.), it emerges as a promising source of biodiesel due to its ideal attributes, wide availability, ease of harvesting and cultivation, frequent harvest cycles, noncompetition with food crops, ease of extraction, cost-effectiveness, and global recognition as an invasive nuisance weed. The low-tech processing involved in harnessing the potential of WH makes it a feasible and economically viable option. It is worth noting that millions of dollars are spent each year to remove and dispose of WH, making its utilization not only environmentally friendly but also economically advantageous [7].

11.2 Review of Literature

11.2.1 Water Hyacinth and Its Uses

WH is known as one of the fastest-growing plants in the world [18]. It has an excellent capacity for reproduction as it easily spreads over a wide area on land or on the sea, particularly on water with high phosphorus and nitrogen levels [19]. According to studies conducted and available in the literature, the WH doubles in weight between 6 and 28 days, and in quantity between 4 and 58 days [20, 21]. The rapid growth or proliferation of WH may have a number of detrimental effects, which include obstructions to navigation, the reduction in the amount of oxygen in the water, irrigation, recreation, generation of power, and ultimately, the possibility of causing harm to the environment, human, and economic growth [21, 22]. Consequently, it is important to manage WH's widespread so as to avert the negative consequences. This can be achieved using different techniques, which can either be biological, mechanical, or chemical technique. The methods mentioned have been used in time past to clear reduce the spread of

WH. Considering the mechanical method, this method requires the use of harvesting machines to remove the WH from water bodies. This approach requires different kinds of harvesting machines, which can either be classical or modern method, it is regarded as the most frequently utilized practice [23]. In tropical and subtropical areas, WH proliferation is known to be a recurrent environmental issue. It is important to have a clear water body since WH is known to sap nutrients out of the eutrophic body of water. The plant has its own advantage of enhancing water quality because it is a phytoremediation plant.

WH has also been used to remove phosphorus and nitrogen from water, along with organic, inorganic, and heavy metal contaminants [17, 24, 25]. It has also been utilized to absorb some metal ions that are challenging to recycle and biodegrade. Moreover, research conducted on dried WH showed that it contains 3.07% of nitrogen and 0.46% of phosphorus, respectively [26]. WH can, therefore, be used to enhance water quality because its scale can be managed, and its assumed environmental value is anticipated to enhance the efficiency of WH in terms of economic value as regards the production of valuable items. WH as a plant is a possible source of hemicellulose and cellulose, and apart from its potential for renewable energy source, it can be used for making office paper, fiberboard for roofing sheets, animal feedstuff for feeding animals, fertilizer for agriculture [27]. In terms of renewable energy, WH biomass can be used to generate energy [28].

11.2.2 Biodiesel as a Renewable Energy Source

Biodiesel, as a sustainable fuel, is produced from organic feedstocks like vegetable oils, recycled cooking oil, castor seed oil, and similar materials. Its application as an alternative to traditional diesel fuel is prevalent in both transportation and industrial sectors. Biodiesel offers various benefits compared to conventional diesel, encompassing lower greenhouse gas emissions, decreased reliance on fossil fuels, and enhanced engine performance. This renewable fuel's production is rooted in biodegradable sources, providing an eco-friendly alternative to address environmental concerns. Its versatility in substituting for conventional diesel makes it a practical and environmentally conscious choice in mitigating the impact of transportation and industrial activities on the environment. In addition, biodiesel contributes to a sustainable energy landscape by reducing greenhouse gas emissions, thereby supporting efforts to combat climate change. Furthermore, its positive impact on engine performance serves as an added advantage in promoting the adoption of biodiesel as a viable and greener fuel option.

WH, an invasive aquatic plant with high oil content, is one such promising feedstock that could help meet the growing demand for renewable fuels while addressing environmental and social challenges [3]. The use of this interesting plant has been demonstrated in previous studies to including production of bioethanol [29, 30], production of biogas [30], and production of hydrogen [31], as its water content is decreased. WH could be employed as the feedstock for second-generation bioethanol due to its resource availability and technical viability [32]. Research conducted by [33, 34] focused on the impacts of various pretreatment methods on the production of ethanol and the effects of various accelerators, enzymes, and strains on the production of ethanol in terms of fermentation and simultaneous saccharification phase [34, 35] are the two primary issues of recent investigations.

11.3 Materials and Method

The process of producing biodiesel from WH involves several steps, including sourcing of the WH plant via harvesting, drying, oil extraction, transesterification, and purification. A flow chart showing the process is presented in Figure 11.2.

As presented in Figure 11.2, the flow chart for the production of biodiesel from WH has been summarized.

11.3.1 Brief Description of Each Step Process

A brief description of each step of the process is further presented below:

Sourcing/harvesting/collection and cleaning: WH is harvested from water bodies using various methods, such as nets, boats, or mechanical harvesters. Remove any impurities like rocks, debris, etc. Clean the plants to remove dirt and unwanted materials. The clean biomass is then transported to a processing facility. The major challenge often encountered during harvesting of WH is the issue of the plant's buoyancy, which can make it difficult to remove from the water. In addition, the plant has a high water content, which can increase transportation costs and reduce the efficiency of subsequent processing steps.

Drying: After cleaning, spread the WH on a flat surface to dry out. It is important to ensure that the WH is fully dried to reduce moisture content, as high moisture content can affect the quality of the biodiesel. The leaf is detached from the stem and roots before being washed with clean water to remove any soil or dust. The specimens are sliced, and harvested biomass is dried to reduce its moisture content and facilitate oil extraction. This is done by spreading the biomass on a flat surface or using a mechanical dryer in the University Laboratory.

Advantages: Drying the harvested biomass reduces its water content and improves the efficiency of oil extraction. This can also help to reduce transportation costs.

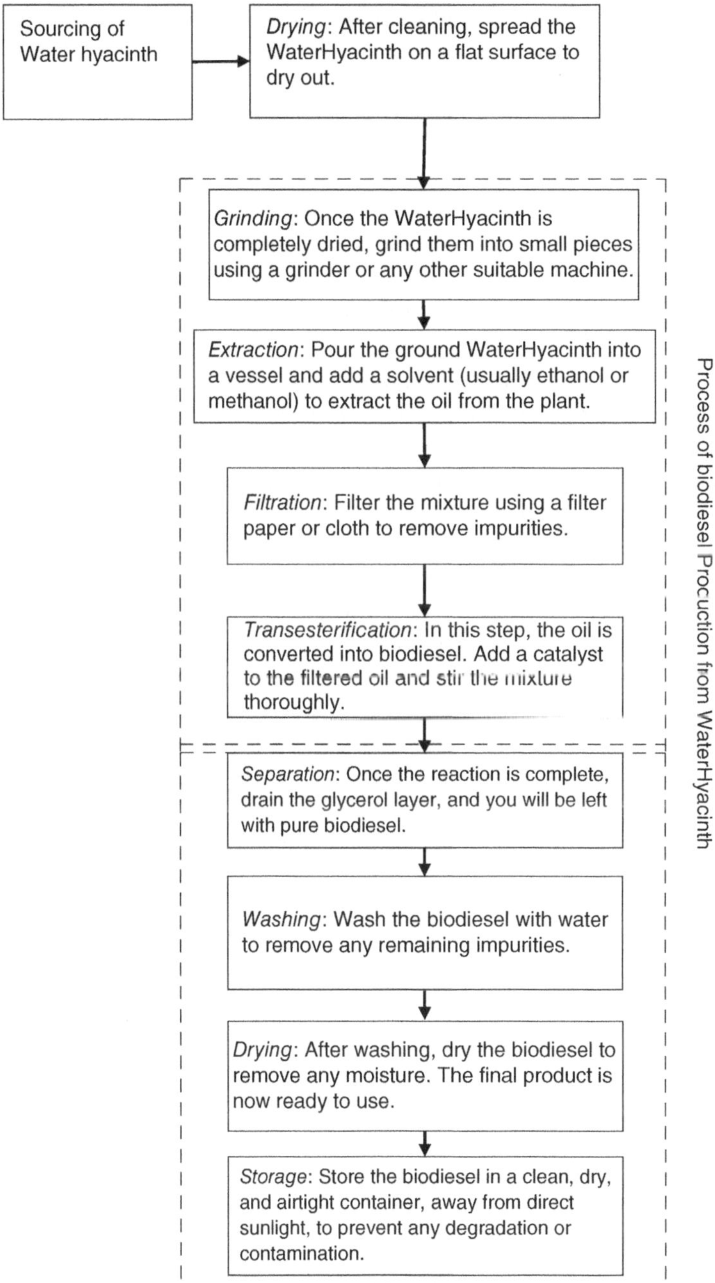

Figure 11.2 Flowchart showing the step-by-step required to produce biodiesel from water hyacinth.

Challenges: Drying WH can be time-consuming and energy-intensive. The plant's high-moisture content can also increase the risk of microbial growth and spoilage.

Grinding and oil extraction: Once the WH is completely dried, the desiccated biomass is crushed or ground to release the WH oil, which is then extracted using a solvent such as methanol or ethanol. An alternative method of extracting the WH oil is the use of a designed mechanical pressing machine available in the workshop. Stir the mixture thoroughly and let it settle down for some time. The oil will float on top of the mixture, and the rest of the impurities will settle down. Filter the mixture using a filter paper or cloth to remove impurities.

Advantages: WH has a high oil content, which can make it a valuable source of feedstock for biodiesel production. The use of solvent extraction methods can also result in high yields of oil.

Challenges: Oil extraction from WH can be challenging due to the tough cell walls of the plant. In addition, the use of solvents can be expensive and may pose environmental and health risks.

Transesterification: This is the pivotal chemical process wherein the extracted oil undergoes conversion into biodiesel. This intricate procedure involves the reaction of the oil with an alcohol, commonly methanol, and a catalyst, typically sodium or potassium hydroxide. The outcome of this reaction is the formation of fatty acid methyl esters (FAME) and glycerol. This reaction occurs efficiently at room temperature, with the entire process extending over an approximate duration of one hour.

During transesterification, the molecular rearrangement of the oil's triglyceride structure takes place, resulting in the creation of biodiesel and glycerol as by-products. This biodiesel, in the form of FAME, holds promise as an eco-friendly and sustainable alternative to traditional diesel fuels. The involvement of alcohol and catalyst in the transesterification process facilitates the conversion of triglycerides into biodiesel, contributing to the broader goal of reducing reliance on nonrenewable energy sources and curbing environmental impacts associated with conventional diesel fuels.

Advantages: Transesterification is a well-established process for biodiesel production, and the use of WH oil as a feedstock can result in high-quality biodiesel that meets established standards.

Challenges: Transesterification requires the use of a catalyst and a large quantity of methanol, which can be expensive and pose environmental risks if not handled properly. In addition, the high free fatty acid in terms of the content of WH oil can make the process more challenging.

Purification: The crude biodiesel is purified to remove impurities such as residual water, glycerol, and catalyst residues. This is done through various methods, such as washing, distillation, or membrane filtration. The purified biodiesel is then ready for use.

Summarily, the production of biodiesel from WH involves several technical and logistical challenges, such as optimizing the harvesting and drying processes, developing efficient oil extraction methods, and ensuring the quality and sustainability of the feedstock. However, the potential benefits of using WH as a feedstock, such as its high oil content, rapid growth, and potential positive environmental and social impacts, make it an attractive option for biodiesel production.

Advantages: Purification removes impurities from the biodiesel, resulting in a high-quality fuel that can be used in existing diesel engines.

Challenges: Purification can be expensive and energy intensive, and the use of certain methods (such as washing) can result in the generation of wastewater that requires treatment. In addition, the use of impure biodiesel can result in engine damage and reduced performance.

11.3.2 Refining of Biodiesel Produced from Water Hyacinth

Refining biodiesel produced from WH is a process that involves removing any remaining impurities and contaminants from the biodiesel to meet industry standards and improve its performance. The refining process typically includes several stages, such as washing, drying, and filtering.

The first step in refining biodiesel is washing: The biodiesel is washed with water to remove any residual catalyst, soap, and glycerol. The washing process is typically done in a settler or separator vessel where the biodiesel is mixed with water and allowed to settle. The water and impurities settle at the bottom, and the biodiesel is removed from the top.

Following the washing phase: The biodiesel undergoes a drying process to eliminate residual water. The presence of water in biodiesel poses risks such as fuel system corrosion, diminished energy content, and the proliferation of microorganisms. To execute the drying process, the biodiesel is commonly directed through a desiccant material bed, often employing substances like silica gel. This method ensures the removal of water content from the biodiesel, safeguarding its quality and mitigating potential issues associated with moisture presence in the fuel.

The next step in the refining process is filtering: The biodiesel is passed through a filter to remove any remaining particles or contaminants. The filter may be a simple mesh screen or a more complex system, such as a filter press or a centrifuge.

In addition to these primary refining steps: Other treatments may be applied to further purify the biodiesel. For example, activated carbon may be used to remove impurities such as pigments, odor compounds, and sulfur compounds. Membrane filtration may also be used to remove contaminants such as salts and metals. The quality of the biodiesel produced from the raw material in the form of WH can be evaluated using various parameters such as acid value, viscosity, density, and flash point. The biodiesel must meet industry standards to be considered appropriate for utilization in diesel engines. Finally, refining biodiesel produced from WH is an important step in ensuring the fuel's quality and performance. The refining process typically includes washing, drying, and filtering, and may also involve other treatments such as activated carbon and membrane filtration. The quality of the biodiesel can be evaluated using various parameters, and it must meet industry standards to be considered appropriate for utilization in diesel engines.

11.3.3 Co-Products of Biodiesel Production from Water Hyacinth

Biodiesel production from WH can result in several co-products that can be utilized for various purposes. The co-products can include glycerol, protein-rich biomass, and residual lignocellulosic material.

The first co-product, glycerol, is produced as a by-product of the transesterification process. Glycerol is a viscous, sweet-tasting liquid that can be used in several applications. For example, it can be used as a feedstock for the production of various chemicals such as propylene glycol, glycerol carbonate, and epichlorohydrin. Glycerol can also be used as a food ingredient, a moisturizing agent in cosmetics, and a component in animal feed.

The second co-product, protein-rich biomass, is produced after the lipid extraction process. WH contains a significant amount of protein, which can be used for animal feed or as a source of nitrogen in fertilizers. The protein-rich biomass can also be used to produce biogas through anaerobic digestion, which can be used as a renewable energy source.

The third co-product, residual lignocellulosic material, is produced after the lipid and protein extraction process. This material consists of the plant's fibrous and woody parts and can be used as a feedstock for the production of cellulosic ethanol or as a source of energy through combustion.

The by-products generated during the production of biodiesel from WH present supplementary income sources, contributing to the enhanced economic feasibility of the procedure. Moreover, the utilization of these by-products not only diminishes waste but also offers sustainable solutions across diverse industries. In conclusion, biodiesel production from WH can result in several co-products, including glycerol, protein-rich biomass, and residual lignocellulosic material. These co-products can be utilized in various applications, such as chemical production, animal feed, fertilizers, and renewable energy sources. The use of co-products can help to improve the overall economic viability of the process and reduce waste.

11.4 Advantages and Challenges of Using Water Hyacinth as Raw Material for Biodiesel Production

11.4.1 Advantages of Water Hyacinth for Biodiesel Production

The use of WH as a feedstock for biodiesel production can be categorized into environmental, economic, and social benefits as discussed:

1) *Environmental benefits*: WH is considered an invasive species and its removal can help to restore ecosystems and improve water quality.

   Biodiesel produced from WH can reduce greenhouse gas emissions compared to fossil fuels, which can help to mitigate climate change.

   Biodiesel from WH is biodegradable and nontoxic, which can reduce pollution and improve air quality.
2) *Economic benefits*: Biodiesel production from WH can provide an additional revenue stream for farmers and rural communities.

   The production of co-products such as glycerol, protein-rich biomass, and residual lignocellulosic material can further enhance the economic viability of the process.

   The process of producing biodiesel from WH holds the potential to diminish reliance on imported fossil fuels, thereby contributing to bolstering energy security. This shift toward biodiesel production from local sources offers a strategic advantage by reducing dependency on external fuel supplies.
3) *Social benefits*: Biodiesel production from WH can create jobs and improve the livelihoods of local communities, particularly in rural areas.

The removal of WH can improve water flow and reduce flooding, which can benefit communities living near water bodies.

The use of biodiesel from WH can improve public health by reducing pollution and improving air quality.

In summary, biodiesel production from WH offers several environmental, economic, and social benefits. It can help restore ecosystems, reduce greenhouse gas emissions, and improve air and water quality. The production of co-products can enhance the economic viability of the process while creating jobs and improving the livelihoods of local communities. Biodiesel production from WH can also enhance energy security and improve public health.

11.4.2 Challenges and Limitations of Biodiesel Produced from the Raw Material (Water Hyacinth)

The challenges faced in the use of WH as a feedstock for biodiesel production can be categorized into technical, economic, and social benefits as discussed:

1) *Technical challenges*: WH contains a significant amount of water, which can reduce the efficiency of lipid extraction and transesterification processes.
   WH can contain high amounts of ash and other impurities, which can reduce the efficiency of the biodiesel production process.
   The cultivation and harvesting of WH require specific knowledge and expertise, which may not be widely available.
2) *Economic challenges*: Biodiesel production from WH may not be economically viable compared to fossil fuels or other renewable energy sources due to the high costs of cultivation, harvesting, and processing.
   The availability and cost of the necessary equipment for lipid extraction and transesterification may also be a barrier to the economic viability of the process.
   The market for biodiesel from WH may be limited, which can affect the profitability of the process.
3) *Social challenges*: WH removal and biodiesel production may conflict with the interests of local communities, particularly if they rely on WH for cultural or economic reasons.
   The introduction of WH cultivation may compete with existing land use and can cause conflicts over land rights.
   The social acceptability of using invasive species for biodiesel production may be a concern for some stakeholders.

Finally, biodiesel production from WH faces several technical, economic, and social challenges. The high water content and impurities in WH can reduce the efficiency of the production process. The high costs of cultivation, harvesting, and processing may limit the economic viability of the process. The use of WH for biodiesel production may also face social challenges, including conflicts with local communities and concerns about the use of invasive species.

11.5 Future Prospects of Water Hyacinth as a Feedstock for Biodiesel Production

WH is a free-floating aquatic plant that is considered a nuisance weed in many parts of the world, including Africa, Asia, and South America. However, it has recently gained attention as a potential feedstock for biodiesel production due to its high oil content and fast growth rate. Despite its promising features, several challenges need to be addressed to make WH biodiesel production economically viable and sustainable. One major challenge is the high cost of harvesting and transporting WH from water bodies to processing plants. The process requires specialized equipment and labor, which can be expensive, especially in remote areas. In addition, WH biomass has a high water content, which reduces its energy content and makes transportation more difficult. Another challenge is the low conversion efficiency of WH oil to biodiesel. WH oil has a high content of free fatty acids, which can lead to soap formation and reduce the yield and quality of biodiesel. Therefore, a pretreatment process is necessary to remove the free fatty acids and improve the quality of WH biodiesel. In response to these challenges, continuous research and development activities are concentrated on enhancing the efficacy of biodiesel production from WH and broadening the utilization of alternative raw materials. One approach is to develop cost-effective and energy-efficient harvesting and processing technologies, such as using biological agents to de-water the biomass and reduce its moisture content. Another approach is to explore new processing technologies, such as enzymatic or microbial biodiesel production, that can reduce the need for chemical catalysts and lower energy consumption and waste generation.

In addition, researchers are investigating the potential of other nonfood feedstocks, such as algae, Jatropha, and waste cooking oil, as alternative sources of biodiesel that can complement WH. This approach can help to reduce the dependence on a single feedstock and improve the sustainability of biodiesel production. WH has the potential to serve as a bioenergy crop and contribute to rural development and poverty reduction. The plant can be cultivated in water bodies that are not suitable for other crops and can provide income and employment opportunities for local communities. In addition, WH biomass can be used for other purposes, such as animal feed, biogas production, and soil amendment.

Finally, while WH holds great promise as a potential feedstock for biodiesel production, several challenges need to be addressed to make it economically viable and sustainable. Ongoing research and development efforts are focused on improving the efficiency of biodiesel production and expanding the use of alternative feedstock. Furthermore, WH has the potential to contribute to rural development and poverty reduction through its use as a bioenergy crop and for other purposes.

11.6 Conclusion

This chapter is an exposition of the possibility of utilizing WH as a raw material for the production of biodiesel. The study also examines the environmental, economic, and social benefits of biodiesel production from WH and the challenges and limitations that need to be addressed. However, several challenges must be addressed to make WH biodiesel production economically viable and sustainable. These include the high cost of harvesting and transporting biomass, the low conversion efficiency of WH oil to biodiesel, and the need for pretreatment to remove free fatty acids. To overcome these challenges, ongoing research and development efforts are focused on improving the efficiency of biodiesel production and expanding the use of alternative feedstocks. It was stated that the plant can be cultivated in water bodies and harvested to provide income and employment opportunities for local communities. In addition, WH biomass can be used for other purposes, such as animal feed, biogas production, and soil amendment. However, it is recommended that further research and development are required to unlock the full potential of WH as a source of biodiesel.

References

1 Rahman, M.A. and Aziz, M.A. (2021). Biodiesel from water hyacinth biomass and its influence on CI engine performance, emission, combust, and heat loss characteristics with the induction of hydrox. *Energy* 24: 120–151.
2 Bote, M.A., Naik, V.R., and Jagadeeshgouda, K.B. (2020). Materials science of energy technologies review on water hyacinth weed as a potential biofuel crop to meet the collective energy needs. *Mater. Sci. Energy Technol.* 3: 397–406.
3 Oyejide, O.J., Okwu, M.O., and Tartibu, L. (2019). Adaptive design and development of a modular water hyacinth briquette stove. In: *Energy Sources Part A Recovery Utilization and Environmental Effects*. Taylor and Francis.
4 Uday, U.S.P., Choudhury, P., Bandyopadhyay, T.K., and Bhunia, B. (2016). Classification, mode of action and production strategy of xylanase and its application for biofuel production from water hyacinth. *Int. J. Biol. Macromol.* 82: 1041–1054.
5 Shanab, S.M.M., Hanafy, E.A., and Shalaby, E.A. (2016). Water hyacinth as non-edible source of biofuel production. Waste Biomass Valor. https://doi.org/10.1007/s12649-016-9816-6.
6 Das, A., Ghosh, P., Paul, T. et al. (2016). Production of bioethanol as useful biofuel through the bioconversion of water hyacinth (*Eichhorniacrassipes*). *3 Biotech* 6 (1): 70.
7 Sagar, C.V. and Kumari, N.A. (2013). Sustainable biofuel production from water hyacinth (*Eicchorniacrassipes*). *Int. J. Eng. Trends Technol.* 4 (10): 4454–4458.
8 Afify, A.M.M.R., Shalaby, E.A., and Shanab, S.M.M. (2010). Enhancement of biodiesel production from different species of algae. *Grasas Aceites* 61 (4): 416–422.
9 Shalaby, E.A. and El-Gendy, N. (2012). Two steps alkaline transesterification of waste cooking oils and quality assessment of produced biodiesel. *Int. J. Chem. Biochem. Sci.* 1: 30–35.

10 Al Ramalli, S.W., Harrington, C.F., Ayub, M., and Haris, P.I. (2005). A biomaterial based approach for arsenic removal from water. *J. Environ. Monit.* 7: 279–282.

11 Shanab, S.M.M., Shalaby, E.A., Lightfoot, D.A., and El Shemy, H.A. (2010). Allelopathic effects of water hyacinth (*Eichhorniacrassipes*). *PLoS One* 5 (10): e13200.

12 Shanab, S.M.M., Hanafy, E.A., and Shalaby, E.A. (2014). Biodiesel production and antioxidant activity of different Egyptian date palm seed cultivars. *Asian J. Biochem.* 9 (3): 119–130.

13 Srinophakun, P., Thanapimmetha, A., Rattanaphanyapan, K. et al. (2017). Feedstock production for third generation biofuels through cultivation of Arthrobacter AK19 under stress conditions. *J. Clean. Prod.* 142: 1259–1266.

14 Lata, N. and Dubey, V. (2010). Preliminary phytochemical screening of Eichhorniacrassipes: the world's worst aquatic weed. *J. Pharm. Res.* 3: 1240–1242.

15 Ruan, T., Zeng, R., Yin, X.Y. et al. (2016). Water hyacinth (*Eichhorniacrassipes*) biomass as a biofuel feedstock by enzymatic hydrolysis. *Bioresources* 11: 2372–2380.

16 Gunnarsson, C.C. and Petersen, C.M. (2007). Water hyacinths as a resource in agriculture and energy production: a literature review. *Waste Manag.* 27: 117–129.

17 Rezania, S., Ponraj, M., FadhilMd Din, M. et al. (2015). The diverse applications of water hyacinth with main focus on sustainable energy and production for new era: an overview. *Renew. Sust. Energ. Rev.* 41: 943–954.

18 Malik, A. (2007). Environmental challenge vis a vis opportunity: the case of water hyacinth. *Environ. Int.* 33: 122–138.

19 Bricker, S.B. (1999). National estuarine eutrophication assessment [electronic resource]. *IEEE Trans. Energy Convers.* 10: 261–267.

20 Epstein, P. (1998). Nairobi weeds bring disease to the east African waterways. *Lancet* 351: 577.

21 Gopal, B. (1987). *Water Hyacinth: Aquatic Plant Studies*. Amsterdam, The Netherlands: Elsevier Science Publishers.

22 Chen, G., Wang, C.Y., and Liu, L. (2007). Consideration of mechanical control over water hyacinth in Fujian. *J. Longyan Univ.* 25 (78–81): 84.

23 Mehra, A., Farago, M.E., Banerjee, D.K., and Cordes, K.B. (1999). The water hyacinth – an environmental friend or pest? A review. *Resour. Environ. Biotechnol.* 2: 255–281.

24 Kang, Z. and Liu, H. (2015). Eco-compensation for removal of nitrogen and phosphorus from Lake Taihu using cost-benefit analysis. *Jiangsu J. Agric. Sci.* 31: 942–945.

25 Lin, H.J., Rong, C.X., Jiu, B.B. et al. (2018). Effects of chromium on pyrolysis characteristic of water hyacinth (*Eichornia crassipes*). *Renew. Energy* 115: 676–684.

26 Zheng, J., Chang, Z., Chen, L. et al. (2008). Feasibility study on treatment of nitrogen and phosphorus pollution in Taihu basin by water hyacinth. *Jiangsu Agric. Sci.* 3: 247–250.

27 Jafari, N. (2010). Ecological and socio-economic utilization of water hyacinth (*Eichhornia crassipes* Mart Solms). *J. Appl. Sci. Environ. Manag.* 14: 43–49.

28 Liu, J., Olsson, G., and Mattiasson, B. (2004). On-line monitoring of a two-stage anaerobic digestion process using a BOD analyzer. *J. Biotechnol.* 109: 263–275.

29 Zhang, Q., Wei, Y., Han, H., and Weng, C. (2017). Enhancing bioethanol production from water hyacinth by new combined pretreatment methods. *Bioresour. Technol.* 251: 358–363.

30 Priya, P., Nikhitha, S.O., Anand, C. et al. (2018). Biomethanation of water hyacinth biomass. *Bioresour. Technol.* 255: 288–292.

31 Wazeri, A., Elsamadony, M., Roux, S.L. et al. (2018). Potentials of using mixed culture bacteria incorporated with sodium bicarbonate for hydrogen production from water hyacinth. *Bioresour. Technol.* 263: 365–374.

32 Kumar, A., Singh, L.K., and Ghosh, S. (2009). Bioconversion of lignocellulosic fraction of water-hyacinth (*Eichhornia crassipes*) hemicellulose acid hydrolysate to ethanol by Pichia stipitis. *Bioresour. Technol.* 100: 3293–3297.

33 Thi, B.T.N., Lu, K.O., Thi, D.T.N., and Ju, Y.H. (2017). Effect of subcritical water pretreatment on cellulose recovery of water hyacinth (*Eichhornia crassipe*). *J. Taiwan Inst. Chem. Eng.* 71: 55–61. Sustainability 2019, 11, 905 19 of 21.

34 Narra, M., Divecha, J., Shah, D. et al. (2017). Cellulase production, simultaneous saccharification and fermentation in a single vessel: a new approach for production of bio-ethanol from mild alkali pre-treated water hyacinth. *J. Environ. Chem. Eng.* 5: 2176–2181.

35 Mohd Azhar, S.H., Abdulla, R., Jambo, S.A. et al. (2017). Yeasts in sustainable bioethanol production: a review. *Biochem. Biophys. Rep.* 10: 52–61.

12

Sustainable Biodiesel Production by Exploring Sunflower as a Potential Feedstock

12.1 Background of the Study

Fossil fuels have recently become humanity's main energy source. Energy has been recognized as the key factor in promoting societal and human growth [1, 2]. The vast use of fossil fuels, however, has stimulated research into renewable energy for sustainable development, with main focus on the production of biofuels [3–5]. Global recognition has been given to the potential of biodegradable feedstock for the manufacture of biodiesel. Oils obtained from seeds and/or plants are used to make biodiesel. One of the most promising feedstocks for the generation of biodiesel is sunflower oil. Certain mutants of sunflower oil display high-stearic characteristics on a background of high oleic acid. Triacylglycerol, which are richer in them and are present in cocoa butter, can serve as a beneficial substitute for palm oil [6].

In this scenario, the recently developed sunflower genotypes with elevated stearic and oleic acid content offer a reliable and novel oil source for the food industry [7]. Sunflower oil is a suitable choice for making biodiesel because it has a high amount of unsaturated fatty acids that are desirable for the process [7]. The feedstock, which makes up around 70% of domestic oilseeds, is essentially regionally accessible. These seeds offer a nutrient-rich combination of minerals, proteins, amino acids, fiber, and unsaturated fats [8]. Due to their carbohydrate, phytosterol, amino acid, protein, and mineral salt content, extracts obtained from various plant parts (stem, leaf, and root) are utilized for both dietary and pharmaceutical applications [8]; the vegetative part of the sunflower could be differently valued and repurposed as natural fibers for biomatrices [9]. One of the classical ways that has been discovered for producing biodiesel is the standardized transesterification process [10].

Feedstocks for Sustainable Biodiesel Production: Characterization, Selection, and Optimization, First Edition. Chinwe P. Okonkwo and Modestus O. Okwu.

12.1.1 Objectives and Scope

The objective of this chapter is focused on the exposition of the potential of sunflower as a feedstock for biodiesel production. The chapter will cover the Oil Extraction Process In Sunflower, the Characteristics of Sunflower as a crop, agronomic practices for cultivation, and methods of oil extraction. The focus will be on the suitability of sunflower oil for biodiesel production, including the transesterification process, catalysts, and quality control; Sunflower for Biodiesel Production; Sunflower oil extraction; Oil quality and yield in Sunflower; Detailed information on Biodiesel Production from Sunflower Oil. This chapter will also explore the challenges and opportunities of sunflower biodiesel production, including economic and environmental considerations, sustainability, and technological challenges. With detailed information on the background of the study, which also includes the oil extraction process in sunflower, other sections will cover the following: Section 12.2 will focus on a review of literature focusing on the Characteristics of Sunflower with detailed information on climate change, growth process and cultivation, sunflower for biodiesel production, sunflower oil extraction, oil quality and yield in Sunflower; Section 12.3 will focus on Detailed information on Biodiesel Production from Sunflower Oil with flow chart showing the step by step production process, including the chemical composition of sunflower oil, the transesterification process, catalysts, and quality control. Other sections will explore the challenges and opportunities of sunflower biodiesel production, including economic and environmental considerations, sustainability, and technological challenges. Highlight the future prospects of sunflower biodiesel production, including potential expansion, government policies, and integration into existing agricultural systems. Finally, this chapter will conclude with a summary of key points and a call for further research and development to unlock the potential of sunflower as a feedstock for biodiesel production.

12.1.2 Oil Extraction Process in Sunflower

The extraction of oil from sunflower seeds can be accomplished through various methods, including mechanical pressing, supercritical fluid extraction, and solvent. While mechanical pressing, whether using traditional or modern equipment, is the most common approach, solvent extraction, and supercritical fluid extraction are more efficient alternatives that require sophisticated and costly equipment. In the context of biodiesel production from sunflower oil, the transesterification process plays a pivotal role. This process involves reacting the fatty acids present in the oil with an alcohol, typically methanol, in the presence of a catalyst, usually sodium or potassium hydroxide, to yield biodiesel (fatty acid methyl esters; FAME) and glycerol [10]. Transesterification can be executed through various methods, including batch, continuous, or supercritical fluid methods, with the selection depending on factors such as the desired yield, quality, and cost-effectiveness [10].

The suitability of sunflower oil for biodiesel production has been confirmed by several studies, which have shown that sunflower biodiesel has similar or better properties than diesel fuel, such as high cetane number, low sulfur content, and good lubricity. Sunflower biodiesel has also been shown to have lower emissions of carbon monoxide, hydrocarbons, and particulate matter compared to diesel fuel. The potential of sunflower biodiesel production has several benefits, such as reducing dependence on fossil fuels, promoting rural development, creating employment opportunities, and reducing greenhouse gas emissions. Nevertheless, there exist challenges that require attention, including potential competition with food crops, environmental consequences of agricultural practices, and the assessment of technological and economic feasibility. Addressing these challenges necessitates additional research and development efforts to fully realize the potential of sunflower as a biodiesel feedstock. The feasibility of biodiesel production is recognized, considering the availability of feedstocks with sufficient oil content, particularly for first-generation biodiesel. Noteworthy edible oils in this context encompass palm [11, 12], coconut [13], olive, mustard, soybean [14], rapeseed [15], and sunflower oil [16]. As shown in Figure 12.1, sunflower has oil content that ranges between 25% and 35% [17, 18].

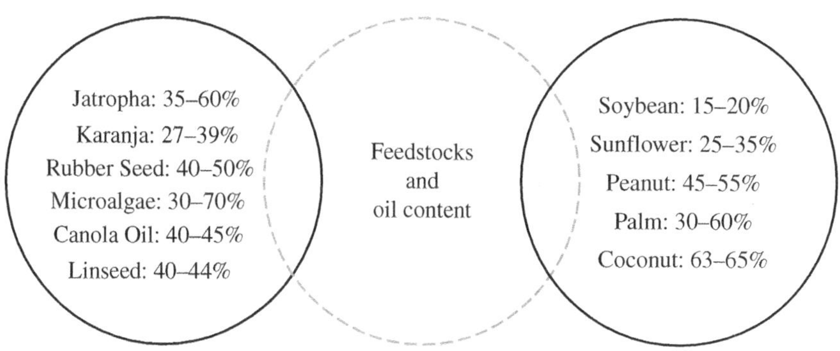

Figure 12.1 The oil content of feedstocks for biodiesel synthesis.

12.2 Review of Literature

12.2.1 Characteristics of Sunflower

Studies on Sunflower cultivation due to an increase in the demand for feedstock as a renewable energy source are detailed in recent studies [19–23]. Sunflower, as an annual plant, can grow up to three meters in height and are mainly harvested under rain-fed conditions as a spring–summer crop [24, 25]. The stem of the sunflower is erect and hairy, with leaves that are alternate, simple, and rough to the touch. The leaves are arranged spirally on the stem and are usually heart-shaped or triangular in shape, with a serrated margin. Sunflower flowers are large and showy, with a characteristic yellow color and a dark center. The flowers are arranged in a head, which is composed of numerous small individual flowers called florets. The seeds of sunflower are covered by a hard outer shell, which is black or gray in color. The seeds contain a high concentration of oil. Furthermore, sunflower relies on water during flowering and achene filling [26]. This interesting flower is adapted to a wide range of soils and climatic conditions. It is a warm-season crop that requires a long growing season. The crop requires moderate amounts of water for effective growth. It can be grown in soils with a moderate pH range and can tolerate reasonable levels of salinity [27, 28]. It has been cultivated for thousands of years and is now widely grown in many parts of the world for various purposes, such as food, feed production, and industrial uses. Numerous studies in the literature have discovered the sowing date of sunflower, which is approximately one month [29, 30]. When selecting sunflower varieties, it is important to consider the climate and growing conditions in your region to ensure that the plants will thrive. Studies where detailed information on climate change, growth process, and cultivation have been discussed in detailare presented in Table 12.1 [31–41].

In terms of feed production, Sunflower is capable of providing many feeds for ruminant animals and monogastrics. Almost every part of this magnificent flower is useful. Sunflower is considered a major feed, and sunflower hulls, meal, seeds, cake, meal, foliage, crop residue, and sunflower oil are considered very useful [42, 43]. Other area where the exposition of sunflower has been detailed for feed production is available in collected works [43–49]. In terms of biodiesel production, Sunflower has been presented as a good source of biofuel. Table 12.2 shows a representation of different crops in terms of oil production.

12.2.2 Sunflower for Biodiesel Production

Sunflower as a crop has gained significant attention as a potential feedstock in recent years, specifically for the production of biodiesel. This is a result of its high oil content and favorable fatty acid composition. It is relatively easy to

Table 12.1 Information on climate change, growth process, and cultivation.

Feedstock	Research	Focus	Citation
Sunflower	Changes in sunflower breeding over the last 50 years	B	[31]
Sunflower	Sunflower ideotyping in changing environments	CS	[32]
Sunflower	Drought stress in sunflower	B	[33]
Sunflower	Responses of sunflower to Irrigation	GP	[34]
Sunflower	Effect of date and final irrigation on yield of Sunflower	GP	[35]
Sunflower	Irrigation timing and relationships between leaf area and yield in sunflower	GP	[36]
Sunflower	Irrigation strategies and sowing dates on sunflower yield	GP	[37]
Sunflower	Biological seed priming mitigates the effects of water stress in sunflower seedlings.	GP	[38]
Sunflower	Agronomic response of sunflower	GP	[39]
Sunflower hybrid	Unraveling the mystery of nonextraction of honey from modern sunflower	B	[40]
Hybrid crop	Increasing crop production	C	[41]

B, breeding; C, cultivation; CS, cropping system; GP, growth process.

grow and harvest due to the botanical characteristics of this interesting sunflower, thereby making it a valuable feedstock for the production of biodiesel [51]. Biodiesel, derived from triglycerides through transesterification, primarily with methanol, stands out as the most widely utilized fuel within the spectrum of oil-refining derivatives. This nontoxic and biodegradable fuel, synthesized from vegetable oils [52], shares a comparable viscosity with diesel, positioning it as an excellent alternative. The versatility of biodiesel is underscored by its potential use with various oil types from different sources [53–56], including by-products from animal processing [51]. Sunflower oil, in particular, distinguishes itself due to its complete biodegradability in contrast to regular biodiesel [57]. Moreover, sunflower oil exhibits superior lubricity characteristics compared to diesel [58]. Given its origin from a crop well-suited to the local terrain, sunflower oil holds significant promise as a substitute for other agricultural products [59].

The promise of sunflower for biodiesel production has been previously unveiled in Italy through the work of Scrosta et al. [59]. Subsequent research emphasized the significance of the hybrid system of sunflower for enhancing productivity [60].

Table 12.2 Liters oil per ha [50].

S/N	Crop/feedstock/raw material	Liters oil/ha	US gal/acre
1	Avocado or Alligator Pear: – Scientific name is *Persea Americana*	2638	282
2	Calendula or Marigold: Scientific name is *Calendula officinalis*	305	33
3	Castor Bean: Scientific name is *Ricinus communis*	1413	151
4	Cocoa: Scientific name for cocoa is *Theobroma cacao*	1026	110
5	Coconut: Scientific name is *Cocos nucifera*	2689	287
6	Coffee: Scientific name is *Coffea* spp.	459	49
7	Corn (Maize): Scientific name is *Zea mays*	172	18
8	Cotton: Scientific name is *Gossypium* spp.	325	35
9	Jatropha: Scientific name is *Jatropha curcas*	1892	202
10	Jojoba: Scientific name is *Simmondsia chinensis*	1818	194
11	Kenaf: Scientific name is *Hibiscus cannabinus*	273	29
12	Macadamia nut: Scientific name is *Macadamia integrifolia*	2246	240
13	Mustard seed: Scientific name is *Brassica juncea*	572	61
14	Oat: Scientific name is *Avena sativa*	217	23
15	Oil palm: Scientific name is *Elaeis guineensis*	5950	635
16	Olive: Scientific name is *Olea europaea*	1212	129
17	Opium poppy: Scientific name is *Papaver somniferum*	1163	124
18	Peanut: Scientific name is *Arachis hypogaea*	1059	113
19	Pecan nut: Scientific name is *Carya illinoinensis*	1791	191
20	Pumpkin seed: Scientific name is *Cucurbita pepo*	534	57
21	Rape seed: Scientific name is *Brassica napus*	1190	127
22	Rice: Scientific name is *Oryza sativa*	828	88
23	Sesame: Scientific name is *Sesamum indicum*	696	74
24	Soybean: Scientific name is *Glycine max*	446	48
25	Sunflower: Scientific name is *Helianthus annuus*	952	102

Notably, sunflower has been cultivated for biodiesel production across various regions globally [61, 62]. The subsequent sections focus on a comprehensive discussion of the biodiesel production process from sunflower, covering key aspects and methodologies.

12.2.3 Sunflower Oil Extraction

Oil content in sunflower seeds: As discussed in the previous section, Sunflower seeds have a high oil content. This makes sunflower oil an ideal feedstock for the production of biodiesel. Traditional and modern methods of oil extraction from sunflower seeds: There are several methods for extracting oil from sunflower seeds, including traditional and modern methods.

Traditional methods include: *Cold-pressing*: This method involves crushing the sunflower seeds with a pestle and mortar or a small-scale oil expeller. The resulting oil is typically of lower quality and yield; *Expeller-pressing*: This method involves using a mechanical expeller press to crush the sunflower seeds and extract the oil. The oil is typically of higher quality and yield than cold-pressing.

Modern methods include *Solvent extraction*: This method involves using a solvent, such as hexane, to extract the oil from the sunflower seeds. The solvent is evaporated, leaving behind the oil. This method is efficient and yields a high-quality oil. Supercritical fluid extraction: This method involves using supercritical carbon dioxide to extract the oil from the sunflower seeds. This method is more expensive than solvent extraction, but it yields a high-quality oil and is more environmentally friendly. After the oil has been extracted, it must be refined to remove impurities and improve its quality. Following oil extraction, a refinement process is imperative to eliminate impurities and enhance the oil's quality. This commonly entails degumming, neutralization, and bleaching. The refined oil can subsequently serve as a valuable feedstock for biodiesel production.

12.2.4 Oil Quality and Yield in Sunflower

Sunflower oil quality and yield depend on various factors, such as the variety of sunflowers, oil extraction methods, and processing conditions. Here are some general factors that influence the quality and yield of sunflower oil:

Sunflower variety: Different sunflower varieties have varying oil content and quality. Some varieties are bred specifically for high oil yield and quality.

Oil extraction method: The method used for extracting oil from sunflower seeds can affect the yield and quality of the oil. Methods such as cold-pressing may result in lower yield but higher quality oil, while solvent extraction can result in higher yield but lower quality oil.

Processing conditions: The conditions under which the oil is processed can also affect its quality. Factors such as temperature, pressure, and the presence of oxygen can influence the oil's oxidative stability and nutrient content.

In general, sunflower oil has a high yield and good quality, making it a popular choice for biodiesel production. The oil contains a high percentage of unsaturated fatty acids, making it a healthier option compared to other vegetable oils. However, it is important to note that the quality and yield of sunflower oil can vary depending on the factors mentioned above. Therefore, it is important to choose the appropriate sunflower variety, extraction method, and processing conditions to obtain the desired quality and yield of sunflower oil.

12.3 Detailed Information on Biodiesel Production from Sunflower Oil

The flow chart showing the biodiesel production process from sunflower oil is presented in Figure 12.2.

Biodiesel serves as a renewable substitute for traditional fossil fuels, and its production from sunflower oil involves a comprehensive process outlined in Figure 12.2. The biodiesel production process from sunflower oil is elucidated below:

Oil extraction: The initial phase of biodiesel production involves extracting oil from sunflower seeds. Various methods, such as cold-pressing, expeller-pressing, solvent extraction, or supercritical fluid extraction, can be employed for this purpose.

Pretreatment: Extracted oil often contains impurities like water, free fatty acids, and gums, which can impact biodiesel quality. Therefore, a pretreatment step is crucial to eliminate these impurities. Common pretreatment methods include degumming, neutralization, and/or bleaching.

Transesterification: The pretreated oil undergoes transesterification, a chemical process converting the oil into biodiesel. This process involves reacting the oil with an alcohol (typically methanol) and a catalyst (such as sodium hydroxide or potassium hydroxide), resulting in the formation of FAME, the primary component of biodiesel.

Separation: Following the transesterification reaction, the mixture is allowed to settle, facilitating the separation of the glycerol by-product from the biodiesel.

Washing: The biodiesel is then washed with water to remove any remaining impurities and traces of the catalyst.

Drying: The washed biodiesel is then dried to remove any water that may have been introduced during the washing process.

Quality control: The final step in biodiesel production is quality control. The biodiesel is tested to ensure that it meets the required specifications for fuel use, including viscosity, flash point, and cetane number.

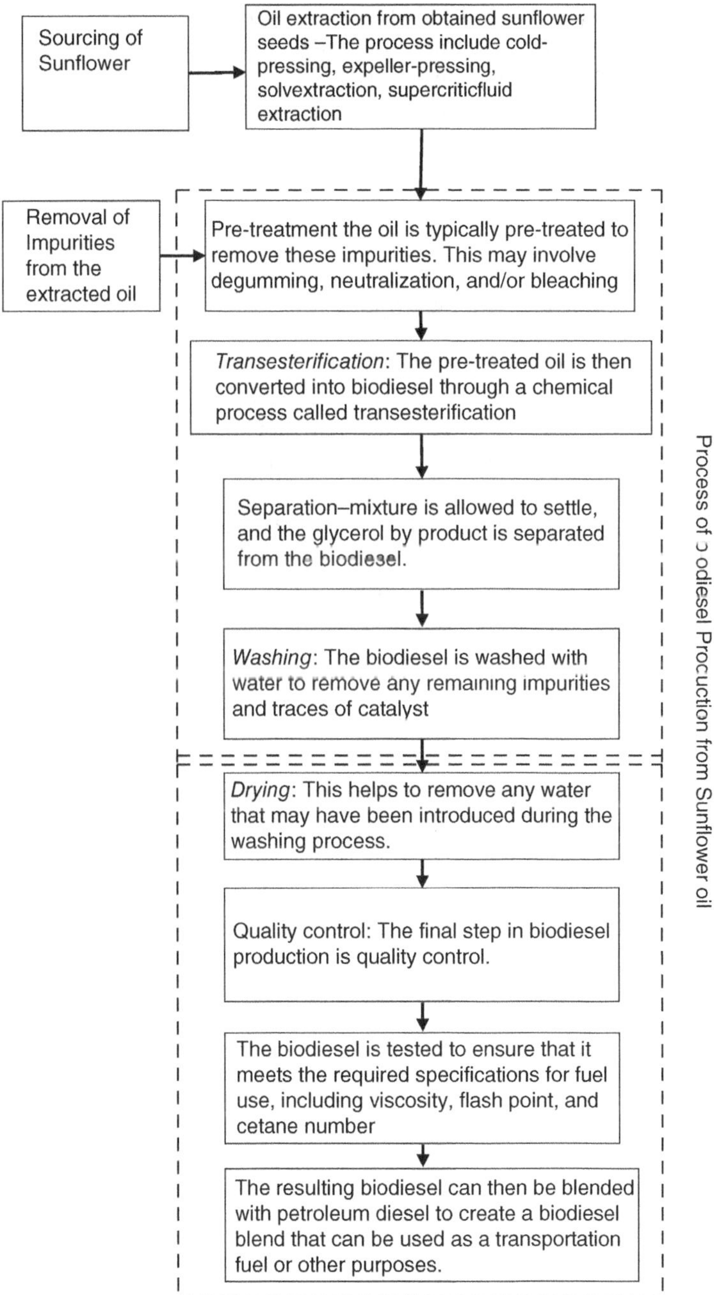

Figure 12.2 Flowchart showing the step-by-step process of producing biodiesel from sunflower.

The resulting biodiesel can then be blended with petroleum diesel to create a biodiesel blend that can be used as a transportation fuel. Sunflower oil is a popular feedstock for biodiesel production due to its high oil content and availability. However, the quality and yield of the biodiesel can depend on various factors, including the quality of the sunflower oil and the transesterification process. Therefore, it is important to use high-quality sunflower oil and optimize the production process to achieve the desired quality and yield of biodiesel.

12.3.1 Biodiesel Production from Sunflower Oil

1) Chemical composition of sunflower oil and its suitability for biodiesel production: Based on the current studies conducted in the lab, sunflower oil, a triglyceride primarily consisting of oleic acid (approximately 20%), linoleic acid (approximately 65%), along with smaller proportions of palmitic, stearic, and other fatty acids, boasts a substantial content of unsaturated fatty acids. This high concentration of unsaturated fatty acids renders sunflower oil an excellent raw material for the production of biodiesel. The fatty acid composition of sunflower oil can affect the properties of the resulting biodiesel, with higher levels of unsaturated fatty acids generally resulting in lower oxidative stability and higher viscosity. Therefore, it is important to choose the appropriate sunflower oil variety and processing conditions to obtain the desired quality of biodiesel.
2) Sunflower oil, composed mainly of oleic acid (around 20%), linoleic acid (around 65%), and smaller amounts of palmitic, stearic, and other fatty acids, possesses a significant concentration of unsaturated fatty acids. This elevated content of unsaturated fatty acids makes sunflower oil a superb feedstock for the biodiesel production process.
3) Catalysts and their Impact on the Transesterification Process: Catalysts play a crucial role in accelerating the transesterification reaction rate and minimizing the required alcohol quantity. Sodium hydroxide (NaOH) and potassium hydroxide (KOH) are the most commonly employed catalysts in biodiesel production. While both catalysts effectively facilitate the reaction, they possess distinct properties influencing biodiesel yield and quality. NaOH, being a strong base, has the potential to cause saponification of free fatty acids, leading to soap formation and a diminished biodiesel yield. On the contrary, KOH, a weaker base, is less prone to saponification but may result in elevated water levels in the reaction mixture. Alternative catalysts, such as acid catalysts and enzymes, are also viable options, each with its specific advantages and drawbacks.

In summary, sunflower oil is a suitable feedstock for biodiesel production due to its high percentage of unsaturated fatty acids. The transesterification process, with the use of catalysts such as NaOH or KOH, is the main method for converting sunflower oil into biodiesel. The choice of catalyst can affect the yield and quality of the biodiesel, and it is important to optimize the reaction conditions to obtain the desired properties of the biodiesel.

12.3.2 Other Conversion Technologies Used in the Production of Biodiesel from Sunflower Oil

In addition to transesterification, there are other conversion technologies that can be used in the production of biodiesel from sunflower oil:

1) *Pyrolysis*: Pyrolysis is a thermochemical process in which organic materials are heated in the absence of oxygen to produce a liquid fuel. In the case of sunflower oil, pyrolysis can be used to produce a bio-oil, which can be further processed into biodiesel. Pyrolysis can be carried out at different temperatures and with or without a catalyst.
2) *Supercritical methanol*: Supercritical methanol is a method in which methanol is heated to a high temperature and pressure above its critical point, which results in a supercritical fluid with unique properties. This method can be used to produce biodiesel from sunflower oil without the need for a catalyst. However, the high pressure and temperature required can make this method more expensive and energy-intensive than traditional transesterification.
3) *Hydrogenation*: Hydrogenation is a process in which hydrogen gas is added to a liquid or solid material to produce a new product. In the case of sunflower oil, hydrogenation can be used to convert the unsaturated fatty acids into saturated fatty acids, which can then be used to produce biodiesel. However, this process requires high pressure and temperature, and it can also result in lower yields of biodiesel.
4) *Enzymatic transesterification*: Enzymatic transesterification is a method in which enzymes are used as catalysts instead of traditional chemical catalysts. This method can be more environmentally friendly and produce higher yields of biodiesel. However, the cost of enzymes and the longer reaction times required can make this method more expensive than traditional transesterification.

In summary, there are various conversion technologies that can be used in the production of biodiesel from sunflower oil, in addition to traditional transesterification. Each method has its own advantages and disadvantages, and the choice of method depends on various factors such as cost, yield, and environmental impact.

12.3.3 Quality Control of Biodiesel and Its Compliance with International Standards

Quality control of biodiesel is essential to ensure that the fuel meets the required specifications and standards for use in diesel engines. The quality control process involves various tests and analyses to determine the fuel's physical and chemical properties, such as viscosity, density, flash point, acid value, and moisture content.

Biodiesel produced from sunflower oil must comply with international standards such as ASTM D6751 and EN 14214 to ensure that it is suitable for use as a fuel. These standards specify the maximum limits for various parameters such as viscosity, flash point, acid value, and sulfur content. The quality control process for biodiesel involves the following steps:

1) *Sampling*: Samples of the biodiesel are taken at various stages of production and storage to ensure that the fuel is consistent and free from contamination.
2) *Physical and chemical testing*: The samples are tested for various physical and chemical properties such as viscosity, density, flash point, acid value, and moisture content. These tests are carried out using standard methods such as ASTM and EN standards.
3) *Performance testing*: The biodiesel is tested for its performance characteristics such as engine power, fuel efficiency, and emissions. These tests are typically carried out in laboratory settings and in real-world driving conditions.
4) *Compliance with standards*: The test results are compared to the requirements of the relevant international standards such as ASTM D6751 and EN 14214. If the results are within the specified limits, the biodiesel is deemed compliant with the standards and suitable for use as a fuel.

In summary, quality control of biodiesel is essential to ensure that the fuel meets the required specifications and standards for use as a fuel. Biodiesel produced from sunflower oil must comply with international standards, which specify the maximum limits for various physical and chemical parameters. Quality control involves various tests and analyses to determine the fuel's properties and performance characteristics, and to ensure compliance with international standards.

12.3.4 Economic and Environmental Considerations in Sunflower Biodiesel Production

Sunflower biodiesel production involves economic and environmental considerations that need to be taken into account. Some of these considerations are:

Economic considerations

1) *Feedstock availability and cost*: The availability and cost of sunflower oil, the primary feedstock for biodiesel production, can have a significant impact on the overall cost of production. Therefore, it is essential to ensure a stable supply of feedstock at a reasonable cost.

2) *Production cost*: The production cost of biodiesel includes the cost of raw materials, labor, equipment, and energy. To make biodiesel production economically viable, it is essential to optimize the production process and reduce production costs wherever possible.
3) *Market demand*: The demand for biodiesel in the market can also influence the profitability of sunflower biodiesel production. It is essential to identify potential markets and ensure that the quality of the biodiesel produced meets the required standards.

Environmental considerations

1) *Land use and biodiversity*: Sunflower biodiesel production can have an impact on land use and biodiversity, particularly if the land is converted from natural habitats to sunflower fields. Therefore, it is essential to ensure that biodiesel production does not lead to the loss of biodiversity and degradation of natural habitats.
2) *Emissions of greenhouse gases*: While biodiesel production presents an opportunity to decrease greenhouse gas emissions in contrast to petroleum diesel, the production process itself may contribute to the release of greenhouse gases, including carbon dioxide and nitrous oxide. It is imperative to minimize these emissions through process optimization and the incorporation of renewable energy sources.
3) *Water use and pollution*: Biodiesel production can also have an impact on water use and pollution. Water is used in various stages of the production process, and if not managed properly, it can lead to water pollution. It is essential to minimize water use and ensure that wastewater is treated appropriately.
4) *Waste management*: Biodiesel production can generate waste products such as glycerol, which must be managed appropriately to avoid environmental pollution.

In summary, sunflower biodiesel production involves economic and environmental considerations that need to be carefully considered. Economic considerations include feedstock availability and cost, production cost, and market demand, while environmental considerations include land use and biodiversity, greenhouse gas emissions, water use and pollution, and waste management. By addressing these considerations, sunflower biodiesel production can be made economically viable and environmentally sustainable.

12.3.5 Competitiveness of Sunflower Biodiesel Compared to Other Feedstocks

Sunflower biodiesel is one of several feedstocks used for biodiesel production, and its competitiveness compared to other feedstocks depends on several factors, including:

1) *Feedstock availability and cost*: The availability and cost of different feedstocks can vary depending on factors such as climate, soil conditions, and market

demand. Sunflower oil may be more readily available and cost-effective in regions where sunflowers are grown, while other feedstocks may be more competitive in other regions.

2) *Biodiesel yield*: The yield of biodiesel that can be obtained from different feedstocks can vary. Sunflower oil has a relatively high yield compared to some other feedstocks, such as soybean oil.
3) *Quality of biodiesel*: The quality of biodiesel produced from different feedstocks can vary. Sunflower biodiesel has properties that make it suitable for use in diesel engines, such as a high cetane number and low sulfur content.
4) *Environmental impact*: The environmental impact of biodiesel production from different feedstocks can vary depending on factors such as land use, greenhouse gas emissions, and water use. Sunflower biodiesel has been found to have a lower environmental impact than some other feedstocks, such as palm oil.
5) *Policy and regulatory support*: Policy and regulatory support can influence the competitiveness of different feedstocks for biodiesel production. For example, government mandates or tax incentives for biodiesel production or use can make sunflower biodiesel more competitive compared to other feedstocks.

In summary, the competitiveness of sunflower biodiesel compared to other feedstocks depends on several factors, including feedstock availability and cost, biodiesel yield, quality of biodiesel, environmental impact, and policy and regulatory support. By considering these factors, producers and policymakers can make informed decisions about the most competitive feedstocks for biodiesel production in different regions and markets.

12.3.6 Sustainability and Social Impacts of Sunflower Cultivation for Biodiesel Production

Sunflower cultivation for biodiesel production can have sustainability and social impacts that need to be considered. Some of these impacts are:

1) *Land use and biodiversity*: Sunflower cultivation can lead to the conversion of natural habitats to sunflower fields, leading to the loss of biodiversity and degradation of natural habitats. Therefore, it is essential to ensure that sunflower cultivation is done sustainably without causing adverse effects on biodiversity.
2) *Soil health*: Intensive sunflower cultivation can lead to soil degradation, such as nutrient depletion, erosion, and compaction. Sustainable cultivation practices, such as crop rotation and intercropping, can help to maintain soil health and fertility.

3) *Water use and pollution*: Sunflower cultivation can lead to high water use, particularly in regions with limited water resources. Water pollution can also result from the use of agrochemicals, such as fertilizers and pesticides. Sustainable practices, such as efficient irrigation and integrated pest management, can help to minimize water use and pollution.
4) *Labor conditions and social equity*: Sunflower cultivation can have social impacts, such as labor exploitation and displacement of small-scale farmers. It is essential to ensure that sunflower cultivation is done in a socially responsible manner, with fair labor conditions and support for small-scale farmers.
5) *Climate change*: Sunflower cultivation can contribute to climate change through greenhouse gas emissions, such as nitrous oxide from the use of fertilizers. Sustainable cultivation practices, such as the use of organic fertilizers and reduced tillage, can help to reduce greenhouse gas emissions and mitigate climate change.

In summary, sunflower cultivation for biodiesel production can have sustainability and social impacts that need to be considered. These impacts include land use and biodiversity, soil health, water use and pollution, labor conditions and social equity, and climate change. By adopting sustainable cultivation practices and promoting social responsibility, sunflower biodiesel production can be made environmentally and socially sustainable.

12.3.7 Technological Challenges in Large-Scale Production of Sunflower Biodiesel

Large-scale production of sunflower biodiesel presents several technological challenges that need to be addressed to ensure efficient and cost-effective production. Some of these challenges are:

1) *Feedstock quality and availability*: The quality and availability of sunflower oil as a feedstock can vary, depending on factors such as weather conditions, pests, and diseases. Therefore, it is essential to ensure a steady supply of high-quality feedstock for biodiesel production.
2) *Efficient oil extraction*: Sunflower oil extraction is a critical step in biodiesel production. Traditional oil extraction methods can be time-consuming and labor-intensive, leading to high production costs. Therefore, efficient oil extraction methods, such as solvent extraction or mechanical pressing, need to be used to ensure high yields of high-quality oil.
3) *Transesterification efficiency*: Transesterification is a chemical reaction that converts sunflower oil into biodiesel. The efficiency of this reaction can vary, depending on factors such as reaction conditions, catalyst type, and feedstock quality. Therefore, optimizing transesterification conditions is essential to ensure high yields of high-quality biodiesel.

4) *Separation and purification*: Separation and purification of the biodiesel from the reaction mixture are crucial steps in large-scale production. This process can be time-consuming and energy-intensive, leading to high production costs. Therefore, efficient separation and purification methods, such as distillation or membrane filtration, need to be used to ensure high yields of high-quality biodiesel.
5) *Waste management*: Biodiesel production generates several waste streams, such as glycerol and spent catalysts, that need to be managed properly to minimize environmental impacts and ensure compliance with regulations.
6) *Infrastructure and logistics*: Large-scale production of biodiesel requires significant investments in infrastructure, such as storage tanks, pipelines, and transport vehicles.

Therefore, efficient logistics planning is essential to ensure the smooth flow of feedstock and biodiesel from production to distribution.

In summary, large-scale production of sunflower biodiesel presents several technological challenges that need to be addressed to ensure efficient and cost-effective production. These challenges include feedstock quality and availability, efficient oil extraction, transesterification efficiency, separation and purification, waste management, and infrastructure and logistics. By addressing these challenges, sunflower biodiesel production can be made more efficient and economically viable.

12.3.8 Opportunities for Research and Development in Sunflower Biodiesel Production

There are several opportunities for research and development in sunflower biodiesel production that can improve the efficiency, sustainability, and cost-effectiveness of the process. Some of these opportunities are:

1) *Improvement of feedstock quality and yield*: Research can be conducted to develop sunflower varieties with high oil content and better resistance to pests and diseases. This can lead to a steady supply of high-quality feedstock for biodiesel production.
2) *Development of efficient oil extraction methods*: Research can be conducted to develop and optimize efficient oil extraction methods, such as ultrasound-assisted extraction or enzymatic hydrolysis. These methods can lead to higher yields of high-quality oil and lower production costs.
3) *Optimization of transesterification conditions*: Research can be conducted to optimize transesterification conditions, such as reaction temperature, catalyst type, and reaction time. This can lead to higher yields of high-quality biodiesel and lower production costs.

4) *Development of waste-to-value strategies*: Research can be conducted to develop and optimize strategies for the conversion of biodiesel waste streams, such as glycerol and spent catalysts, into value-added products, such as bio-based chemicals or fertilizers.
5) *Implementation of sustainable production practices*: Research can be conducted to develop and implement sustainable production practices, such as intercropping, crop rotation, or organic farming. These practices can lead to reduced environmental impacts and increased social and economic benefits.
6) *Integration with other bioenergy production systems*: Research can be conducted to explore the integration of sunflower biodiesel production with other bioenergy production systems, such as biogas or bioethanol. This can lead to increased energy efficiency and reduced production costs.

In summary, research and development in sunflower biodiesel production can lead to significant improvements in feedstock quality and yield, oil extraction efficiency, transesterification optimization, waste-to-value strategies, sustainable production practices, and integration with other bioenergy production systems. These improvements can make sunflower biodiesel production more efficient, sustainable, and cost effective.

12.4 Future Prospects of Sunflower Biodiesel Production

Sunflower biodiesel production has promising future prospects due to its potential for expansion, government policies and incentives, integration with existing agricultural systems, and its role in energy security and climate change mitigation.

1) *Potential for expansion of sunflower biodiesel production*: Sunflower biodiesel production has the potential for expansion as sunflower is a versatile crop that can be grown in various regions of the world. In addition, sunflower biodiesel has the potential to replace a significant portion of petroleum-based diesel fuel. The expansion of sunflower biodiesel production can also create new job opportunities and contribute to rural development.
2) *Government initiatives and incentives for sunflower biodiesel promotion*: Numerous governments globally have adopted policies and incentives aimed at fostering the production and utilization of biodiesel, specifically emphasizing sunflower biodiesel. These policies include tax credits, subsidies, and mandates that require a certain percentage of diesel fuel to come from renewable sources. These policies and incentives can provide a favorable market environment for sunflower biodiesel production.

3) *Integration of sunflower biodiesel production into existing agricultural systems*: Sunflower biodiesel production can be integrated into existing agricultural systems by using sunflower as a rotational crop. This can provide farmers with an additional source of income and can promote soil health and fertility. In addition, the use of biodiesel in agricultural machinery can reduce the environmental impact of farming practices.
4) *Role of sunflower biodiesel in climate change mitigation and energy security*: Sunflower biodiesel can play an important role in climate change mitigation and energy security. It can reduce dependence on petroleum-based diesel fuel, which is often imported from politically unstable regions of the world. In addition, sunflower biodiesel has lower greenhouse gas emissions compared to petroleum-based diesel fuel, which can help mitigate climate change.

In summary, the future prospects of sunflower biodiesel production are promising due to its potential for expansion, government policies and incentives, integration with existing agricultural systems, and its role in energy security and climate change mitigation. However, further research and development are needed to optimize the production process and to improve the economic and environmental sustainability of sunflower biodiesel production.

12.4.1 Challenges and Opportunities of Sunflower Biodiesel Production

Sunflower biodiesel production faces several challenges and opportunities, which are outlined thus:

Challenges

1) *Feedstock availability*: One of the main challenges of sunflower biodiesel production is the availability of sunflower oil as a feedstock. The production of sunflower oil is subject to weather conditions, pests and diseases, and fluctuations in market prices, which can affect the availability and price of the feedstock.
2) *Production costs*: The production costs of sunflower biodiesel can be higher than those of petroleum diesel, which can make it less competitive in the market. This is due to factors such as the cost of feedstock, equipment, and labor, as well as the cost of complying with environmental and safety regulations.
3) *Infrastructure*: The production and distribution infrastructure for biodiesel may not be well-established in some regions, which can limit the availability and use of the fuel.
4) *Policy and regulatory support*: The development of sunflower biodiesel production may require policy and regulatory support from governments, such as tax incentives, subsidies, and mandates for the use of biodiesel in transportation.

Opportunities

1) *Sustainable energy resource*: Sunflower biodiesel stands as a renewable energy source capable of diminishing reliance on fossil fuels, thereby aiding in the mitigation of greenhouse gas emissions.
2) *Rural development*: Sunflower biodiesel production can provide opportunities for rural development by creating jobs and income for farmers and small-scale producers.
3) *Local production*: Sunflower biodiesel can be produced locally, reducing dependence on imported fossil fuels and promoting energy independence.
4) *Sustainability*: Sunflower biodiesel production can contribute to the development of sustainable agricultural practices, such as crop rotation and conservation tillage, which can improve soil health and reduce environmental impacts.

In summary, sunflower biodiesel production faces several challenges related to feedstock availability, production costs, infrastructure, and policy and regulatory support. However, there are also opportunities related to its renewable nature, potential for rural development, local production, and sustainability. The development of sunflower biodiesel production requires a comprehensive approach that addresses these challenges and promotes these opportunities.

12.5 Conclusion

This chapter is focused on the exposition of Sunflower for Biodiesel Production. It highlights the potential of sunflower as a feedstock for biodiesel production. The key points discussed include the chemical composition of sunflower oil, methods of oil extraction, transesterification process for biodiesel production, catalysts used in the process, quality control measures, and economic and environmental considerations. The implications for the future of sunflower biodiesel production are promising. There is potential for expansion of sunflower biodiesel production due to its versatility and potential to replace petroleum-based diesel fuel. Government policies and incentives, integration into existing agricultural systems, and the role of sunflower biodiesel in energy security and climate change mitigation all provide a favorable market environment for sunflower biodiesel production. Further research and development are needed to optimize the production process, improve economic and environmental sustainability, and address challenges such as technological limitations and social impacts. Therefore, there is a call for further research and development to fully harness the potential of sunflower as a feedstock for biodiesel production. In conclusion, sunflower biodiesel production presents significant potential as a sustainable and renewable alternative to petroleum-based diesel fuel. With further research and development, it can provide economic, social, and environmental benefits and play a crucial role in addressing global energy and environmental challenges.

References

1 Chilakamarry, C.R., Sakinah, A., Zularisam, A. et al. (2022). Bioconversion of glycerol into biofuels – opportunities and challenges. *BioEnergy Res.* 15: 46–61.

2 Andreo-Martínez, P., Ortiz-Martínez, V.M., Salar-García, M.J. et al. (2022). Waste animal fats as feedstock for biodiesel production using non-catalytic supercritical alcohol transesterification: a perspective by the PRISMA methodology. *Energy Sustain. Dev.* 69: 150–163.

3 Ennetta, R., Soyhan, H.S., Koyunoglu, C., and Demir, V.G. (2022). Current technologies and future trends for biodiesel production: areview. *Arab. J. Sci. Eng.* 47: 15133–15151.

4 Rincón-Pérez, J., Celis, L.B., Morales, M. et al. (2021). Improvement of methane production at alkaline and neutral pH from anaerobic co-digestion of microalgal biomass and cheese whey. *Biochem. Eng. J.* 169: 107972.

5 Soudagar, M.E.M., Khan, H.M., Khan, T.Y. et al. (2021). Experimental analysis of engine performance and exhaust pollutant on a single-cylinder diesel engine operated using *Moringa oleifera* biodiesel. *Appl. Sci.* 11: 7071.

6 Anushree, S., André, M., Guillaume, D., and Frédéric, F. (2017). Stearic sunflower oil as a sustainable and healthy alternative to palm oil: a review. *Agron. Sustain. Dev.* 37: 1–10.

7 Parsons, S., Raikova, S., and Chuck, C.J. (2020). The viability and desirability of replacing palm oil. *Nat. Sustain.* 3: 412–418.

8 Adeleke, B.S. and Babalola, O.O. (2020). Oilseed crop sunflower (*Helinathus annuus*) as a source of food: nutritional and healths benefits. *Food Sci. Nutr.* 8: 4666–4684.

9 Mathias, J.D., Alzina, A., Grédiac, M. et al. (2015). Upcycling sunflower stems as natural fibers for biocomposite applications. *Bioresources* 10: 8076–8088.

10 Çamur, H. and Alassi, E. (2021). Physicochemical properties enhancement of biodiesel synthesis from various Feedstocks of waste/residential vegetable oils and palm oil. *Energies* 14: 4928.

11 Peer, M.S., Kasimani, R., Rajamohan, S., and Ramakrishnan, P. (2017). Experimental evaluation on oxidation stability of biodiesel/diesel blends with alcohol addition by rancimat instrument and FTIR spectroscopy. *Mech. Sci. Technol.* 31: 455–463.

12 Kafuku, G. and Mbarawa, M. (2010). Biodiesel production from *Croton megalocarpus* oil and its process optimization. *Fuel* 89: 2556–2560.

13 Silitonga, A.S., Atabani, A., Mahlia, T. et al. (2011). A review on prospect of *Jatropha curcas* for biodiesel in Indonesia. *Sustain. Energy Rev.* 15: 3733–3756.

14 Atabani, A., Silitonga, A., Badruddin, I.A. et al. (2012). A comprehensive review on biodiesel as an alternative energy resource and its characteristics. *Sustain. Energy Rev.* 16: 2070–2093.

15 Janaun, J. and Ellis, N. (2010). Perspectives on biodiesel as a sustainable fuel. *Sustain. Energy Rev.* 14: 1312–1320.

16 Cao, P., Tremblay, A.Y., and Dubé, M.A. (2009). Kinetics of canola oil transesterification in a membrane reactor. *Eng. Chem. Res.* 48: 2533–2541.

17 Kılıç, M., Uzun, B.B., Pütün, E., and Pütün, A.E. (2013). Optimization of biodiesel production from castor oil using factorial design. *Fuel Process. Technol.* 111: 105–110.

18 Cremonez, P.A., Feroldi, M., Nadaleti, W.C. et al. (2015). Biodiesel production in Brazil: current scenario and perspectives. *Sustain. Energy Rev.* 42: 415–428.

19 Dauguet, S., Labalette, F., Fine, F. et al. (2016). Genetic impact on protein content and hullability of sunflower seeds, and on the quality of sunflower meal. *OCL Oilseeds Fats Crops Lipids* 23: D205.

20 Qi, L.L., Foley, M.E., Cai, X.W., and Gulya, T.J. (2016). Genetics and mapping of a novel downy mildew resistance gene, Pl 18, introgressed from wild *Helianthus argophyllus* into cultivated sunflower (*Helianthus annuus* L.). *Theor. Appl. Genet.* 129: 741–752.

21 Fusari, C.M., Di Rienzo, J.A., Troglia, C. et al. (2012). Association mapping in sunflower for sclerotinia head rot resistance. *BMC Plant Biol.* 12: 1–13.

22 Zollinger, R.K. (2004). Advances in sunflower weed control in the USA. In: *Proceedings of the 16th International Sunflower Conference*, 435–439, Fargo, ND, USA (29 August–2 September 2004).

23 Tuemmler, C. and Schroeder, G. (2013). Study on chemical control of common ragweed (*Ambrosia artemisiifolia*) in sunflower and grain legumes. In: *Proceedings of the International Multidisciplinary Scientific GeoConference: Ambrosia in Germany—Can the Invasion be Halted?*, 105–110, Berlin, Germany (10–12 September 2013).

24 García-Vila, M., Fereres, E., Prieto, M.H. et al. (2012). Sunflower. In: *Crop Yield Response to Water*, 66. Rome, Italy: FAO Irrigation and Drainage.

25 IPCC, Pachauri, R.K., and Meyer, L.A. (eds.) (2014). Climate change 2014: synthesis report. In: *Contribution of Working Groups I, II and III to the Fifth Assessment Report of the Intergovernmental Panel on Climate Change*, 151. Core Writing Team, IPCC: Geneva, Switzerland.

26 Flagella, Z., Rotunno, T., Tarantino, E. et al. (2002). Changes in seed yield and oil fatty acid composition of high oleic sunflower (*Helianthus annuus* L.) hybrids in relation to the sowing date and the water regime. *Eur. J. Agron.* 17: 221–230.

27 Ebrahimian, E., Seyyedi, S.M., Bybordi, A., and Damalas, C.A. (2019). Seed yield and oil quality of sunflower, safflower, and sesame under different levels of irrigation water availability. *Agric. Water Manag.* 218: 149–157.
28 Alberio, C., Izquierdo, N.G., and Aguirrezábal, L.A.N. (2015). Sunflower crop physiology and agronomy. In: *Sunflower*, 53–91. Champaign, IL, USA: AOCS Press.
29 Ferreira, A.M. and Abreu, F.G. (2001). Description of development, light interception and growth of sunflower at two sowing dates and two densities. *Math. Comput. Simul.* 56: 369–384.
30 Balalic, I., Zoric, M., Brankovic, G. et al. (2012). Interpretation of hybrid× sowing date interaction for oil content and oil yield in sunflower. Field. *Crop. Res.* 137: 70–77.
31 Vear, F. (2016). Changes in sunflower breeding over the last fifty years. *OCL Oilseeds Fats Crops Lipids* 23: 1–8.
32 Debaeke, P., Casadebaig, P., and Langlade, N. (2021). New challenges for sunflower ideotyping in changing environments and more ecological cropping systems. *OCL Oilseeds Fats Crops Lipids* 28: 1–23.
33 Hussain, M., Farroq, S., Hasan, W. et al. (2018). Drought stress in sunflower: physiological effects and its management through breeding and agronomic alternatives. *Agric. Water Manag.* 201: 152–166.
34 Göksoy, A.T., Demir, A.O., Turan, Z.M., and Dagüstü, N. (2004). Responses of sunflower (*Helianthus annuus* L.) to full and limited irrigation at different growth stages. *Field Crop Res.* 87: 167–178.
35 Browne, C.L. (1997). Effect of date and final irrigation on yield and yield components of sunflowers in a semi-arid environment. *Aust. J. Exp. Agric.* 17: 482–488.
36 Rawson, H.M. and Turner, N.C. (1983). Irrigation timing and relationships between leaf area and yield in sunflower. *Irrig. Sci.* 4: 167–175.
37 Giannini, V., Mula, L., Carta, M. et al. (2022). Interplay of irrigation strategies and sowing dates on sunflower yield in semi-arid Mediterranean areas. *Agric. Water Manag.* 260: 107287.
38 Singh, N.B., Singh, D., and Singh, A. (2015). Biological seed priming mitigates the effects of water stress in sunflower seedlings. *Physiol. Mol. Biol. Plants* 21: 207–214.
39 Langeroodia, A.S., Tedeschi, P., Allevato, E. et al. (2022). Agronomic response of sunflower subjected to biochar and arbuscular mycorrhizal fungi application under drought conditions. *Ital. J. Agron.* 17: 2086.
40 Poonia, R. and Chaudhary, O.P. (2022). Unraveling the mystery of non extraction of honey from modern sunflower hybrids. *J. Apic. Res.* 61: 557–566.
41 Deppermann, A., Balkovi˘c, J., Bundle, S.C. et al. (2018). Increasing crop production in Russia and Ukraine – regional and global impacts from intensification and recultivation. *Environ. Res. Lett.* 13: 025008.

42 M.O. Okwu, O. Omonigho, O.D. Samuel, E. Eyitimi (2022). Production and charaterization of briquettes produced from blend of rice-husk and water-hyacinth. "Advances in Biofeedstocks and Biofuels" Scrivener-Wiley. https://doi.org/10.1002/9781119785842.ch4

43 Heuzé, V., Tran, G., and Vittet, M.A. (2015). Sunflower (general). In: *Feedipedia, a Programme by INRAE, CIRAD, AFZ and FAO*. https://www.feedipedia.org/node/6225 (accessed 28 May 2015).

44 Putnam, D.H., Oplinger, E.S., Hicks, D.R. et al. (1990). Sunflower. In: *Alternative Field Crop Manual*. Madison, WI, USA: University of Wisconsin-Exension, Cooperative Extension.

45 Dohme-Meier, F. and Bee, G. (2012). Feeding unprotected CLA methyl esters compared to sunflower seeds increased milk CLA content level but inhibited milk fat synthesis in cows. *Asian Australas. J. Anim. Sci.* 25: 75–85.

46 Shingfield, K.J., Reynolds, C.K., Hervás, G. et al. (2006). Examination of the persistency of milk fatty acid composition responses to fish oil and sunflower oil in the diet of dairy cows. *J. Dairy Sci.* 89: 714–732.

47 INRA (2018). *INRA Feeding System for Ruminants*, 640p. Wageningen, The Netherlands: Wageningen Academic Publishers.

48 Finoli, C., Vecchio, A., Sirtori, C., and Vercesi, A. (2004). Fungal contamination and mycotoxin occurrence in raw materials and finished feeds. *Tecn. Molit.* 55: 223–234.

49 National Research Council (2001). *Nutrient Requirements of Dairy Cattle*, 8th revised ed. Washington, DC, USA: The National Academies Press.

50 Johnston, M. (2007). *Redefining Biodiesel Crop Yields*. Center for Sustainability and the Global Environment, University of Wisconsin.

51 Tashtoush, G.M., Al Widyan, M.I., and Al Jarrah, M.M. (2004). Experimental study on evaluation and optimization of conversion of waste animal fat into biodiesel. *Energy Convers. Manag.* 45: 2697–2711.

52 Omoregbee, H., Okwu, M.O., Tartibu, L.K., and Edward, B. (2022). Effect of process parameter on biogas yield: a review. In: *Advances in Biofeedstocks and Biofuels*. Scrivener-Wiley.

53 Frohlich, A. and Rice, B. (2005). Evaluation of *Camelina sativa* oil as a feedstock for biodiesel production. *Ind. Crop. Prod.* 21: 25–31.

54 Cardone, M., Mazzoncini, M., Menini, S. et al. (2003). *Brassica carinata* as an alternative oil crop for the production of biodiesel in Italy: agronomic evaluation, fuel production by transesterification and characterization. *Biomass Bioenergy* 25: 623–636.

55 Okonkwo C.P., V. Ajiwe, M. Obiadi, M. Okwu, J. Oyogu (2023). Production of biodiesel from the novel non-edible seed of *Chrysobalanus icaco* using natural heterogeneous catalyst: modeling and prediction using artificial neural network. *J. Clean. Prod.* https://doi.org/10.1016/j.jclepro.2022.135631.

56 Auld, D.L., Peterson, C.L., and Korus, R.A. (1989). Production, processing, and utilization of rapeseed oil as a diesel fuel substitute. In: *Energy from Biomass and Wastes XII*, 1187–1207. Chicago, IL, USA: Institute of Gas Technology.

57 Krahl, J., Munack, A., Stein, H. et al. (2005). Fuel economy and environment characteristics of biodiesel and low sulfur fuels in diesel engines. *Landbauforsch. Volkenrode* 55: 99–106.

58 Goodrum, J.W. and Geller, D.P. (2005). Influence of fatty acid methyl esters from hydroxylated vegetable oils on diesel fuel lubricity. *Bioresour. Technol.* 96: 851–855.

59 Scrosta, V. (2003). Buone prospettive per il biodiesel ma solo a certe condizioni. *Inf. Agrar.* 59: 29–31.

60 Chiu, C.W., Schumacher, L.G., and Suppes, G.J. (2004). Impact of cold flow improvers on soybean biodiesel blend. *Biomass Bioenergy* 27: 485–491.

61 Szeto, W. and Leung, D.Y. (2022). Is hydrotreated vegetable oil a superior substitute for fossil diesel? A comprehensive review on physicochemical properties, engine performance and emissions. *Fuel* 327: 125065.

62 Riello, L. and Bona, S. (2006). Life cycle assessment for evaluating on-farm energy production: the case of sunflower oil. *Ital. J. Agron.* 1: 705–709.

13

Sustainable Biodiesel Production Using *Moringa oleifera* as Biodegradable Feedstock

13.1 Introduction

13.1.1 Description of *Moringa oleifera*

The previous chapters of this book focused on the paradigm shift from fossil fuel to renewable energy. It is a known fact that fossil fuels, including natural gas, coal, and oil, have served as a major energy source for the economic development of many countries [1]. However, the combustion of these fuels has presented significant environmental and health concerns. Biofuels have, in no doubt, emerged as a promising alternative to fossil fuels, as they offer several potential benefits, including improved energy security, reduced greenhouse gas emissions, and increased rural development. Among the various biofuel feedstocks, *Moringa oleifera* (*M. oleifera*), a fast-growing tree that is native to India and widely cultivated in tropical and subtropical regions, Eastern and Southern Asia has garnered increasing attention as a potential source of oil for biodiesel production [2]. *M. oleifera* is a highly versatile plant that has been used for centuries for food, medicine, and other applications [3]. Popoola and Obembe [3] further exposed the significant use of *M. oleifera* in agro-ecological regions of Africa, especially in the pharmaceuticals. Rockwood et al. [4] emphasized the substantial healing potential of *M. oleifera*, attributing it to the abundance of essential bioactive compounds and functional ingredients found in its leaves, seeds, and pods. This characteristic renders *M. oleifera* a valuable and nutritionally rich resource with multifaceted health benefits. The diverse array of bioactive compounds contributes to its reputation as a plant with significant therapeutic properties, making it a noteworthy source of nourishment and potential applications in various health-related contexts.

Feedstocks for Sustainable Biodiesel Production: Characterization, Selection, and Optimization, First Edition. Chinwe P. Okonkwo and Modestus O. Okwu.

Figure 13.1 Pictorial view of the tree and seed pods of *Moringa oleifera* (*Sources:* Krish Dulal/Wikimedia Commons/CC BY-SA 3.0 and Prof. Chen Hualin/Wikimedia Commons/CC BY-SA 4.0).

Subsequent investigations reveal that *M. oleifera* has been cultivated and embraced globally, with a notable presence in regions including Sub-Saharan Africa, the Caribbean and Pacific Islands, Indonesia, Sudan, the Middle East, Florida, and North and South America [5]. The *M. oleifera* tree has garnered significant attention for its pivotal role in the treatment of various diseases, including but not limited to herpes, cardiovascular disorders, cancer, cramps, diabetes, sexually transmitted diseases, gout, and rheumatism [6, 7]. Particularly in tropical and subtropical regions where the plant thrives abundantly [8], the leaves of Moringa stand out as the most extensively utilized plant part. This preference is attributed to their remarkable efficacy in addressing malnutrition-related issues [9]. The widespread cultivation and diverse applications of Moringa underscore its global importance in addressing health challenges and nutritional deficiencies across different geographical regions. Figure 13.1 represents the pictorial view of the Tree and Seed Pods of *M. oleifera,* while Figure 13.2 represents the Pollen and Flowers of *M. oleifera.*

13.1.2 *Moringa oleifera* Seed

In recent years, there has been a surge in interest surrounding the diverse applications of Moringa. Notably, *M. oleifera* seeds have gained attention for their rich content of bioactive compounds [10]. These bioactive compounds, including alkaloids, glucosinolates, isothiocyanates, oil, proteins, carbohydrates, and thiocarbamates, exhibit significant pharmacological potential, offering numerous health benefits [11]. While a plethora of reviews have explored the remarkable properties of this consequential and sustainable plant, the majority has traditionally focused on the bioactive compounds present in the leaves [12–16].

Figure 13.2 Pictorial view of *Moringa oleifera* pollen and flowers (*Sources:* Harvey McDaniel/Flickr/CC BY 2.0).

However, recent studies have begun to shed light on the distinct value of Moringa seeds in various applications, extending beyond traditional leaf-centric analyses. Some of these applications include their use in animal feed, medicinal purposes, and water treatment [17–19]. Moreover, Moringa seeds have been identified as rich sources of proteins, lipids, and carbohydrates, comprising 33%–60% of their dry weight [20]. This multifaceted exploration into Moringa seeds broadens our understanding of their potential contributions to animal nutrition, medical advancements, and water purification initiatives.

The plant is known to produce oil-rich seeds that can be processed into biodiesel. Subsequent sections of this book chapter provide a comprehensive assessment of the potential of *M. oleifera* as a feedstock for biofuel production, highlighting the advantages and challenges associated with its use and discussing the future prospects of this promising feedstock. By further exploring the potential of *M. oleifera* for biofuel production, this manuscript contributes to the ongoing efforts to develop sustainable, environmentally friendly alternatives to fossil fuels.

To the best of the authors' knowledge, only a few studies have demonstrated the use of *M. oleifera* for biodiesel production. Some recent studies as presented in Table 13.1 shows clearly the fact that Moringa has got more attention on the use of the plant for food products, pharmaceuticals, water treatment and other generic use. This chapter is an exposition of Moringa for biodiesel production. The next section will focus more on the procedure for biodiesel production using the versatile and unique plant.

Table 13.1 Studies conducted using *Moringa oleifera* in diverse applications.

Citations	Authors	Title	Applications	Year
[17]	Falowo et al.	Multi-functional application of *Moringa oleifera* Lam. in nutrition and animal food products	FP	2018
[13]	Alain Mune et al.	A comparison of the nutritional quality of proteins from *Moringa oleifera* leaves and seeds	FP	2016
[21]	Oyeyinka and Oyeyinka	*Moringa oleifera* as a food fortificant: recent trends and prospects	FP	2018
[22]	Saucedo-Pompa et al.	Moringa plants: bioactive compounds and promising applications in food products	FP	2018
[23]	Jaja-Chimedza et al.	A dietary isothiocyanate-enriched moringa (*Moringa oleifera*) seed extract improves glucose tolerance in a high-fat diet.	FP	2018
[24]	Ma et al.	Evaluation of phytochemical and medicinal properties of Moringa (*Moringa oleifera*) as a potential functional food	FP	2020
[12]	Vergara-Jimenez et al.	Bioactive components in *Moringa oleifera* leaves protect against chronic disease.	Ph	2017
[25]	Amina et al.	Exploiting the potential of *Moringa oleifera* oil/polyvinyl chloride polymeric bionanocomposite film enriched with silver nanoparticles for antimicrobial activity	AA	2019
[16]	Khor et al.	The *in vitro* and *in vivo* anticancer properties of *Moringa oleifera*	Ph	2020
[19]	Yamaguchi et al.	*Moringa oleifera* seeds in water treatment: trends and future challenges.	WT	2020
[26]	Zulfiqar et al.	An overview of plant-based natural biostimulants for sustainable horticulture with a particular focus on moringa leaf extracts	GP	2019
[27]	Bancessi et al.	Present and potential uses of *Moringa oleifera* as a multipurpose plant in Guinea-Bissau. South African	GP	2019
[28]	Niju et al.	A comprehensive review of biodiesel production using *Moringa oleifera* oil	B	2019
[29]	Nijuet al.	Process optimization for biodiesel production from *Moringa oleifera* oil using conch shells as heterogeneous catalyst	B	2019

AA, antimicrobial activity; B, biodiesel; FP, food products; GP, general purpose; Ph, pharmaceuticals; WT, water treatment.

13.2 *Moringa oleifera* for Biodiesel Production

13.2.1 Stages of Biodiesel Production from *Moringa oleifera*

The production of biodiesel from *M. oleifera* classically involves two main steps: Oil extraction is the first step, while the second step is the transesterification process.

13.2.2 Oil Extraction

The first step is to extract oil directly from the *M. oleifera* plant (with a focus on the seed). This can be achieved by using a variety of methods; the two major methods include:

- Mechanical pressing
- Solvent extraction

Mechanical pressing is the commonly used method and it appears as the simplest to use method especially for small-scale production. The procedure is such that the seeds are sourced for and later dried. The next process is the crushing of the dried seed into a fine powder. The *M. oleifera* powder is then mixed with water and pressed so as to extract the oil.

For solvent extraction, this method involves using a solvent, such as hexane, to get the oil extracted from the seeds. The solvent is then separated from the oil through distillation. This is a more advanced method that uses supercritical carbon dioxide to extract the oil. The carbon dioxide is heated and pressurized to become a supercritical fluid, which can then be used to extract the oil.

13.2.3 Transesterification

The second step required for the production of biodiesel from the *M. oleifera* plant is the transesterification method, this technique typically involves converting the available moringa oil into biodiesel. The process begins by adding a catalyst, typically potassium hydroxide or sodium, to the oil. The catalyst reaction with the oil results in the formation of fatty acid methyl esters (FAMEs), the primary constituents of biodiesel. In addition, glycerol is generated as a by-product of this reaction and needs to be separated from the FAMEs. After the transesterification reaction is complete, the mixture is allowed to settle so that the glycerol can be separated from the FAMEs. The glycerol can be used for other applications, such as soap-making or as a feedstock for biogas production. The FAMEs are then washed with water to remove any remaining impurities before being dried and filtered. The resulting biodiesel can then be used as a fuel for diesel engines. Figure 13.3 represents a flowchart showing the step-by-step process of biodiesel production from the *M. oleifera* plant.

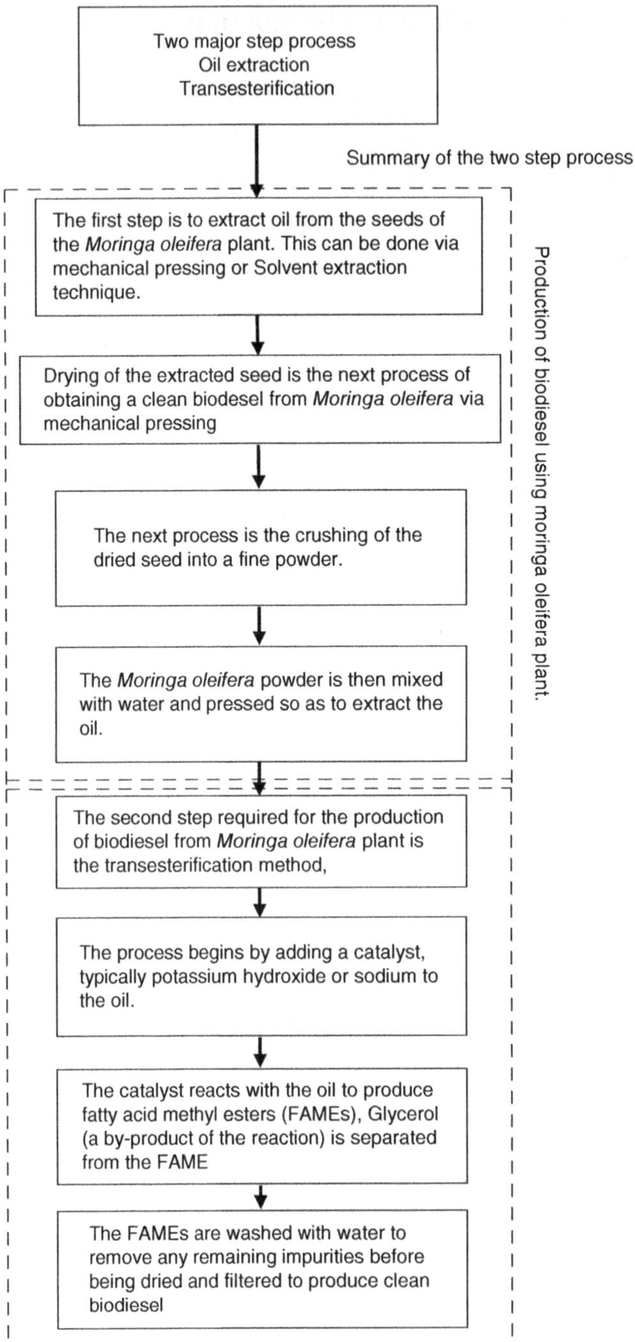

Figure 13.3 Flowchart showing the step-by-step process of biodiesel production from the *Moringa oleifera* plant.

13.3 Overview of the Chemical Reactions Involved in Transesterification

Transesterification is the process of converting triglycerides, such as those found in plant oils, into FAMEs, generally known as biodiesel. *M. oleifera* oil is a promising source of biodiesel due to its high content of unsaturated fatty acids.

The transesterification reaction involves the following chemical reactions.

13.3.1 Saponification

In the first step, the *M. oleifera* oil is mixed with an alcohol, typically methanol, and a catalyst, such as sodium hydroxide or potassium hydroxide, to form methyl esters and glycerol. This process is known as saponification, and it breaks down the triglycerides into their component fatty acids.

13.3.2 Transesterification

The methyl esters are then separated from the glycerol and washed to remove any residual catalyst or impurities. The methyl esters are then mixed with a small amount of acid, such as sulfuric acid, and heated to drive the transesterification reaction. This reaction involves the transfer of the methyl group from the alcohol to the fatty acid, resulting in the formation of FAMEs and glycerol as a by-product.

13.3.3 Purification

The FAMEs are then purified to remove any remaining impurities, such as water or methanol. This is typically done by washing the biodiesel with water or using a drying agent such as silica gel. The final product of this process is high-quality biodiesel, which can be used as a renewable and sustainable alternative to petroleum-based diesel fuel.

13.4 Benefits of Utilizing *M. oleifera* as a Raw Material for Biodiesel

There are potential economic and environmental benefits to utilizing *M. oleifera* as a sustainable, biodegradable raw material for biofuel production. *M. oleifera* is known to be a popular and fast-growing tree, which is native to tropical and subtropical regions of the world. In the literature review section, it has been shown that this plant is commonly used in the pharmaceutical and food industries. In recent times, the attention of researchers has been drawn to the benefit of Moringa beyond the current use in medicine, pharmaceutical industries and food industries. The global recognition of this potential plant or feedstock has been extended to the production of large quantities of biofuel. This is a result of the high oil content of the raw material, which can easily be extracted and converted into biodiesel. There is no doubt that the use of this raw material as a feedstock for the production of biofuel will bring both economic and environmental benefits.

13.4.1 Economic Benefits of *M. oleifera*

Some of the economic benefits of *M. oleifera* include:

1) *Sustainable income generation*: The abundance of this plant in Sub-Sahara Africa, the Caribbean, and other parts of the globe has made it much more interesting for generic use. The cultivation of *M. oleifera* for biofuel production can provide a sustainable income to young people, strategists in industry, farmers, movement and other stakeholders. Farmers will benefit greatly by investing in cultivation since Moringa is a hardy plant that can grow in a wide range of soils, making it suitable for cultivation in many areas.

   In addition, the oil extracted from *M. oleifera* is highly valued for its use in biofuels, making it a lucrative crop for farmers.
2) *Cost-effective production:* Compared to other biofuel feedstocks such as corn, soybeans, and palm oil, *M. oleifera* is relatively inexpensive to produce. This is because it grows naturally and requires fewer inputs, such as fertilizers, pesticides, and water.
3) *Job creation:* The production of *M. oleifera* for biofuel production can create jobs for young people in various sectors. An inflow of income is possible during farming of *M. oleifera* especially on a large scale, during harvesting so much sales is made since the use is beyond biodiesel production, during processing supply chain is required at the inbound and outbound stage, and finally is the transportation from source to destinations.

13.4.2 Environmental benefits of *M. oleifera*

Some of the environmental benefits of *M. oleifera* include:

1) *Reduced greenhouse gas emissions:* The use of *M. oleifera* as a feedstock for biofuel production will significantly reduce greenhouse gas emissions. Biodiesel produced from *M. oleifera* has been found to emit fewer greenhouse gases than petroleum diesel.

2) *Land use efficiency: M. oleifera* is a highly productive crop that requires less land than other biofuel crops. This means that more land can be used for other purposes, such as food production or conservation.
3) *Reduced soil erosion*: The deep root system of *M. oleifera* can help to reduce soil erosion, which is a common problem in many agricultural areas. This can help to protect the soil and improve its fertility.

In summary, the use of *M. oleifera* as a feedstock for biofuel production has the potential to provide economic benefits for farmers, create jobs for young people, create opportunities for investors and reduce greenhouse gas emissions. It can also help to improve land use efficiency and reduce soil erosion, leading to environmental benefits.

13.4.3 Rapid Growth of *Moringa oleifera* in Areas with Little Water and Marginal Soils

The process of growth of the plant – *M. oleifera* in areas with little water and marginal soils has made it a potential source of income for farmers in developing countries.

M. oleifera as a hardy tree can grow in a wide range of soils, including those that are considered marginal or of poor quality. It can also tolerate drought and is therefore well-suited to areas with limited water resources. These characteristics make it a potential source of income for farmers in developing countries, where access to fertile land and water is often limited.

Some of the noticeable ways in which *M. oleifera* can be grown in areas with marginal soils and little water include the process of soil improvement, water conversation, income generation, and agro-forestry. This is expanded further thus:

1) *Soil improvement*: *M. oleifera* has the ability to improve soil fertility and structure. Its leaves are rich in nutrients such as nitrogen, potassium, and phosphorus, which can be used as a natural fertilizer to improve the quality of the soil. By planting *M. oleifera*, farmers can improve the quality of their soil, making it more productive and enabling them to grow other crops alongside it.
2) *Water conservation*: *M. oleifera* is a drought-tolerant tree that can survive with minimal water. It has a deep taproot that allows it to access water from deeper soil layers, making it less dependent on surface water. By planting *M. oleifera*, farmers can conserve water resources and reduce their reliance on irrigation.
3) *Income generation*: *M. oleifera* is a valuable crop that can provide a source of income for farmers. Its leaves, seeds, and pods can be harvested and sold for a variety of uses, including food, medicine, and cosmetic products. The oil extracted from its seeds can also be used as a feedstock for biofuel production, as mentioned earlier.

4) *Agroforestry*: *M. oleifera* can be integrated into agroforestry systems, where it can be planted alongside other crops such as maize or beans. By planting *M. oleifera* as a border crop, it can help to improve soil fertility, provide shade, and conserve water, which can benefit the other crops in the system.

In summary, *M. oleifera* can be grown in areas with marginal soils and little water by improving soil fertility, conserving water, generating income, and integrating it into agroforestry systems. By doing so, farmers in developing countries can benefit from this hardy and valuable crop and improve their livelihoods.

13.5 Challenges of Using *Moringa oleifera* as a Feedstock

13.5.1 Potential Environmental Challenges Associated with *Moringa oleifera* Cultivation

Although *M. oleifera* has many potential environmental benefits, there are also some potential environmental challenges associated with its cultivation. These include habitat loss, water use, soil degradation, pesticide use, and others:

1) *Habitat loss*: The cultivation of *M. oleifera* on a large scale could lead to habitat loss and deforestation if natural forests are cleared to make way for plantations. This could have negative impacts on biodiversity and ecosystem services.
2) *Water use*: Although *M. oleifera* is a drought-tolerant crop, large-scale cultivation of this crop could still lead to increased water use in areas where water resources are already limited. This could exacerbate water scarcity in these regions and have negative impacts on aquatic ecosystems.
3) *Soil degradation*: Although *M. oleifera* can improve soil fertility and structure, monoculture cultivation of this crop could lead to soil degradation over time. This is because monoculture can lead to soil erosion, nutrient depletion, and increased susceptibility to pests and diseases.
4) *Pesticide use*: Pests and diseases can pose a threat to *M. oleifera* cultivation, and farmers may resort to using pesticides to protect their crops. However, the overuse of pesticides can lead to environmental pollution, harm nontarget organisms, and pose health risks to humans.
5) *Genetic erosion*: The widespread cultivation of a single variety of *M. oleifera* could lead to genetic erosion and loss of genetic diversity. This could make the crop more vulnerable to pests and diseases and limit its potential for adaptation to changing environmental conditions.

In summary, although *M. oleifera* has many potential environmental benefits, the large-scale cultivation of this crop could also lead to environmental challenges, as mentioned above. These challenges need to be carefully managed to ensure that the environmental benefits of *M. oleifera* cultivation are maximized and its negative impacts are minimized.

13.5.2 Potential Cost Challenges of Producing Biodiesel from *Moringa oleifera*

Producing biodiesel from *M. oleifera* can be challenging due to several cost factors. Some of the potential cost challenges associated with producing biodiesel from *M. oleifera* include low seed yield, labor-intensive harvesting and processing, transportation costs, infrastructure costs, and competition with food production:

1) *Seed yield*: The seed yield of *M. oleifera* is relatively low, which can make it more expensive to produce biodiesel compared to other feedstocks with higher yields. This means that more seeds are needed to produce the same amount of biodiesel, which can increase production costs.
2) *Harvesting and processing*: Harvesting and processing *M. oleifera* seeds can be labor-intensive, which can add to production costs. The seeds need to be collected, cleaned, dried, and pressed to extract the oil, which requires specialized equipment and expertise.
3) *Transportation*: The transportation of *M. oleifera* seeds and biodiesel can also add to production costs, especially if the production site is located far from the processing plant or distribution center.
4) *Infrastructure*: The infrastructure needed for biodiesel production, such as processing plants and distribution networks, can be expensive to build and maintain. This can add to the overall cost of producing biodiesel from *M. oleifera*.
5) *Competition with food production*: There is concern that the use of *M. oleifera* as a feedstock for biodiesel production could compete with food production, which could drive up the cost of food and have negative social impacts.

In summary, producing biodiesel from *M. oleifera* can be challenging due to several aforementioned cost factors. These cost challenges need to be carefully considered when evaluating the economic viability of using this beneficial raw material as a feedstock for the production of biodiesel.

13.5.3 Need for Further Research on Optimal Growing Conditions and Cultivation Practices for *Moringa oleifera*

From the study so far, it is a noticeable fact that *M. oleifera* has many potential benefits. This has been demonstrated in research focused on the medical benefits of Moringa [2]. Also, another studies revealed the relevance of Moringa in the food products [13, 17, 21–24]. The benefit of the useful moringa plant has also been showcased in pharmaceuticals [12, 16, 25]; water treatment is another area where moringa has been considered useful in previous studies [19]. Recent studies were equally conducted showcasing the significance of moringa for general purpose use [26, 27]. The importance of the production of biodiesel is equally encouraging [28, 29]. However, based on previous studies, it has become important to highlight other areas of need for further research on optimal growing conditions, cultivation practices, and the significance of *M. oleifera* as a general-purpose plant. The motives are further highlighted thus:

1) *Variability in growing conditions*: *M. oleifera* can grow in a range of different climatic and soil conditions, but optimal growing conditions for maximum yield and quality have not yet been fully identified. Further research is needed to identify the best growing conditions for different varieties of *M. oleifera*.
2) *Disease and pest management*: Although *M. oleifera* is known to be resistant to pests and diseases, there is still a need for research to develop effective and sustainable disease and pest management strategies that do not harm the environment or human health.
3) *Water management*: *M. oleifera* is known to be drought-tolerant but there is still a need for research to identify optimal irrigation strategies that conserve water while ensuring maximum crop yield and quality.
4) *Soil management*: *M. oleifera* can improve soil fertility, but there is still a need for research to identify optimal soil management practices that maintain soil health and prevent soil degradation.
5) *Nutrient management*: *M. oleifera* requires nutrients such as nitrogen, phosphorus, and potassium to grow, but there is still a need for research to identify optimal nutrient management practices that ensure maximum crop yield and quality while minimizing environmental impacts.

In summary, further research is needed to optimize growing conditions and cultivation practices for *M. oleifera*. This study is an exposition of the generic use of moringa and specific use for biodiesel production. This research has equally provided details on maximizing the potential economic and environmental benefits of this crop while minimizing its potential negative impacts.

13.6 Conclusion

This research focuses on exploring the potential of *M. oleifera* as a viable raw material for the production of biodiesel, primarily due to its impressive oil content. The chapter covers an in-depth examination of *M. oleifera*'s capabilities as a biodiesel feedstock, shedding light on the method of production of biodiesel, its advantages and challenges. One of the noteworthy advantages is *M. oleifera*'s adaptability to thrive in marginal soils with minimal water availability, making it a resilient option for biodiesel production. In addition, its cultivation provides an income source for farmers in developing countries, contributes to the reduction of greenhouse gas emissions, and enhances soil fertility. Despite these positive aspects, challenges associated with *M. oleifera* as a feedstock include its relatively low seed yield, the labor-intensive nature of harvesting and processing, and the potential competition with food production. To address these challenges and maximize the potential of *M. oleifera* as a biodiesel feedstock, further research is imperative. This includes optimizing growing conditions and cultivation practices, devising sustainable pest and disease management strategies, and implementing measures to reduce production costs. Despite the existing challenges, the future outlook for *M. oleifera* as a biofuel feedstock appears promising, offering a sustainable and eco-friendly alternative to conventional fossil fuels.

References

1 Mofijur, M., Rasul, M.G., Hassan, N.M.S., and Nabi, M.N. (2019). Recent development in the production of third generation biodiesel from microalgae. *Energy Procedia* 156: 53–58. https://doi.org/10.1016/j.cgypro.2018.11.088.

2 Padayachee, B. and Baijnath, H. (2012). An overview of the medicinal importance of Moringaceae. *J. Med. Plant Res.* 6 (48): 5831–5839.

3 Popoola, J.O. and Obembe, O.O. (2013). Local knowledge, use pattern, and geographical distribution of *Moringa oleifera* Lam. (Moringaceae) in Nigeria. *J. Ethnopharmacol.* 150 (2): 682–691.

4 Rockwood, J., Anderson, B., and Casamatta, D. (2013). Potential uses of *Moringa oleifera* and an examination of anti-biotic efficacy conferred by *M. oleifera* seed and leaf extracts using crude extraction techniques available to underserved indigenous populations. *Int. J. Phytother. Res.* 3 (2): 61–71.

5 Fahey, J.W. (2005). *Moringa oleifera*: a review of the medical evidence for its nutritional, therapeutic, and prophylactic properties. Part 1. *Trees Life J.* 1 (5): 1–15.

6 Nair, S. and Varalakshmi, K. (2011). Anticancer, cytotoxic potential of *Moringa oleifera* extracts on HeLa cell line. *J. Nat. Pharm.* 2 (3): 138–142.

7 Sutalangka, C., Wattanathorn, J., Muchimapura, S. et al. (2013). *Moringa oleifera* mitigates memory impairment and neurodegeneration in animal model of age-related dementia. *Oxidative Med. Cell. Longev.* 2013: 695936.

8 Leone, A., Spada, A., Battezzati, A. et al. (2015). Cultivation, genetic, ethnopharmacology, phytochemistry and pharmacology of *Moringa oleifera* leaves: an over-view. *Int. J. Mol. Sci.* 16 (6): 12791–12835.

9 Saa, R.W., Fombang, E.N., Ndjantou, E.B. et al. (2019). Njintang NYJFs, nutrition. Treatments and uses of *Moringa oleifera* seeds in human nutrition: a review. *Food Sci. Nutr.* 7 (6): 1911–1919.

10 Saini, R.K., Sivanesan, I., and Keum, Y.-S. (2016). Phytochemicals of *Moringa oleifera*: a review of their nutritional, therapeutic and industrial significance. *3 Biotech* 6 (2): 203.

11 Leone, A., Spada, A., Battezzati, A. et al. (2016). *Moringa oleifera* seeds and oil: characteristics and uses for human health. *IJMS* 17 (12): 2141.

12 Vergara-Jimenez, M., Almatrafi, M.M., and Fernandez, M.L. (2017). Bioactive components in *Moringa oleifera* leaves protect against chronic disease. *Antioxidants* 6 (4): 91.

13 Alain Mune Mune, M., Nyobe, E.C., Bakwo Bassogog, C. et al. (2016). A comparison on the nutritional quality of proteins from *Moringa oleifera* leaves and seeds. *Cogent Food Agric.* 2 (1): 1–8.

14 Mutiara, T., Titi, E., and Estiasih, W. (2013). Effect lactagogue moringa leaves (*Moringa oleifera* Lam) powder in rats white female wistar. *J. Basic Appl. Sci. Res.* 3 (4): 430–434. Critical Reviews in Biotechnology 287.

15 Richter, N., Siddhuraju, P., and Becker, K.J.A. (2003). Evaluation of nutritional quality of moringa (*Moringa oleifera* Lam.) leaves as an alternative protein source for Nile tilapia (*Oreochromis niloticus* L.). *Aquaculture* 217 (1–4): 599–611.

16 Khor, K.Z., Lim, V., Moses, E.J. et al. (2018). The in vitro and in vivo anticancer properties of *Moringa oleifera*. *Evid. Based Complement. Alternat. Med.* 2018: 1071243.

17 Falowo, A.B., Mukumbo, F.E., Idamokoro, E.M. et al. (2018). Multi-functional application of *Moringa oleifera* Lam. innutrition and animal food products: a review. *Food Res. Int.* 106: 317–334.

18 Ma, Z., Ahmad, J., Zhang, H. et al. (2020). Evaluation of phyto-chemical and medicinal properties of moringa (*Moringa oleifera*) as a potential functional food. *S. Afr. J. Bot.* 129: 40–46.

19 Yamaguchi, N.U., Cusioli, L.F., Quesada, H.B. et al. (2020). A review of *Moringa oleifera* seeds in water treatment: trends and future challenges. *Process. Saf. Environ. Prot.* 147: 405–420.

20 Vats, S. and Gupta, T. (2017). Evaluation of bioactive compounds and antioxidant potential of hydroethanolic extract of *Moringa oleifera* Lam. from Rajasthan, India. *Physiol. Mol. Biol. Plants* 23 (1): 239–248.

21 Oyeyinka, A.T. and Oyeyinka, S.A. (2018). *Moringa oleifera* as a food fortificant: recent trends and prospects. *J. Saudi Soc. Agric. Sci.* 17: 127–136. https://doi.org/10.1016/j.jssas.2016.02.002.

22 Saucedo-Pompa, S., Torres-Castillo, J.A., Castro-López, C. et al. (2018). Moringa plants: bioactive compounds and promising applications in food products. *Food Res. Int.* 111: 438–450. https://doi.org/10.1016/j.foodres.2018.05.062.

23 Jaja-Chimedza, A., Graf, B.L., Simmler, C. et al. (2017). Biochemical characterization and anti-inflammatory properties of an isothiocyanate-enriched moringa (*Moringa oleifera*) seed extract. *PLoS One* 12 (8): e0182658.

24 Ma, Z., Ahmad, J., Zhang, H. et al. (2020). Evaluation of phytochemical and medicinal properties of moringa (*Moringa oleifera*) as a potential functional food. *S. Afr. J. Bot.* 129: 40–46.

25 Amina, M., Al Musayeib, N.M., Alarfaj, N.A. et al. (2019). Exploiting the potential of *Moringa oleifera* oil/polyvinyl chloride polymeric bionanocomposite film enriched with silver nanoparticles for antimicrobial activity. *Int. J. Polym. Sci.* 2019: 1–11.

26 Zulfiqar, F., Casadesús, A., Brockman, H., and Munné-Bosch, S. (2019). An overview of plant-based natural biostimulants for sustainable horticulture with a particular focus on moringa leaf extracts. *Plant Sci.* 110194. https://doi.org/10.1016/j.plantsci.2019.110194.

27 Bancessi, A., Bancessi, Q., Baldé, A., and Catarino, L. (2019). Present and potential uses of *Moringa oleifera* as a multipurpose plant in Guinea-Bissau. *S. Afr. J. Bot.* https://doi.org/10.1016/j.sajb.2019.06.013.

28 Niju, S., Balajii, M., and Anushya, C. (2019). A comprehensive review on biodiesel production using *Moringa oleifera* oil. *Int. J. Green Energy* 16: 702–715. https://doi.org/10.1080/15435075.2019.1619565.

29 Niju, S., Anushya, C., and Balajii, M. (2019). Process optimization for biodiesel production from *Moringa oleifera* oil using conch shells as heterogeneous catalyst. *Environ. Prog. Sustain. Energy* 38. https://doi.org/10.1002/ep.13015.

14

Sustainable Biodiesel Production with Hemp as a Renewable Feedstock

14.1 Introduction

As the global population continues to expand, the demand for energy generation experiences a steady rise [1]. Presently, fossil fuels serve as the primary feedstock in various processes within today's energy production systems. It is undeniable that the majority of countries heavily depend on fossil fuels, with many provinces deriving almost all of their energy from these nonrenewable sources [2]. Therefore, it becomes imperative to consider alternative energy production techniques, aiming to reduce the energy production sector's reliance on fossil fuels, ensuring sustainability, and mitigating environmental threats.

One significant alternative is bioenergy, derived from biomass, which holds the potential to substitute for fossil fuels, offering an environmentally sustainable energy source. Various bioenergy products, including biodiesel, biogas, bio-hydrogen, and bioethanol, can be produced using biomass as a feedstock [3]. Embracing these bioenergy alternatives not only diversifies the energy portfolio but also contributes to reducing greenhouse gas emissions and promoting environmental sustainability. As the world grapples with the challenges of climate change and the depletion of fossil fuel resources, the exploration and adoption of bioenergy solutions become increasingly crucial for fostering a more sustainable and environmentally friendly energy future.

As an alternative to traditional fossil fuels, biomass stands out with significant advantages, primarily due to its abundant availability [4]. The majority of biomass resources are in the form of biodegradable waste products, making them accessible at minimal or no cost [5]. These distinctive characteristics make biomass a unique and highly potent energy source. Research studies indicate a growing trend toward the increased abundance of biomass and biofuels, with projections suggesting that by 2035, they could emerge as dominant alternatives in the energy

Feedstocks for Sustainable Biodiesel Production: Characterization, Selection, and Optimization, First Edition. Chinwe P. Okonkwo and Modestus O. Okwu.

space [6]. One noteworthy contributor to the biomass sector is industrial hemp, a crop with exceptional potential. *Cannabis sativa* L., commonly known as industrial hemp, has a rich history of cultivation [7]. It represents a unique variety of species that can readily interbreed [8]. The cultivation of industrial hemp is multifaceted, serving as a source of various materials [9–11].

The appeal of biomass, and particularly industrial hemp, lies not only in its availability and cost-effectiveness but also in its versatility as a renewable resource for a wide range of applications. As the world increasingly seeks sustainable and eco-friendly alternatives to traditional energy sources, the potential of biomass, including industrial hemp, becomes a focal point in the quest for a more environmentally conscious and resilient energy future.

With its diverse components, including medicinal ingredients, oil, and fiber, among others, industrial hemp proves to be a valuable resource with applications in pharmaceuticals, medical purposes, and as a bioenergy fuel [12]. The versatile nature of industrial hemp positions it as a promising asset with significant potential across various industries.

Despite being cultivated in several countries, the hemp market encounters challenges, primarily stemming from societal and legal pressures associated with its close relation to a plant of the same species known for its psychotropic compound commonly referred to as marijuana. This creates a complex system where companies and manufacturing industries face competition and hurdles in fully exploiting the potential of industrial hemp. To maximize its benefits, there is a pressing need for rapid development in the production of industrial hemp. The challenges posed by regional economic and legal conditions underscore the importance of highlighting the potential and advantages of industrial hemp. Remarkably, research on industrial hemp has not received as much attention in the literature as other biodegradable feedstocks, especially in the field of biofuel production and process optimization. This exposition aims to shed light on the untapped potential of industrial hemp, emphasizing the need for increased research and development efforts to overcome existing hurdles and fully harness the benefits that industrial hemp can offer to diverse industries, ranging from medicine to bioenergy.

Many countries view the negative aspects of this crop.As a result, it is highly prohibited to cultivate industrial hemp as a feedstock for use in many countries; this makes it difficult to access the crop for research purposes [13], and the use of hemp is seen as taboo in most countries [14]. The application is not only in the field of renewable energy; previous research has focused on the medical application of Indian Hemp, and this has constrained the potential of Indian hemp as a plant. The good side is that hemp is considered legal in Canada, and research is ongoing to find its usefulness in diverse areas [15, 16]. New research prospectshave opened up as a result of this development.

Figure 14.1 Pictorial view of hemp seed. *Source:* Michelle/Adobe Stock.

Moreover, the popularity of hemp in developed nations is creating new areas of applications of this crop; there is no doubt that the potential for success in the years to come is quite high, and the importance will be felt in areaslike pyrolysis [17], insulation [18], and concrete [19]. Hemp's efficiency in conversion procedures holds out a lot of hope. This is quite true for the production of ethanol, where hemp is more lucrative than other crops like Kenaf and switchgrass [20], and also for the production of biodiesel, where hemp has a far greater conversion rate than the industry standard [20]. While most research is focusing on hybrid feedstocks [21], industrial hemp which is not so popular will require such comparative studies. The benefit from numerous studies in terms of their performance in conversion processes will serve as a benchmark. Industrial hemp has been considered an excellent feedstock for bioenergy purposes, given its high biomass [22]. A pictorial view of Hemp seed is presented in Figure 14.1.

14.2 Brief Overview of Hemp as a Potential Feedstock for Biodiesel Production

Hemp is a versatile plant that has the potential to serve as a feedstock for biodiesel production. The oil content of hemp seeds can range from 25% to 35%, making it a viable source of feedstock for biodiesel production. In addition, hemp grows quickly and can be cultivated in a variety of climates, making it a reliable and renewable source of feedstock. Hemp-based biodiesel also has the potential to be biodegradable, which is an important environmental advantage over traditional fossil fuels. Despite these advantages, there are also limitations to the use of hemp as a feedstock for biodiesel production, including the limited availability of hemp as a feedstock and the high cost of production. Nevertheless, ongoing research and development efforts are focused on optimizing the production process and improving the quality of hemp-based biodiesel, which could help to overcome these challenges and promote the use of this promising feedstock.

14.2.1 Review of Fuels Obtained from Industrial Hemp

Industrial hemp and other crops are often converted into liquid fuels via pyrolysis, anaerobic digestion, and transesterification [23]. To create effective fuel mixes, these biofuels are frequently combined with traditional fuels like diesel and gasoline [24]. Biodiesels are produced by trans esterifying natural oils, and they can be used as a direct replacement for some fossil fuels with a renewable substitute. With a reported 75.9% conversion rate from hempseed oil to biodiesel using a potassium hydroxide catalyst [25], hemp appears as a valuable source of fuel, while certain oils can achieve conversion rates of over 95% [26]. Yet, in other studies, it was revealed that hempseed oil may achieve conversion rates of up to 97%, indicating that it can be utilized successfully to produce biodiesel [27]. Several important characteristics of biodiesel from various crops, including hemp, are listed in a study for comparative analysis [28–31]. Desirable characteristics of diesel are described by the ASTM specifications, which include low sulfur content for low sulfur emissions [32], a high flash point for increased fuel stability and safety [33], as well as kinematic viscosity within the ASTM range for practical use in internal combustion engines.

According to a study [28], hemp biodiesel has a sulfur concentration that is 53% lower than the average sulfur content of the other biodiesels listed, which is substantially below ASTM criteria. It also has a 10% higher flash point and kinematic viscosity that is comfortably within the acceptable range. Despite having cheap growing costs, hemp has a yield that is 62% greater per hectare than the average of the other crops listed, however, it is still lower than some feedstocks. As hemp biodiesel production requires hempseed oil, but hemp straw and the remaining seed can be used for other purposes, hemp biodiesel has the potential to be economically competitive with other diesel by utilizing its by-products [25]. Hemp bio-oil is another way to produce liquid fuel from industrial hemp. Hemp bio-oil, produced by pyrolysis, has the potential to be upgraded by hydro deoxygenation to produce liquid fuels high in hydrocarbons [34].

14.3 Steps Involved in Producing Biodiesel from Hemp Oil

A flowchart representing the steps required to produce biodiesel from hemp oil is presented in Figure 14.2.

As presented in Figure 14.2, further explanation of the process requirement for the production of biodiesel is presented thus.

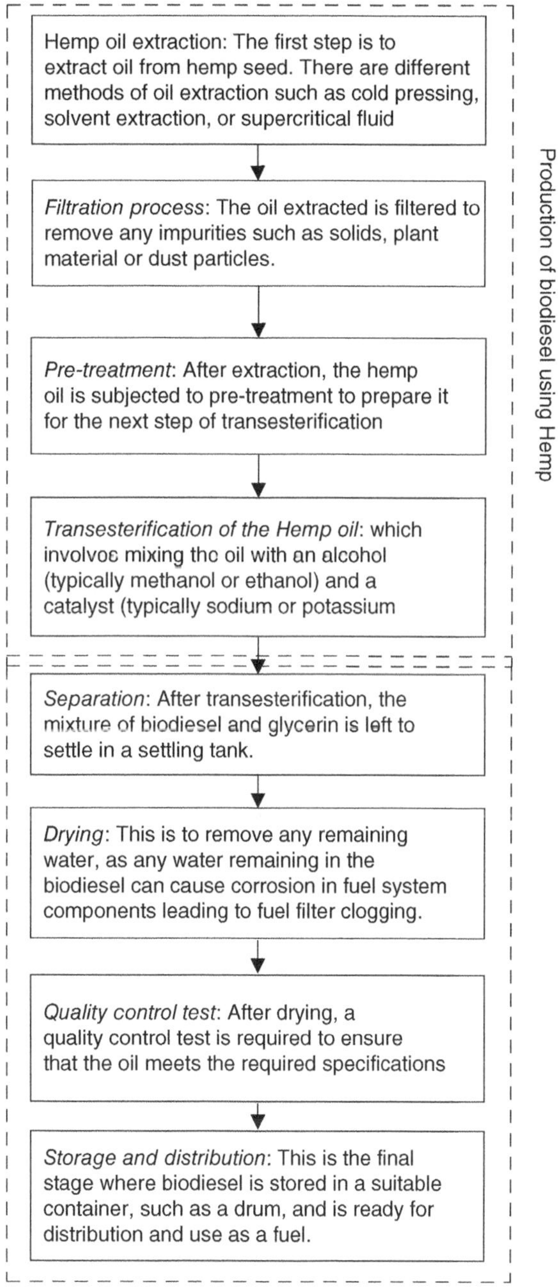

Figure 14.2 Flowchart showing the step-by-step process of biodiesel production from Hemp.

14.3.1 Extraction of Hemp Oil

The first step required in producing biodiesel from hemp oil is to extract the oil from the hemp seeds. There are several methods required for the production of oil, including cold pressing, solvent extraction, or supercritical fluid extraction. The extracted oil is further filtered to remove any impurities such as solids, plant material, or dust particles.

14.3.2 Pretreatment

After extraction of hemp oil, the oil is further subjected to a pretreatment step to remove any impurities and to prepare it for the next step of transesterification. In this step, the oil is mixed with a small amount of methanol or ethanol and an acid catalyst to neutralize any free fatty acids present in the oil.

14.3.3 Transesterification

The next step is the transesterification of the hemp oil, which involves mixing the oil with an alcohol (typically methanol or ethanol) and a catalyst (typically sodium or potassium hydroxide) in the presence of heat. This reaction breaks down the oil molecules and recombines them with the alcohol to form fatty acid esters (i.e. biodiesel) and glycerine.

14.3.4 Separation

After transesterification, the mixture of biodiesel and glycerine is then left to settle in a settling tank. The glycerine sinks to the bottom and is removed from the top layer of biodiesel. The biodiesel is then washed with water to remove any remaining impurities and any excess catalyst.

14.3.5 Drying

After washing, the biodiesel is typically dried to remove any remaining water. This is an important step, as any water remaining in the biodiesel can cause corrosion in fuel system components and can lead to problems with fuel filter clogging.

14.3.6 Quality Control

Once the biodiesel is dried, it is then subjected to quality control testing to ensure that it meets the required specifications. Testing may include analysis of the biodiesel's viscosity, flash point, and acid number. Any biodiesel that fails to meet the required specifications may need to be reprocessed.

14.3.7 Storage and Distribution

Finally, the biodiesel is stored in a suitable container, such as a tanker or drum, and is ready for distribution and use as a fuel. It can be blended with petro diesel in various proportions or used as a pure fuel in diesel engines that are compatible with biodiesel.

It is important to note that the specific process for producing biodiesel from hemp oil may vary depending on the equipment, feedstock, and resources available. However, the basic steps of oil extraction, pretreatment, transesterification, separation, drying, quality control, and storage and distribution are common to most biodiesel production processes.

14.4 Benefits, Limitations, Efficiency, and Potential Yield of Hemp-Based Biodiesel

14.4.1 Benefits of Hemp-Based Biodiesel

Hemp-based biodiesel is an alternative fuel that is derived from hemp oil, a renewable and sustainable resource. Here are some of the advantages of using hemp-based biodiesel:

Renewable and sustainable: Hemp is a renewable resource that can be grown and harvested annually, making it a sustainable source of fuel. Unlike fossil fuels, which are finite resources that are being depleted, hemp can be grown and used indefinitely.

Reduced carbon emissions: Hemp-based biodiesel has lower carbon emissions compared to petroleum-based diesel. The production and combustion of hemp-based biodiesel emits up to 80% less carbon dioxide compared to petroleum diesel.

Improved air quality: Hemp-based biodiesel produces fewer harmful emissions than petroleum diesel, including particulate matter, nitrogen oxides, and sulfur dioxide. This can lead to improved air quality and reduced pollution levels.

Biodegradable: Hemp-based biodiesel is biodegradable, which means it breaks down naturally and does not pose a threat to the environment.

Energy independence: Hemp-based biodiesel can be produced domestically, reducing dependence on foreign oil and increasing energy independence.

Versatile: Hemp can be grown in a variety of climates and soil types, making it a versatile crop that can be used to produce a range of products, including biodiesel.

Job creation: Hemp-based biodiesel production can create jobs in farming, processing, and manufacturing, helping to stimulate local economies.

Generally, the advantages of hemp-based biodiesel make it a promising alternative to petroleum-based diesel and a sustainable solution to meet the energy needs of today and the future.

14.4.2 Challenges and Limitations

Biodiesel produced from hemp has the potential to be a sustainable and environmentally friendly alternative to fossil fuels. However, there are several challenges and limitations associated with this type of biodiesel. Some of the key ones are presented thus:

Feedstock limited availability: In countries where hemp cultivation is legal, the availability of hemp as a feedstock for biodiesel production is limited. This is because hemp is also used for other purposes, such as textile production, and there may not be enough hemp available for both purposes.

Cost: The cost of producing hemp-based biodiesel is currently higher than that of petroleum-based diesel. This is due to the relatively low yield of hemp oil per acre, as well as the high cost of processing and refining the oil.

Infrastructure: Hemp-based biodiesel requires a different infrastructure for production, storage, and distribution than petroleum-based diesel. This means that significant investment is needed to build the necessary infrastructure to support its use.

Land use: Growing hemp for biodiesel production requires land, water, and other resources that may compete with other uses such as food production. This can create environmental and social challenges, such as deforestation, soil degradation, and displacement of communities.

Regulatory challenges: Hemp production is regulated in many countries, and there may be legal and regulatory challenges to its use as a fuel source. In some places, hemp production is still illegal.

Hemp cultivation: Hemp cultivation is still illegal or heavily restricted in some countries, which limits its availability as a source of biodiesel.

Energy balance: While biodiesel from hemp is renewable, its overall energy balance is not clear. This is because the energy required to grow, harvest, and process hemp into biodiesel may be greater than the energy obtained from the biodiesel itself.

Sustainability concerns: Hemp cultivation may require large amounts of water and fertilizers, and there are concerns about the impact of intensive agriculture on the environment.

Generally, while biodiesel produced from hemp has the potential to be a sustainable and environmentally friendly alternative to fossil fuels, there are still several challenges and limitations that need to be addressed.

14.4.3 Discussion of the Efficiency and Potential Yield of the Process

The efficiency and potential yield of the process for producing biodiesel from hemp can vary depending on several factors, such as the type and quality of the hemp feedstock, the extraction and transesterification methods used, and the equipment and resources available. However, some general information on the efficiency and potential yield of the process is presented thus:

Feedstock quality: The quality of the hemp feedstock can affect the efficiency of the process. Hemp seeds with high oil content and low impurities will produce higher yields of biodiesel. It is important to use clean and high-quality hemp oil as impurities can affect the efficiency of the transesterification process.

Extraction efficiency: The efficiency of the oil extraction process can also affect the yield of biodiesel. Different methods of extraction can result in different yields of hemp oil. For example, cold pressing is a popular method for producing hemp oil, but it may result in a lower yield compared to solvent extraction or supercritical fluid extraction.

Transesterification efficiency: The efficiency of the transesterification process can also affect the yield of biodiesel. The use of high-quality catalysts and optimization of process parameters such as temperature, pressure, and reaction time can improve the efficiency of the process and increase the yield of biodiesel.

Yield potential: The yield of biodiesel from hemp oil can vary but typically ranges from 75% to 85% of the oil volume. This means that for every 100 L of hemp oil, 75–85 L of biodiesel can be produced [21]. However, the yield can be affected by various factors, such as the quality of the feedstock, the extraction and transesterification methods used, and the efficiency of the process. In conclusion, the efficiency and potential yield of the process for producing biodiesel from hemp can vary depending on several factors. However, with proper feedstock selection, extraction, and transesterification methods, it is possible to achieve high yields of biodiesel from hemp oil.

14.5 Opportunities for Hemp as a Viable Feedstock for Biodiesel Production

The utilization of hemp as a feedstock for biodiesel production presents a range of opportunities, aligning with the broader goals of sustainability and environmental consciousness. Here are detailed opportunities associated with using hemp as a feedstock for biodiesel production:

High oil content: Hemp seeds are rich in oil content, particularly in the form of fatty acids. This high oil content enhances the potential yield of biodiesel per unit of cultivation, making hemp an efficient and productive feedstock.

Rapid growth and high yield: Hemp is known for its fast growth rate and high biomass yield. This characteristic allows for multiple harvests in a single growing season, contributing to increased feedstock availability for biodiesel production.

Adaptability to diverse climates: Hemp is a resilient plant that can thrive in various climates and soil conditions. This adaptability enables widespread cultivation, making it suitable for biodiesel production in diverse geographical regions.

Reduced environmental impact: Hemp cultivation typically requires minimal pesticide and fertilizer usage. Its ability to grow without relying on harmful agrochemicals contributes to reduced environmental impact, aligning with sustainable and eco-friendly agricultural practices.

Biodegradability: Biodiesel produced from hemp is inherently biodegradable. This property ensures that the end product is environmentally friendly and can be integrated into ecosystems without causing long-term harm.

Carbon sequestration: Hemp plants are known for their capacity to sequester carbon dioxide from the atmosphere. Biodiesel production from hemp contributes to a carbon-neutral or even carbon-negative fuel cycle, helping mitigate greenhouse gas emissions.

Crop rotation and soil health: The cultivation of hemp can be integrated into crop rotation systems, improving soil health by preventing soil degradation and reducing the need for synthetic fertilizers. This sustainable farming practice enhances the overall environmental footprint of biodiesel production.

Economic opportunities for farmers: The cultivation of hemp for biodiesel production can provide economic opportunities for farmers. Diversifying crops and participating in the growing bioenergy sector may contribute to increased income and stability for agricultural communities.

Research and development potential: Ongoing research and development efforts in hemp-based biodiesel production present opportunities for innovation. Optimization of cultivation techniques, genetic improvements, and advanced processing methods can further enhance the efficiency and economic viability of hemp as a biodiesel feedstock.

Legislative support and market demand: Legislative support and growing market demand for sustainable fuels can create favorable conditions for hemp biodiesel. As governments and industries increasingly prioritize renewable energy sources, hemp-based biodiesel may benefit from supportive policies and market incentives.

Integration with the circular economy: Utilizing hemp waste or by-products for additional purposes, such as animal feed, building materials, or other bio-based products, contributes to a circular economy. This integrated approach maximizes resource use and minimizes waste.

In summary, the opportunities associated with using hemp as a feedstock for biodiesel production extend beyond its high oil content. Hemp's adaptability, environmental benefits, and potential for economic development position it as a valuable and versatile contributor to the sustainable bioenergy system.

14.5.1 The Potential for Hemp in the Context of the Bio-Economy and Sustainable Development

The potential for hemp in the context of the bio-economy and sustainable development is considerable, encompassing various sectors and contributing to a more environmentally conscious and economically viable future. Here's an exploration of the potential for hemp in the broader context of the bio-economy and sustainable development:

Bio-based products: Hemp serves as a versatile raw material for the production of bio-based products. Beyond biodiesel, it can be utilized in the creation of textiles, construction materials, bioplastics, and other sustainable alternatives. This diversification contributes to the bio-economy by reducing dependence on nonrenewable resources.

Carbon sequestration and soil health: Hemp's ability to sequester carbon dioxide from the atmosphere and enhance soil health makes it a valuable component of sustainable agriculture. As part of a circular economy, hemp cultivation can contribute to carbon-neutral or even carbon-negative practices, mitigating the impacts of climate change.

Renewable energy source: Biodiesel production from hemp contributes to the generation of renewable energy. The integration of hemp into the bio-economy as a renewable energy source helps meet energy demands while reducing reliance on fossil fuels and minimizing environmental degradation.

Crop rotation and sustainable farming: Hemp's adaptability and rapid growth make it suitable for crop rotation systems. Integrating hemp into agricultural practices promotes soil fertility, reduces the need for synthetic fertilizers, and supports sustainable farming methods that contribute to overall ecosystem health.

Job creation and economic development: The cultivation and processing of hemp create opportunities for job creation and economic development, particularly in rural areas. The establishment of hemp-related industries, such as processing facilities and manufacturing plants, fosters local economies and contributes to sustainable development goals.

Eco-friendly construction materials: Hemp fibers can be used to produce sustainable and eco-friendly construction materials, such as hempcrete. This not only reduces the environmental impact of construction but also aligns with the principles of sustainable building practices.

Health and wellness products: Hemp has diverse applications in the health and wellness sector. The extraction of cannabidiol (CBD) from hemp contributes to the production of natural remedies and supplements. This aligns with the growing demand for plant-based and natural health products in the bio-economy.

Biodiversity conservation: Hemp cultivation, when practiced sustainably, can contribute to biodiversity conservation. The adoption of agroecological practices in hemp farming enhances ecosystem resilience, protects natural habitats, and promotes the overall health of agricultural systems.

Textiles and fashion industry: Hemp fibers are known for their durability and sustainability, making them suitable for textile production. Integrating hemp into the fashion industry provides eco-friendly alternatives to conventional fabrics and contributes to sustainable practices in the textile sector.

Regenerative agriculture practices: Hemp cultivation can be part of regenerative agriculture practices that focus on enhancing soil health, reducing water usage, and promoting biodiversity. This aligns with the principles of sustainable development by fostering resilient and adaptive agricultural systems.

Legislative and regulatory frameworks: The development of supportive legislative and regulatory frameworks for hemp cultivation and processing is crucial for maximizing its potential in the bio-economy. Governments can play a pivotal role in creating an enabling environment for sustainable hemp-related industries.

In summary, the potential for hemp in the context of the bio-economy and sustainable development is multifaceted. From renewable energy to sustainable products and agricultural practices, hemp's versatility positions it as a valuable resource in building a more sustainable and resilient future.

14.6 Conclusion

In conclusion, the production of biodiesel from hemp has the potential to be a sustainable and environmentally friendly alternative to traditional sources of energy. Hemp is a renewable feedstock that can be grown in a variety of climates, and its use as a feedstock for biodiesel production can reduce greenhouse gas emissions and dependence on fossil fuels. Despite the challenges and limitations associated with hemp-based biodiesel, including limited availability of feedstock and concerns about sustainability and energy balance, research and development in this area continue to progress. Improvements in cultivation methods, processing technologies, and government regulations could help to overcome these challenges and promote the growth of the hemp-based biodiesel industry. As we continue to face the challenges of climate change and the need for sustainable energy sources, hemp-based biodiesel represents a promising avenue for reducing our reliance on fossil fuels and promoting a more sustainable future. With continued investment in research and development, the potential of hemp-based biodiesel as a sustainable and environmentally friendly alternative to traditional sources of energy can be realized.

References

1 Khan, I., Hou, F., and Le, H.P. (2021). The impact of natural resources, energy consumption, and population growth on environmental quality: fresh evidence from the United States of America. *Sci. Total Environ.* 754: 142–222.

2 Natural Resources Canada (2017). Electricity facts. Nat Resourc Canada. https://www.nrcan.gc.ca/energy/facts/electricity/20068#L3 (accessed 1 April 2023).

3 Ozturk, M., Saba, N., Altay, V. et al. (2017). Biomass and bioenergy: an overview of the development potential in Turkey and Malaysia. *Renew. Sust. Energ. Rev.* 79: 1285–1302.

4 Okonkwo, C.P., Ajiwe, V., Obiadi, M. et al. (2023). Production of biodiesel from the novel non-edible seed of *Chrysobalanus icaco* using natural heterogeneous catalyst: modeling and prediction using Artificial Neural Network. *J. Clean. Prod.* https://doi.org/10.1016/j.jclepro.2022.135631.

5 Liu, C.-M. and Wu, S.-Y. (2016). From biomass waste to biofuels and biomaterial building blocks. *Renew. Energy* 96: 1056–1062.

6 Matzenberger, J., Kranzl, L., Tromborg, E. et al. (2015). Future perspectives of international bioenergy trade. *Renew. Sust. Energ. Rev.* 43: 926–941.

7 Fike, J. (2016). Industrial hemp: renewed opportunities for an ancient crop. *Crit. Rev. Plant Sci.* 35 (5–6): 406–424.

8 Sandler, L.N., Beckerman, J.L., Whitford, F., and Gibson, K.A. (2019). Cannabis as conundrum. *Crop Prot.* 117: 37–44.

9 Correspondent DJS (2023). New York, "early research on industrial hemp looks promising. Lancaster Farming. https://www.lancasterfarming.com/news/northern_edition/early-research-on-industrial-hemp-looks-promising/article_ead3fb74-1e4b-11e8-9409-4f9e6181d7ca.html (accessed 1 April 2023).

10 Kumar, I.M.K.V., Mazhar, M.S., and Nawaz, S. (2024). Potential of establishing industrial Hemp value chains in Northern Australia. *Electron. J. Struct. Eng.* 24: 17–22.

11 Visković, J., Zheljazkov, V.D., Sikora, V. et al. (2023). Industrial hemp (Cannabis sativa L.) agronomy and utilization: a review. *Agronomy* 13 (3): 931.

12 Schluttenhofer, C. and Yuan, L. (2017). Challenges towards revitalizing hemp: a multifaceted crop. *Trends Plant Sci.* 22 (11): 917–929.

13 Bonini, S.A., Premoli, M., Tambaro, S. et al. (2018). Cannabis sativa: a comprehensive ethnopharmacological review of a medicinal plant with a long history. *J. Ethnopharmacol.* 227: 300–315.

14 Zolotov, Y., Vulfsons, S., Zarhin, D., and Sznitman, S. (2018). Medical cannabis: an oxymoron? Physicians' perceptions of medical cannabis. *Int. J. Drug Policy* 57: 4–10.

15 Salentijn, E.M.J., Zhang, Q., Amaducci, S. et al. (2015). New developments in Fiber hemp (*Cannabis sativa* L.) breeding. *Ind. Crop. Prod.* 68: 32–41.

16 Cinnamon Bidwell, L., York Williams, S.L., Mueller, R.L. et al. (2018). Exploring cannabis concentrates on the legal market: user profiles, product strength, and health-related outcomes. *Addict. Behav. Rep.* 8: 102–106.

17 Branca, C., Di Blasi, C., and Galgano, A. (2017). Experimental analysis about the exploitation of industrial hemp (cannabis sativa) in pyrolysis. *Fuel Process. Technol.* 162: 20–29.

18 Barth, M. (2015). *Carbon Footprint and Sustainability of Different Natural Fibres for Biocomposites and Insulation Material.* Hürth, Germany: Nova Institute.

19 Florentin, Y., Pearlmutter, D., Givoni, B., and Gal, E. (2017). A life-cycle energy and carbon analysis of hemp-lime bio-composite building materials. *Energy Build.* 156: 293–305.

20 Das, L., Liu, E., Saeed, A. et al. (2017). Industrial hemp as a potential bioenergy crop in comparison with kenaf, switchgrass and biomass sorghum. *Bioresour. Technol.* 244: 641–649.

21 Okwu, M.O., Omonigho, O., Samuel, O.D., and Eyitimi, E. (2022). Production and charaterization of briquettes produced from blend of rice-husk and water-hyacinth. In: *Advances in Biofeedstocks and Biofuels*. Scrivener-Wiley. doi: 10.1002/9781119785842.ch4.

22 Danielewicz, D. and Surma-Slusarska, B. (2017). Properties and fibre characterisation of bleached hemp, birch and pine pulps: a comparison. *Cellulose* 24 (11): 5173–5186.

23 Enagi, I.I., Al-attab, K.A., and Zainal, Z.A. (2018). Liquid biofuels utilization for gas turbines: a review. *Renew. Sust. Energ. Rev.* 90: 43–55.

24 Chakraborty, R. and Mukhopadhyay, P. (2019). Green fuel blending: a pollution reduction approach. In: *Reference Module in Materials Science and Materials Engineering*. Elsevier.

25 Ahmad, M., Ullah, K., Khan, M.A. et al. (2011). Physicochemical analysis of hemp oil biodiesel: a promising non edible new source for bioenergy. *Energy Sources Part A* 33 (14): 1365–1374.

26 Goh, B.H.H., Ong, H.C., Cheah, M.Y. et al. (2019). Sustainability of direct biodiesel synthesis from microalgae biomass: a critical review. *Renew. Sust. Energ. Rev.* 107: 59–74.

27 Veljkovic, V.B., Biberdzic, M.O., Bankovic-Ilic, I.B. et al. (2018). Biodiesel production from corn oil: a review. *Renew. Sust. Energ. Rev.* 91: 531–548.
28 Nautiyal, P., Subramanian, K.A., and Dastidar, M.G. (2014). Production and characterization of biodiesel from algae. *Fuel Process. Technol.* 120: 79–88.
29 Verma, P., Sharma, M.P., and Dwivedi, G. (2016). Evaluation and enhancement of cold flow properties of palm oil and its biodiesel. *Energy Rep.* 2: 8–13.
30 Brown, L.R. (2006). *Plan B 2.0: Rescuing a Planet under Stress and a Civilization in Trouble*. New York, NY: W.W. Norton & Co.
31 Mat Yasin, M.H., Mamat, R., Najafi, G. et al. (2017). Potentials of palm oil as new feedstock oil for a global alternative fuel: a review. *Renew. Sust. Energ. Rev.* 79: 1034–1049.
32 Krishna, S.M., Abdul Salam, P., Tongroon, M., and Chollacoop, N. (2019). Performance and emission assessment of optimally blended biodiesel-diesel-ethanol in diesel engine generator. *Appl. Therm. Eng.* 155: 525–533.
33 Fu, J. (2019). Flash points measurements and prediction of biofuels and biofuel blends with aromatic fluids. *Fuel* 241: 892–900.
34 (2018). In situ hydrodeoxygenation upgrading of pine sawdust bio-oil to hydrocarbon biofuel using Pd/C catalyst. *J. Energy Inst.* 91 (2): 163–171.

15

Exploration of the Potential Benefits, Challenges, and Opportunities Associated with Utilizing Rapeseed Oil in Biodiesel Production

15.1 Introduction

Rapeseed oil has emerged as a promising feedstock in the field of biodiesel production, offering a sustainable and versatile alternative to traditional fossil fuels. Its unique characteristics make it an attractive option for researchers, engineers, and policymakers seeking to enhance the sustainability and efficiency of biodiesel production. Biodiesel is frequently regarded as a highly favorable substitute for conventional diesel fuel, owing to its numerous potential advantages. It is recognized for its environmentally friendly nature, renewable origin, uncomplicated production process, positive impact on air quality, contributions to enhanced energy security, and safety benefits [1]. This has sparked considerable enthusiasm for biodiesel as a promising alternative to fossil fuel in the realm of diesel, which is prevalent. This enthusiasm is primarily fueled by the nonrenewable and environmentally detrimental nature of fossil fuels. The cost-effective and uncomplicated large-scale production of biodiesel involves employing alkali-catalyzed transesterification on freshly extracted vegetable oils [2, 3]. Biodiesel stands out as a distinctive form of biofuel crafted from alcohol, animal fats, or vegetable oils. Its distinguishing features, such as renewability, low toxicity, and innate biodegradability, position it as a compelling substitute for conventional petroleum-based fuels. Notably, the seamless integration of biodiesel into contemporary engines with minimal or no modification requirements facilitates the straightforward introduction of blended diesel/biodiesel fuels or pure biodiesel (B100) fuels [4]. A pivotal characteristic of biodiesel lies in its sourcing from replenishable feedstocks, with a significant share, approximately 70%, attributed to rapeseed feedstocks in the global biodiesel production setting [5]. The productivity of rapeseed seeds, yielding up to one ton of oil per hectare, establishes them as the

Feedstocks for Sustainable Biodiesel Production: Characterization, Selection, and Optimization, First Edition. Chinwe P. Okonkwo and Modestus O. Okwu.

highest-yielding oil crop. Beyond sheer productivity, the resulting biodiesel exhibits a low gel point, indicating the temperature at which diesel ceases to flow through fuel lines [6].

The composition of biodiesels primarily comprises mono-alkyl esters of long-chain fatty acids. In the case of rapeseed oil, a noteworthy 90% of these fatty acids consist of C-18 or longer chains, imparting a high cetane number to the biodiesel. This attribute is particularly desirable as it ensures optimal combustion quality, a critical factor in the efficacy of biodiesels [7–9]. It is essential to recognize that the specific fatty acid composition can vary depending on the rapeseed cultivar. Nevertheless, the general makeup typically includes oleic, linoleic, linolenic, stearic, and palmitic acids [8]. This diversified fatty acid profile contributes to the versatility and potential applications of rapeseed-derived biodiesel in various contexts. The production of biodiesel from rapeseed presents notable challenges on a commercial scale, characterized by its energy-intensive and inefficient nature. The intricacies of this process begin with the collection of raw seeds in the agricultural phase, followed by the time-consuming steps of drying and storage before extraction. Extraction itself can be achieved through physical means, such as pressing or chemical methods involving conditioning, solvent extraction, and distillation. This initial phase demands substantial energy and resources, contributing to the overall inefficiency of rapeseed biodiesel production [9].

The primary outputs of this extraction process are raw oil and rapeseed meal co product. To refine the raw oil, it undergoes a series of physical and chemical treatments for purification. Transesterification of the refined oil is then conducted to yield fatty acid methyl esters, commonly known as biodiesel, along with a glycerol by-product [10]. As the transesterification reaction is reversible, an excess of alcohol is typically required to drive the reaction toward high product yields. A catalyst, commonly a base like sodium hydroxide or potassium hydroxide, is an essential component, facilitating faster reaction rates and reduced corrosion [11]. However, the use of a catalyst introduces challenges, necessitating intense and costly product cleanup and additional processes for recycling the catalyst. Homogenous catalysts are often favored due to their high biodiesel conversion rates, but they come with drawbacks such as high energy consumption, the need for intricate catalyst separation, and the generation of substantial wastewater during separation and cleaning processes [12]. Notably, the wastewater from biodiesel production, created at a staggering volume of 28 million cubic meters in 2011 alone, poses environmental concerns. It exhibits high pH levels and concentrations of oil and phosphorus, coupled with low nitrogen concentrations, making it resistant to natural degradation under standard conditions, as microorganism growth is hindered [13–15] (Figure 15.1).

Figure 15.1 Rapeseed crop with seed (*Source:* Aneriksson/Adobe Stock Photos and Vincent van Zeijst/Wikimedia Commons/CC BY 3.0).

15.2 Further Review of Rapeseed as a Potential Feedstock

Biofuels can be derived from various vegetable oils, so it is significant to note that rapeseed oil stands out as the most practical choice. Rapeseed, belonging to the cruciferous family, is an annual oil plant. The seeds typically contain between 30% and 50% oil, with the exact percentage depending on the specific variety and cultivation conditions. The composition of rapeseed oil includes linoleic acid of approximately 21%, oleic acid of 44%, linolenic acid of 8.6%, eruca acid of 15.4%, palmitic acid of approximately 5%, stearic acid of approximately 2%, and nicotine of approximately 5% [16]. Despite being a valuable dietary product due to its high content of polyunsaturated fatty acids, rapeseed oil is predominantly utilized for technical purposes.

Due to the advantages offered by rapeseed, folks leverage these benefits for profitable opportunities within the agricultural sector. Agricultural operations are linked to considerable energy expenditures, particularly concerning fuel and lubricants. In the context of crop cultivation technology, the average fuel consumption per hectare is estimated to be around 50 kg of diesel fuel. Itis essential to underscore that the portion of fuel-related expenses within the overall production costs of cultivated agricultural products surpasses 20%. Consequently, any surge in fuel prices directly translates to an upswing in the overall cost of agricultural products. Conversely, efforts to restrain prices within the market can result in a concomitant reduction in the overall profitability of the agricultural sector [17]. This intricate relationship between energy costs, specifically in terms of fuel consumption, and the economic dynamics of agricultural production underscores the vulnerability of the sector to fluctuations in fuel prices.

The significance of fuel in the cost structure of agricultural products accentuates the need for strategic planning and sustainable practices to mitigate the impact of such external factors on the economic viability of agricultural enterprises. In addition, this interplay emphasizes the delicate balance that must be maintained between cost containment measures and ensuring the financial health and sustainability of the agricultural economy. A critical concern revolves around substituting conventional fossil fuels with environmentally sustainable alternatives. In succeeding chapters, different biodegradable feedstocks have been exposed, and biodiesel, specifically derived from vegetable oils, emerges as a promising option, particularly in the context of agricultural practices [18]. In certain regions, the cultivation area for oil crops, encompassing varieties like rapeseed, requires 1000 ha [19].

In ancient times, rapeseed oil served as a source of illumination for streets through lanterns and was utilized to lubricate metal components exposed to water and steam, preventing corrosion. In contemporary settings, it finds application in the hydraulic systems of tractors and agricultural machinery. Moreover, it stands out as a valuable raw material for producing environmentally friendly fuels for heat engines and biodegradable plastics. In contrast to mineral oil, which can lead to plant death and hinder germination in contaminated areas for several years, rapeseed oil poses no harm to plants or living organisms when it enters the soil or water. Furthermore, it lacks a benzene smell and undergoes rapid decomposition in approximately four weeks. During the combustion of rapeseed oil, the emitted carbon dioxide matches the amount absorbed by the plant from the atmosphere, highlighting its carbon-neutral nature. The oil's commendable lubricating properties contribute to extending the lifespan of both the fuel pump and the engine [20]. Rapeseed oil and diesel fuel are different substances with distinct properties and purposes. Rapeseed oil and diesel fuel differ in their sources, composition uses, energy content, and properties. While rapeseed oil can be converted into biodiesel for certain applications, straight rapeseed oil is not directly interchangeable with diesel fuel without modifications.Some of the key differences between rapeseed oil and diesel fuel are presented in Table 15.1.

The investigation carried out by [21] focused on the analysis of the physical and chemical attributes of both rapeseed oil and diesel fuel, with the detailed findings conveniently presented in Table 15.2. This comparative examination proves instrumental in unraveling the distinctions between rapeseed oil and diesel fuel concerning their physical and chemical compositions. Such disparities bear significant implications, particularly in comprehending how these variations might impact their functionalities across diverse applications, including engine performance and other industrial uses. By scrutinizing the data provided in Table 15.2, researchers and industry experts can gain valuable insights into the nuanced characteristics of rapeseed oil and diesel fuel. These insights are crucial for making informed decisions regarding the selection of suitable fuels for specific applications. Whether considering combustion efficiency, environmental impact, or overall compatibility with engines, understanding the intricate details of physical and chemical properties becomes pivotal.

Table 15.1 Difference between rapeseed oil and diesel fuel.

S/N	Consideration	Rapeseed oil	Diesel fuel
1	Source and composition	This is a vegetable oil extracted from the seeds of the rapeseed plant, which is a member of the Brassicaceae family. Rapeseed oil is mainly composed of triglycerides, similar to other vegetable oils, and it contains fatty acids such as oleic acid, linoleic acid, and erucic acid.	Diesel fuel, on the other hand, is a hydrocarbon-based fuel derived from crude oil through a refining process. It consists of a mixture of hydrocarbons, and its composition can vary depending on the refining process and the type of crude oil used.
2	Usage	Rape seed oil is commonly used as a cooking oil, and it is also utilized in the production of biodiesel. Biodiesel is a renewable fuel made from various vegetable oils, including rapeseed oil, through a process called transesterification.	Diesel fuel is primarily used as a fuel for internal combustion engines in vehicles, industrial machinery, and generators.
3	Energy content	While rapeseed oil can be converted into biodiesel, it has a lower energy density compared to conventional diesel fuel.	Diesel fuel has a higher energy density, which means it provides more energy per unit of volume than biodiesel.
4	Properties	It has a higher viscosity and a lower cetane number compared to diesel fuel. These properties may require modifications in engines or the use of additives when using straight rapeseed oil as a fuel.	Diesel fuel has specific characteristics, such as a defined cetane number and viscosity, optimized for efficient combustion in diesel engines.
	Engine compatibility	It may require engine modifications or the use of a blend with conventional diesel fuel to be compatible with standard diesel engines.	Diesel engines are designed to run on conventional diesel fuel, and using alternative fuels like biodiesel or straight vegetable oil may require engine modifications or the use of specific technologies.

The findings of this study not only contribute to the existing body of knowledge but also offer practical applications, aiding engineers, scientists, and industries in optimizing their use of rapeseed oil and diesel fuel. As the demand for sustainable and efficient energy sources continues to grow, such studies play a pivotal role in shaping the future setting of fuel utilization and technological advancements.

Table 15.2 Comparison of physical and chemical properties of rapeseed oil and diesel fuel.

No.	Parameter	Rapeseed oil	Diesel fuel
1	Content, % C; H; O	78.0; 11.5; 10.5	85.2; 13.7; 1.1
2	Density at 15 °C, kgm^3	917	800–845
3	Kinematic viscosity at 40 °C: mm^2/s	42.1	1.5–4.0
4	Dynamic viscosity at 20 °C, Pa/s	68.7×10^{-3}	3.15×10^{-3}
5	Surface tension, n/m	33.2×10^{-3}	27.1×10^{-3}
6	Lower combustion heat, mJ/m^3	36 992	42 437
7	Cetane number	36. . .55	46–49
8	Flash temperature, not lower, °C	100	55
9	Solidification temperature, °C	−23	−10; −35; −45; −55
10	Sulfur content, %	0.005	0.5
11	Cokingcharacteristics of 10%-th residue, notmore than, %	0.4	0.3

From the table presented, the differences in the physical–chemical properties of rapeseed oil and diesel fuel is exposed (Table 15.2), rapeseed oil is currently being used more and more as a main component in producing biodiesel fuels [22]. These two components mix well, and the resulting mixture has properties that help it to be burned in a diesel engine without changing its composition. Rapeseed oil significantly differs from diesel fuel in viscosity and other properties. It has been shown that in terms of viscosity, Rapeseed is more viscous than diesel fuel. The next section presents the process required for the production of Rapeseed into biodiesel.

15.3 Biodiesel Production from Rapeseed

Rapeseed, a versatile and high-yielding oilseed crop, has become a focal point in the quest for renewable energy sources. Its oil-rich composition makes it an ideal candidate for biodiesel production, offering a sustainable alternative to conventional fossil fuels. The production of biodiesel from rapeseed involves several key stages, encompassing the cultivation of the crop, extraction of oil, and the conversion of that oil into biodiesel through a process known as transesterification. Figure 15.2 shows a flowchart with step-by-step procedure for producing biodiesel from Rapeseed.

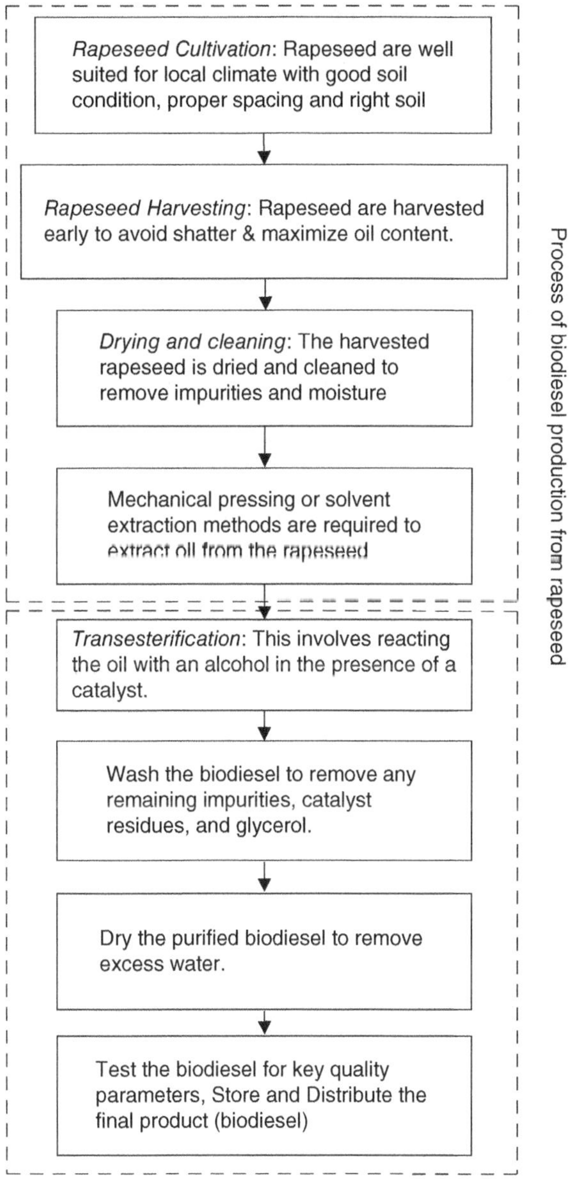

Figure 15.2 Flowchart showing the step-by-stepprocess of producing biodiesel from rapeseed.

15.3.1 Detailed Explanation of the Production Process

Summarily, based on Figure 15.2, the main stages involved in biodiesel production from rapeseed are summarized as cultivation of rapeseed; planting, harvesting and preparation or sourcing; harvesting and preparation; drying and cleaning; oil extraction; oil refining via transesterification; drying, quality testing, and storage. Further explanation of the process is detailed below:

Cultivation of rapeseed: It is important to carefully and systematically select varieties of rapeseed. Making a favorable choice of rapeseed for cultivation requires going for high-yielding and well-suited local climate and soil conditions.

Planting: The appropriate planting season for rapeseed is usually during the rainy season in Nigeria; this is necessary to ensure soil preparation and proper spacing during planting.

Cultivation practices: A good agricultural practices is required for implementation. This includes pest control, irrigation, and fertilization, to maximize yields.

Harvesting and preparation: The timing of harvest is very important to maximize the oil content in rapeseed. It is best to harvest rapeseed when the seeds have matured but before they smash to smithereens to maximize oil content.

Drying and cleaning: After harvesting the rapeseed, the feedstock is dried and cleaned to remove impurities and moisture, this is to ensure a good quality of the oil.

Oil extraction: The extraction of oil from the rapeseed can be achieved by using either the mechanical pressing or solvent extraction method. Mechanical pressing involves using a hydraulic press to squeeze oil out, while solvent extraction uses chemical solvents to dissolve and separate the oil.

Oil refining: The extracted oil can be purified through processes such as degumming, neutralization, and deodorization to remove impurities and enhance quality.

Transesterification: This is the core process where the rapeseed oil is converted into biodiesel. The heart of biodiesel production lies in the transesterification process. Rapeseed oil reacts with an alcohol (typically methanol or ethanol) in the presence of a catalyst (such as sodium hydroxide or potassium hydroxide). This chemical reaction results in the formation of biodiesel (methyl or ethyl esters) and glycerol, a valuable by-product.

Formation of biodiesel and glycerol: The reaction results in the formation of biodiesel (methyl or ethyl esters) and glycerol. The glycerol is separated from the biodiesel.

Biodiesel purification and quality assurance: Post-transesterification, the biodiesel undergoes purification processes to remove impurities, residual catalysts, and glycerol. Rigorous quality testing ensures that the biodiesel meets industry standards, addressing concerns related to fuel efficiency and environmental impact.

Drying: The purified biodiesel is dried to remove excess water.

Quality testing: The produced biodiesel is tested for key quality parameters, including acidity, moisture content, viscosity, and other specifications to ensure it meets regulatory standards.

Storage and distribution: The obtained biodiesel is stored in appropriate conditions to prevent degradation and maintain its quality. Produced biodiesel can thus be distributed for use in various applications, such as transportation, industrial processes, or as a heating fuel.

By-product utilization: The by-product glycerol can be further processed for various applications, such as in the pharmaceutical or cosmetic industries.

It is important to note that each stage requires careful attention to detail to ensure the efficiency and sustainability of the biodiesel production process. In addition, adherence to quality standards and regulations is essential to produce biodiesel that meets industry and environmental requirements.

15.3.2 Key Features that Position Rapeseed Oil as a Favorite Feedstock

In the pursuit of sustainable and economically viable biofuels, rapeseed oil stands out as an exceptional feedstock due to its key features that contribute to both environmental friendliness and economic feasibility. Some of the key features are further discussed:

1) *The abundance and renewability*: Rapeseed, a member of the Brassicaceae family, is a high-yielding crop that thrives in various climates. Its seeds contain a substantial amount of oil, making it a prolific and renewable source of biodiesel feedstock. The plant's adaptability contributes to its widespread cultivation and availability as a consistent source of raw material.
2) *Versatility in end-use applications*: Beyond its application in biodiesel production, rapeseed oil finds use in various industries. Its versatility extends to the culinary world, where it is used as cooking oil, and to industrial applications such as lubricants and bio-based plastics. This multifaceted utility enhances the overall economic viability of rapeseed cultivation. Potential for sustainable agriculture: Rapeseed cultivation can be integrated into sustainable agricultural practices. The crop has the ability to break pest and disease cycles, reducing the need for chemical inputs. In addition, its deep root system enhances soil structure and reduces erosion, promoting overall agricultural sustainability.
3) *Economic impact*: The cultivation of rapeseed for biodiesel production has the potential to stimulate rural economies. Job creation, income generation for farmers, and the establishment of local processing facilities contribute to the economic viability of rapeseed oil as a biodiesel feedstock.
4) *Optimal fatty acid composition*: Rapeseed oil has favorable fatty acid composition, particularly high levels of monounsaturated and polyunsaturated fats.

This composition makes it well-suited for the transesterification process, a key step in biodiesel production. The resulting biodiesel exhibits desirable properties such as good flow characteristics and improved cold-weather performance.
5) *Reduced greenhouse gas emissions*: Biodiesel derived from rapeseed oil offers a substantial reduction in greenhouse gas emissions compared to conventional diesel. The closed carbon cycle of rapeseed, where the carbon dioxide released during combustion is offset by the carbon dioxide absorbed during its growth, contributes to its environmentally friendly profile.

15.4 Potential Benefits, Challenges, and Opportunities of Utilizing Rapeseed Oil in Biodiesel Production

15.4.1 Potential Benefits

The utilization of rapeseed oil in biodiesel production offers several potential benefits, which include the following:

1) *Abundant and renewable source*: Rapeseed is a widely cultivated oilseed crop, providing a sustainable and abundant source of feedstock for biodiesel production. Its availability helps in meeting the growing demand for renewable energy.
2) *High oil content*: Rapeseed oil is characterized by a high oil content, making it an efficient and productive feedstock for biodiesel. The oil-rich nature of rapeseed contributes to a higher biodiesel yield during the transesterification process.
3) *Reduced greenhouse gas emissions*: Biodiesel derived from rapeseed oil typically exhibits lower carbon dioxide emissions compared to conventional fossil fuels. Its use in diesel engines can contribute to a reduction in overall greenhouse gas emissions, promoting environmental sustainability.
4) *Improved lubricity*: Rapeseed biodiesel often possesses enhanced lubricating properties compared to petrodiesel. This can lead to better engine performance and reduced wear and tear on engine components.
5) *Domestic production and energy security*: Cultivating and processing rapeseed for biodiesel can contribute to domestic energy production, reducing dependence on imported fossil fuels. This enhances energy security and supports local economies.
6) *Compatibility with existing infrastructure*: Rapeseed biodiesel can be seamlessly integrated into existing diesel infrastructure without significant modifications. It can be used in conventional diesel engines and distributed through existing fuel supply chains.

7) *Biodiversity and crop rotation*: The cultivation of rapeseed promotes biodiversity and can be part of a sustainable crop rotation system. Rotating rapeseed with other crops helps maintain soil fertility, reduces pests, and minimizes environmental impact.
8) *Job creation*: The production and processing of rapeseed for biodiesel create opportunities for employment in agriculture, bioenergy industries, and associated sectors, contributing to rural development.
9) *Potential for by-product utilization*: By-products generated during rapeseed oil extraction, such as meal, can be utilized for various purposes, including animal feed or even as a source of additional bioenergy.
10) *Government incentives*: Many governments offer incentives and subsidies to encourage the production and utilization of biodiesel, further supporting the economic viability of rapeseed biodiesel. It is essential to consider factors such as land use, cultivation practices, and the overall life cycle analysis to ensure the sustainability and environmental benefits of rapeseed biodiesel production.

15.4.2 Potential Challenges Associated with Utilizing Rapeseed Oil in Biodiesel Production

While there are potential benefits to using rapeseed oil in biodiesel production, there are also challenges associated with its utilization. Some of the potential challenges are discussed below:

1) *Competition with food production*: Rapeseed is also used for culinary purposes, and the competition between biodiesel production and food production can raise concerns about food security. Striking a balance between food and bioenergy needs is crucial.
2) *Land use change and deforestation*: The expansion of rapeseed cultivation for biodiesel production may lead to land use change, including deforestation. Clearing land for rapeseed cultivation can result in habitat loss, biodiversity decline, and increased greenhouse gas emissions.
3) *Resource intensity*: Rapeseed cultivation requires resources such as water, fertilizers, and pesticides. The environmental impact of these inputs, including water usage and potential runoff of chemicals, needs to be carefully managed to ensure sustainable practices.
4) *Climate sensitivity*: Rapeseed cultivation is sensitive to climatic conditions, and variations in temperature and precipitation can impact yields. Climate change could potentially affect the reliability and productivity of rapeseed crops.
5) *Biodiesel yield variability*: The biodiesel yield from rapeseed oil can be variable, influenced by factors such as the specific cultivar, soil conditions, and

agronomic practices. Achieving consistent and high yields may require careful crop management.

6) *High production costs*: The costs associated with cultivating, harvesting, and processing rapeseed into biodiesel can be relatively high. This economic aspect may pose challenges in making rapeseed biodiesel competitive with other forms of energy.
7) *Energy intensive processing*: The conversion of rapeseed oil into biodiesel involves energy-intensive processes, particularly during transesterification. This raises questions about the overall energy balance and whether the energy gained from biodiesel offsets the energy invested.
8) *Transportation and storage challenges*: The logistics of transporting rapeseed oil from farms to processing facilities and then distributing biodiesel to end-users can pose challenges. Efficient transportation and storage infrastructure are essential for the success of the supply chain.
9) *Social impacts*: Large-scale rapeseed cultivation for biodiesel production may have social implications, including changes in rural landscapes, impacts on local communities, and potential conflicts over land use.
10) *Technological advancements*: Continuous research and development are necessary to improve the efficiency of biodiesel production from rapeseed. This includes exploring advanced technologies to enhance yields, reduce processing costs, and mitigate environmental impacts.

Addressing these challenges requires a holistic approach that considers environmental, social, and economic factors. Sustainable practices, responsible land use planning, and ongoing technological advancements are essential to maximize the benefits of rapeseed biodiesel while minimizing its drawbacks.

15.4.3 Potential Opportunities Associated with Utilizing Rapeseed Oil in Biodiesel Production

The utilization of rapeseed oil in biodiesel production presents several potential opportunities, encompassing economic, environmental, and agricultural aspects. Exploration of the opportunities associated with utilizing rapeseed oil for biodiesel is further discussed below:

1) *High biodiesel yield*: Rapeseed oil is known for its high oil content, which translates to a relatively high biodiesel yield. This efficiency in conversion can contribute to increased biofuel production and reduced dependency on fossil fuels.
2) *Crop rotation and soil health*: Integrating rapeseed cultivation into crop rotation systems can enhance soil health. Rapeseed is known for its deep taproots, which can help break up compacted soil, reduce erosion, and contribute to overall soil fertility.

3) *Reduced greenhouse gas emissions*: Biodiesel derived from rapeseed oil has the potential to significantly reduce greenhouse gas emissions compared to traditional fossil fuels. This contributes to efforts to mitigate climate change and improve overall environmental sustainability.
4) *Diversification of agriculture*: Including rapeseed in agricultural rotations provides farmers with a diversified income source. This diversification can contribute to more resilient farming systems and reduce the economic risks associated with reliance on a single crop.
5) *Job creation*: The expansion of rapeseed cultivation for biodiesel production can create job opportunities across the supply chain, from farming and harvesting to processing and distribution. This can have positive economic impacts on rural communities.
6) *Bioenergy security*: Utilizing rapeseed oil for biodiesel production enhances bioenergy security by diversifying the sources of renewable energy. This reduces dependence on finite fossil fuel resources and supports the development of a more sustainable energy mix.
7) *Improved energy independence*: Countries that invest in rapeseed biodiesel production can enhance their energy independence by relying on domestically produced biofuels. This reduces vulnerability to fluctuations in global oil markets.
8) *Rural development*: The cultivation of rapeseed for biodiesel can contribute to the development of rural areas. This includes infrastructure development, increased economic activity, and the establishment of local processing facilities.
9) *Feedstock availability*: Rapeseed is a widely cultivated and abundant crop, providing a readily available source of feedstock for biodiesel production. This can help meet the growing demand for renewable fuels.
10) *Research and innovation*: The exploration of rapeseed oil for biodiesel production encourages ongoing research and innovation. This includes developing new crop varieties, optimizing cultivation practices, and improving processing technologies, fostering advancements in sustainable agriculture.
11) *Co-products and by-products*: The by-products of rapeseed processing, such as meal, can be used for various purposes, including animal feed. This maximizes resource utilization and minimizes waste.
12) *Bioenergy policy support*: Government policies that support the production and use of biodiesel, including incentives and subsidies, can create a conducive environment for the expansion of rapeseed-based biodiesel.

While rapeseed biodiesel presents significant opportunities, it is important to address challenges and ensure sustainable practices to maximize the positive impacts on agriculture, the economy, and the environment.

15.5 The Future of Rapeseed Oil in Biodiesel Production

In the 21st century, the future of rapeseed oil in biodiesel production holds promise and is influenced by various factors, including technological advancements, environmental considerations, economic dynamics, and policy developments. The key aspects that shape the future outlook for rapeseed oil in biodiesel production in Industry 4.0 include:

1) *Technological advancements*: Ongoing research and technological innovations in biodiesel production processes, such as improved transesterification methods and catalyst development, may enhance the efficiency and cost-effectiveness of converting rapeseed oil into biodiesel.
2) *Genetic engineering and crop improvement*: Continued efforts in genetic engineering and crop breeding can lead to the development of rapeseed varieties with enhanced oil content, improved resistance to pests and diseases, and increased adaptability to diverse climatic conditions. This can positively impact biodiesel yield and crop productivity.
3) *Sustainability practices*: The adoption of sustainable agricultural practices, such as precision farming and organic cultivation, can contribute to reducing the environmental footprint of rapeseed cultivation. Sustainable practices align with growing consumer and regulatory preferences for environmentally friendly biofuels.
4) *Policy support*: Government policies and regulations play a crucial role in shaping the future of rapeseed oil in biodiesel production. Supportive policies, such as incentives, subsidies, and mandates for biofuel blending, can stimulate investment and market growth.
5) *Market demand*: The increasing global demand for renewable and sustainable energy sources, coupled with the desire to reduce greenhouse gas emissions, creates a favorable market environment for rapeseed oil biodiesel. Consumer awareness and preferences for eco-friendly fuels can further drive demand.
6) *Global economic dynamics*: Economic factors, including oil prices, geopolitical events, and market dynamics, can influence the competitiveness of rapeseed biodiesel compared to conventional diesel. Economic stability and incentives for renewable energy can contribute to market growth.
7) *Research and development*: Ongoing research initiatives focused on optimizing biodiesel production from rapeseed oil, addressing challenges, and exploring innovative solutions will be instrumental in shaping the industry's future.
8) *Integration with other renewable sources*: The integration of rapeseed biodiesel production with other renewable energy sources, such as solar and wind energy, can create synergies and contribute to a more diversified and resilient energy landscape.

9) *Environmental considerations*: Increasing emphasis on sustainability and environmental considerations may drive the development of certification schemes and standards for sustainable rapeseed cultivation and biodiesel production. Meeting these standards can enhance the marketability of rapeseed biodiesel.
10) *International trade*: The globalization of biofuel markets and international trade agreements can impact the accessibility and competitiveness of rapeseed biodiesel on a global scale.
11) *Circular economy initiatives*: As part of circular economy initiatives, the utilization of by-products and co-products from rapeseed processing for various applications, such as animal feed or bio-based chemicals, can contribute to a more resource-efficient and circular biofuel production system.

In summary, the future of rapeseed oil in biodiesel production in Industry 4.0 is likely to be shaped by a combination of technological advancements, sustainability practices, policy support, market dynamics, and evolving consumer preferences. As the biofuel industry continues to evolve, rapeseed oil remains a significant contributor to the pursuit of renewasable and environmentally friendly energy solutions.

15.6 Conclusion

The exploration of rapeseed oil as a potential feedstock for sustainable biodiesel production represents a significant stride toward addressing the pressing challenges in the global energy landscape. The analysis of its physicochemical properties and its inherent suitability for biodiesel production underscores the viability of rapeseed oil as a promising alternative to traditional fossil fuels. The environmental benefits, including the reduction of greenhouse gas emissions and decreased dependence on nonrenewable resources, position rapeseed oil as a sustainable contributor to the bioenergy sector. As our findings illuminate the challenges and opportunities associated with this feedstock, it becomes evident that a holistic approach is essential for its successful integration into biodiesel production. Considerations of crop yield, land use efficiency, and socioeconomic implications provide a nuanced understanding of the broader impacts of adopting rapeseed oil in biodiesel production. The ongoing research and development initiatives aimed at refining production processes and improving the quality of rapeseed oil-based biodiesel signal a commitment to continuous innovation in the pursuit of sustainable energy solutions. In conclusion, this study not only contributes valuable insights to the scientific community but also offers pragmatic considerations for policymakers and industry stakeholders. The potential of rapeseed oil as a feedstock for sustainable biodiesel production represents a step toward a more resilient and environmentally conscious energy future, inviting collaboration and concerted efforts in realizing the promise of this renewable resource.

References

1 Okonkwo, C.P., Ajiwe, V.I.E., Obiadi, M.C. et al. (2021). Kinetics and transesterification of the oil obtained from *Cussoniabateri* (Jansa seed) as a step in biodiesel production using natural heterogeneous catalyst. *Am. J. Appl. Chem.* 9 (5): 164–171.

2 Gebremariam, S.N. and Marchetti, J.M. (2018). Economics of biodiesel production: review. *Energy Convers. Manag.* 168: 74–84. https://doi.org/10.1016/j.enconman.2018.05.002.

3 Mansir, N., Teo, S.H., Rashid, U. et al. (2018). Modified waste egg shell derived bifunctional catalyst for biodiesel production from high FFA waste cooking oil: a review. *Renew. Sust. Energ. Rev.* 82: 3645–3655. https://doi.org/10.1016/j.rser.2017.10.098.

4 Karmakar, B. and Halder, G. (2019). Progress and future of biodiesel synthesis: advancements in oil extraction and conversion technologies. *Energy Convers. Manag.* 182: 307–339.

5 Kim, D.S., Hanifzadeh, M., and Kumar, A. (2018). Trend of biodiesel feedstock and its impact on biodiesel emission characteristics. *Environ. Prog. Sustain.* 37: 7–19.

6 Rapier, R. (2008). Renewable diesel. In: *Biofuels, Solar and Wind as Renewable Energy Systems*, 1e (ed. D. Pimentel), 153–157. Dordrecht, The Netherlands: Springer.

7 Demirbas, A. (2005). Biodiesel production from vegetable oils via catalytic and non-catalytic supercritical methanol transesterification methods. *PECS* 31: 466–487.

8 Balat, M. and Balat, H. (2010). Progress in biodiesel processing. *Appl. Energy* 87: 1815–1835.

9 Knutsen, H.K., Alexander, J., Barregård, L. et al. (2016). Erucic acid in feed and food. *EFSA* 14: e04593.

10 Yang, F., Hanna, M.A., and Sun, R. (2012). Value added uses for crude glycerol –roduct of biodiesel production. *Biotechnol. Biofuels* 5: 13.

11 Samuel, O.D., Okwu, M.O., Semiu, A., and Adeniran, S.A. (2019). Production of fatty acid ethyl esters from rubber seed oil in hydrodynamic cavitation reactor: study of reaction parameters and some fuel properties. *Ind. Crops Prod.* 141: 1–13. https://doi.org/10.1016/j.indcrop.2019.111658.

12 Vicente, G., Martinez, M., and Aracil, J. (2004). Integrated biodiesel production: a comparison of different homogenous catalysts systems. *Bioresour. Technol.* 92: 297–305.

13 Veljkovic, V.B., Stamenkovic, O.S., and Tasic, M.B. (2014). The wastewater treatment in the biodiesel production with alkali-catalyzed transesterification. *Renew. Sust. Energ. Rev.* 32: 40–60.

14 Srirangsan, A., Ongwandee, M., and Chavalparit, O. (2009). Treatment of biodiesel wastewater by electrocoagulation process. *Environ. Asia* 2: 15–19.

15 Kolesarova, N., Hutnan, M., Bodik, I., and Spalkova, V. (2011). Utilization of biodiesel by-products for biogas production. *J. Biotechnol. Biomed.* 2011: 126798.

16 Sagan, A., Blicharz-Kania, A., Szmigielski, M. et al. (2019). Assessment of the properties of rapeseed oil enriched with oils characterized by high content of α-linolenic acid. *Sustainability* 11 (20): 5638. https://doi.org/10.3390/su11205638.

17 Zybalov, V.S., Sergeev, N.S., and Zapevalov, M.V. (2019). The rational use of rapeseed in agricultural production. *AIC Russ.* 26 (2).

18 Zybalov, V.S. and Kozhamkulova, Y.S. (2013). *Analysis of Chemical Composition and Physical Properties of Sunflower and Rapeseed Oil*, 33–38. Chelyabinsk: ChGAA.

19 Zybalov, V.S. (2019). Spring rapeseed is a culture of great opportunities in the southern Urals. *AIC Russ.* 26 (5): 755–762.

20 Markov, V.A., Gaivoronskiy, A.I., Devyanin, S.N., and Ponomarev, E.G. (2006). Rapeseed oil as an alternative fuel for diesel. Automotive Industry, No. 2.

21 Zapevalov, M.V., Sergeev, N.S., and Redreev, G.V. (2020). Rapeseed oil is the base for biodiesel fuel. *IOP Conf. Ser.: Earth Environ. Sci* 688 (012013): 1–6.

22 Józsa V. (2020). Experimental Comparison of Diesel and Crude Rapeseed Oil Combustion in a Swirl Burner. *Applied Sciences.* 10(14): 4907.

16

Sustainable Biodiesel Production by Exploring Novel Nonedible Aidan Seed as a Potential Feedstock

16.1 Introduction

The increasing rate of energy use has made energy demand and security one of the biggest challenges for all economies in the globe. In addition, the worldwide commitment to ecologically friendly and sustainable energy consumption has increased awareness of the use of biofuels in place of traditional petroleum-based fuels. When used as diesel engine fuel, biodiesel dramatically reduces toxic and other hazardous emissions, making it a promising biodegradable and nontoxic fuel. The primary benefits of biodiesel comprise its elevated cetane number, enhanced combustion efficiency, reduced emissions of sulfurous and aromatic compounds, and elevated flash point [1]. Extensive and innovative research is being conducted on the production, investigation, and application of biodiesel in various diesel engines [2]. Researchers are looking more closely for alternative fuel sources as a result of the depletion of fossil fuels, environmental protection concerns, and a constant rise in the market value of these fuels. It is well acknowledged that biodiesel is a significant source of diesel for transportation, and demand for this diesel is rising. They have a high oxygen content, minimal to no pollution, regulated toxicity, and biodegradability. It can also be utilized to run an engine without having to be adaptable [3]. Regarding industrial rankings, transport has taken the third spot when it comes to overall world energy consumption. Because of industrialization, population growth, and rising living standards, it is projected that consumption will increase by 60% by 2030 [4]. Due to the fact that biodiesels are very sustainable and fossil fuels are scarce, they stand out from other fuels. Regardless of the state of the atmosphere, biodiesels have the ability to run

Feedstocks for Sustainable Biodiesel Production: Characterization, Selection, and Optimization, First Edition. Chinwe P. Okonkwo and Modestus O. Okwu.

multiple diesel engines. Increasing the amount of biodiesel produced will require cheap, accessible energy sources that differ significantly from fuels derived from fossil fuels.

16.1.1 Biodiesel Transesterification Process

The transesterification process has gained a lot of recognition as a biodiesel-producing technique. Biodiesel is produced by reacting alcohol with oil triglycerides in the presence of an acidic or basic catalyst. In addition, it requires the use of three consecutive, reversible methods to react with triglycerides to create diglycerides, monoglycerides, and glycerol [5]. The primary five factors that determine the generation of biodiesel are the proportion of free fatty acids (FFA), the type and concentration of the catalyst, the type of alcohol employed and its molar ratio (alcohol: oil), reaction temperature, and reaction time [6]. Different transesterification procedures determine the characteristics of the feedstock. Certain methods can handle both highly FFA and extremely sensitive feedstocks with low FFA levels [7]. Monoalkyl esters of long-chain fatty acids, such as those found in vegetable oil, leftover cooking oil, animal fat, and microalgae, are used to make biodiesel [8]. Utilizing nonedible oil as a second-generation feedstock has proven to be an excellent way to produce biodiesel. Any seed's oil content fluctuates according to several conditions, such as climate and geographic location. Higher cetane number, increased combustion efficiency, reduced emissions of sulfurous and aromatic compounds, and a high flash point are among the main benefits of biodiesel.

16.1.2 Aidan Fruit and Benefits

The tropical fruit known as Aidan fruit, also known as Tetrapleura tetraptera or Aidan pods, has become more well-known due to rumors of potential health benefits. Western Africa is home to the Aidan fruit (Tetraptera), a kind of covering plant in the pea family. The Aidan fruit tree is useful for a variety of tasks and is prized for its pleasant scent. The fruit's bark has therapeutic uses and is used to flavor food. The fruit's essential oil concentration is responsible for its unique scent [9]. It is widely used in nursing mothers' soup to avoid digestive diseases, particularly stomach ulcers and postpartum contractions. The fruit is widely used in Nigerian traditional medicine to treat a variety of conditions, such as diabetes mellitus, asthma, rheumatoid arthritis, hypertension, and epilepsy [10]. It is an abundant source of vital nutrients. Vitamins C, A, and B-complex vitamins, such as thiamine and niacin, are abundant in it. Aidan fruit also includes nutrients like potassium, iron, and calcium. Maintaining general health and well-being depends heavily on these nutrients [11]. Figure 16.1 presents a novel nonedible fruit and leaves of Aiden feedstock.

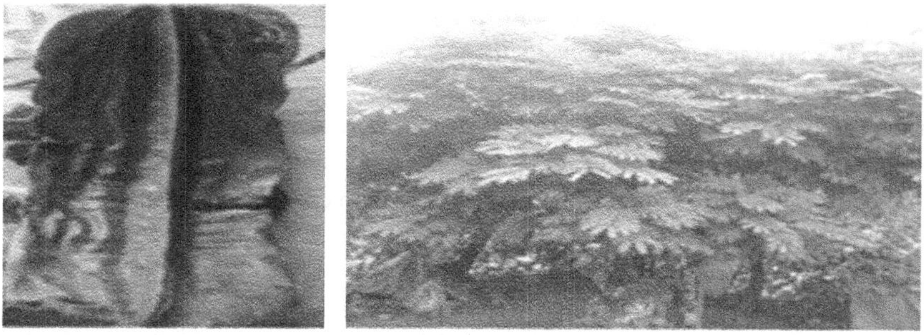

Figure 16.1 Novel Aidan fruit seed and leaves.

16.2 Materials and Method

16.2.1 Materials

Freshly dried Aidan seeds, scientifically known as Tetrapleura tetraptera, were procured for this study. The acquisition took place at Eke Awka market, located in Anambra state, Nigeria. The choice of Aidan seeds as the primary material for biodiesel production is grounded in their nonedible nature, making them a potential and sustainable feedstock for biofuel exploration. The seeds were carefully selected to ensure freshness and uniformity, as the quality of the starting material significantly influences the outcomes of oil extraction and subsequent biodiesel production processes. The geographical specificity of the procurement from the Eke Awka market is noted as favorable, as the regional variations in climate and soil conditions can affect the composition and properties of the Aidan seeds. The procurement and preparation of the Aidan seeds marked the initial step in the research, laying the foundation for subsequent analyses aimed at evaluating their potential as a viable and sustainable feedstock for biodiesel production.

16.2.2 Preparation of Aidan Seeds

Prior to the experimental procedures, the procured Aidan seeds were thoroughly sorted and cleaned with water, and then the wet seeds underwent a meticulous drying process to remove excess moisture, ensuring the stability of the seeds during storage and experimentation, this was done using an oven at 60 °C for 12 hours before being processed using a hammer mill and packaged. The extracts were kept at 4 °C until they were needed for analysis.

The extraction of oil from the sample employed the Soxhlet extraction technique, a process renowned for its efficiency. Initially, the sample was accurately weighed and placed within the thimble of a 500 mL Soxhlet extractor, strategically positioned within a semi-permeable cotton material. Simultaneously, a 500 mL flat-bottom round flask was loaded with 400 mL of *n*-hexane, chosen as the extraction solvent due to its favorable properties. The Soxhlet extraction apparatus was then subjected to controlled heating on a hot plate, facilitating the circulation of water at the condenser's outer jacket. This meticulously controlled system ensured the optimal extraction of oil from the sample. Upon complete extraction, the process was halted, leading to the disposal of the de-fatted sample.

To separate the oil and *n*-hexane amalgamation in the flat-bottom flask, a rotary evaporator was employed, employing a process of evaporation under reduced pressure. This step ensured the isolation of the extracted oil while eliminating the solvent, thereby enhancing the purity of the obtained oil. The quantification of the extraction's success was determined through the calculation of percentage yields. This multi-step process, incorporating Soxhlet extraction, solvent separation, and yield calculation, exemplifies a comprehensive and precise methodology for extracting oil from the sample, ensuring accuracy and reliability in the experimental outcomes.

16.2.3 Biodiesel Production (Transesterification)

The synthesis of methyl ester through transesterification reactions took place in a 500 mL spherical, round-bottom glass reactor equipped with three necks. The reactor underwent a prestirring phase lasting 10 minutes to ensure thorough dissolution before introducing a molar ratio of alcohol to oil at 12:1. A reflux condenser, with cold water circulating around its exterior, was affixed to the mid-neck of the reactor. In addition, a mercury thermometer, housed within a plastic bung, was inserted into the right neck of the reactor to monitor and regulate the reaction temperature effectively.

A specialized stirring knob, designed for capsular stirring, was introduced into the reactor through its left neck and securely fastened using a plastic bung. The entire assembly was activated and positioned on a magnetic heating mantle. Simultaneously, the stirring system was initiated, maintaining a consistent speed and temperature. This particular moment marked the initiation of the reaction, referred to as time zero. Operating at a controlled temperature of 65 °C, each reaction was permitted to proceed for a duration of two hours. This meticulous setup and precise regulation of conditions ensured the uniformity and reliability of the transesterification reactions, optimizing the production of methyl ester with enhanced efficiency.

Following the completion of the reaction, the resultant product underwent a 12-hour settling period in a separating funnel to facilitate the separation of glycerol. During this time, the crude glycerol was drawn out through the funnel tap, leaving behind the methyl ester. Subsequently, the biodiesel underwent a hot water rinse before being transferred to a 250 mL beaker. To eliminate residual water molecules from the biodiesel, it underwent a heating process, reaching a temperature of 105 °C. After cooling, the biodiesel was carefully stored in calibrated specimen vials. The upper layer of the biodiesel was then transferred to a rotary evaporator to facilitate the evaporation of methanol. The resulting purified biodiesel was collected and stored for subsequent analysis. This methodical process ensured the removal of impurities and water content, enhancing the overall quality of the biodiesel for further examination (Table 16.1).

The acid value (AV), alternatively referred to as the neutralization number or acid number, quantifies the quantity of potassium hydroxide (KOH) in milligrams needed to neutralize the acidic components present in 1 g of the sample. A lower AV suggests that the oils can potentially undergo direct conversion into biodiesel through the base-catalyzed transesterification process [8].

High levels of FFA in oil can deactivate the base catalyst, leading to the generation of soap and a subsequent reduction in the overall yield of biodiesel. The acidic compounds found in biodiesel may stem from various sources, including

Table 16.1 Physiochemica characterization of Aidan seed oil.

Properties	***Chrysobalanus icaco***
Color	Light brown
Yield (%)	±32.700
Moisture (%)	±2.500
Kinematic viscosity at 40 °C (mm^2/s)	±22.140
Refractive index at 29 °C	±1.327
Energy value (kJ/kg)	41 729
Acid value (mgKOH/g)	1.250
Saponification (mgKOH/kg)	175.000
Peroxide value (meq/kg)	0.940
Iodine value (g/100 g of oil)	102.000
Molecular weight (g/mol)	871.660
Flashpoint (°C)	240.000
Cloud point (°C)	10.000
Pour point (°C)	4.000
Fire point (°C)	260.000
Density ($g \times mL^{-1}$)	0.964

lingering mineral acids from the manufacturing process, remaining FFA resulting from ester hydrolysis or post-hydrolysis of esters, and ultimately from the oxidation of by-products manifested as other organic acids [12].

The nondrying nature of the oils is indicated by their iodine value. This measure serves to assess the extent of unsaturation in fatty acids, implying a higher propensity for polymerization and oxidation at elevated temperatures. A direct correlation exists between the iodine value and the degree of unsaturation: oils with higher iodine values are more unsaturated, thereby exhibiting faster oxidation and polymerization tendencies [13]. The heightened unsaturation of fatty acids also renders them more vulnerable to lipid peroxidation or rancidity; however, the introduction of antioxidants can effectively shield fats from this oxidative process. A high iodine content Biodiesel may tend to polymerize and produce deposits on piston rings, injector nozzles, and piston ring grooves. The tendency of polymerization increases with the degree of unsaturation of the fatty acids [14].

This implies that biodiesel possessing an elevated iodine content is more susceptible to reduced stability at higher temperatures. The European standard stipulates an iodine value of 120 in biodiesel (EN 14214), while the German standard under DIN 5160 sets it at 115 [15]. The iodine value is notably influenced by the nature and ester composition of the feedstocks utilized in biodiesel production. The kinematic viscosity of the oil surpasses that of petroleum diesel, primarily attributed to the oil's fatty acid composition and molecular weight [16]. The kinematic viscosity plays a direct role in fuel performance, impacting aspects such as atomization and combustion quality [17] (Table 16.2).

Table 16.2 Mineral composition of Aidan seed oil.

Minerals	Composition (Mg/100 g)
Calcium	185.11
Magnesium	98.56
Copper	9.22
Sodium	20.95
Potassium	271.22
Manganese	332.00
Iron	17.11
Zinc	13.75
Phosphorus	38.22

16.2.4 Impact of Biodiesel on the Environment

The automotive sector or industry stands as a leading contributor to greenhouse gas (GHG) emissions, experiencing more rapid growth than other industries. In addition, the transportation sector is responsible for 23% of all energy-related carbon dioxide (CO_2) emissions and contributes to 15% of global GHG emissions [18]. Conventional vehicles predominantly rely on fuels derived from petroleum, the combustion of which adversely affects both humans and the environment. This combustion process results in the release of various harmful pollutants, including carbon monoxide (CO), unburned hydrocarbons (HC), nitric oxide (NO_x), and smoke emissions, among others [19, 20].

In the presence of sunlight and moisture, these primary pollutants form secondary pollutants such as photochemical haze. Due to its ecologically benign, nontoxic, biodegradable, and cleaner-burning qualities, biodiesel has the potential to be employed as a diesel alternative without requiring extensive engine modifications. Biodiesel, in reality, reduces GHG emissions by up to 86% due to the fact that CO_2 is collected by plants throughout their growing period via photosynthesis [21].

Biodiesel utilization has the potential to decrease emissions of hydrocarbons (HC) by 67% and particulate matter by 47%. In contrast to diesel, the biodiesel production process can result in a remarkable reduction of 79% in wastewater and 96% in hazardous waste. This preference for biodiesel is attributed to its higher cetane number and increased oxygen content, making it the preferred choice when blended with any fuels derived from waste sources [22, 23].

The greater emissions of nitrogen oxides (NO_x) in biodiesel-powered engines are primarily attributed to the presence of oxygen molecules and unsaturated fatty acids [24]. Numerous endeavors have been undertaken to explore technologies for reducing NO_x, including approaches like water injection, emulsification, simultaneous technology, and fuel injection timing retardation [25, 26]. In addition, nanoparticles have been utilized in various studies to enhance engine performance and mitigate detrimental emissions [27–29]. According to reports, the combustion of biodiesel results in lower pollutant emissions.

16.3 Experimental Result

Table 16.3 displays the results of biodiesel obtained during the experiment. It equally presents the input variables for the FAME yield from Aidan seed oil which include the Concentration of the catalyst, mole ratio = 1000:1. Others are constant such as temperature = 550 °C, time = 3 hours, speed = 300 rpm. The values correspond to those obtained during the laboratory experiment.

Table 16.3 Experimental results to illustrate the FAME yield from Aidan seed oil. Temperature = 55 °C, mole ratio = 10:1, time = 3 hours, speed = 300 rpm.

Std run	A: A	Input variable Catalyst. Conc. wt. % of oil	Exp. value Yield 1 vol. %
1	23	0.75	51
2	14	3.75	51
3	12	0.75	61
4	30	3.75	61
5	20	0.75	51
6	13	3.75	51
7	5	0.75	61
8	7	3.75	61
9	9	0.75	51
10	2	3.75	51
11	16	0.75	61
12	4	3.75	61
13	18	0.75	51
14	3	3.75	51
15	11	0.75	61
16	26	3.75	61
17	6	−0.75	56
18	25	5.25	56
19	22	2.25	46
20	24	2.25	66
21	15	2.25	56
22	21	2.25	56
23	29	2.25	56
24	28	2.25	56
25	27	2.25	56
26	10	2.25	56
27	19	2.25	56
28	17	2.25	56
29	8	2.25	56
30	1	2.25	56

The presented table showcases the outcomes of the biodiesel obtained in the experimental process. It also detailed the input variables influencing the FAME yield derived from Aidan seed oil, with factors such as concentration of the catalyst and a mole ratio set at 10:1. Meanwhile, certain parameters remained constant throughout the experiment, including a temperature of 55 °C, a duration of three hours, and a stirring speed of 300 rpm. The values recorded in the table align with the results gathered during the laboratory experiment.

16.4 Opinion and Analysis of Biodiesel for the Future

In order to achieve such fits, the economic value of biodiesel should be combined with the environmental benefits. By examining the dual benefits of economic viability and environmental sustainability, more light is shed on the potential transformative impact of Aidan seed-based biodiesel.

Economic value: Aidan seeds, with their nonedible nature, offer a promising avenue for biodiesel production, presenting economic advantages at various stages of the supply chain. The cultivation and processing of Aidan seeds can create employment opportunities in local communities, contributing to economic development. In addition, the biodiesel produced from Aidan seeds has the potential to reduce dependency on imported fossil fuels, fostering energy security and minimizing foreign exchange expenditures. Furthermore, the economic value extends to waste reduction. Aidan seeds can be sourced locally, reducing transportation costs and minimizing the environmental footprint associated with fuel production. The by-products of the biodiesel manufacturing process, such as glycerol, can also find applications in various industries, adding value to the overall biodiesel production chain.

Environmental value: From an environmental standpoint, biodiesel derived from Aidan seeds holds considerable promise. The growth of Aidan plants requires minimal inputs, and their cultivation can contribute to carbon sequestration, mitigating GHG emissions. In addition, Aidan seed biodiesel is known for its lower carbon footprint compared to traditional diesel, offering a potential avenue for reducing overall emissions in the transportation sector.

The utilization of Aidan seeds also aligns with principles of biodiversity conservation, as their cultivation does not compete with food crops. This reduces the pressure on arable land and promotes sustainable agricultural practices. Furthermore, Aidan seed biodiesel exhibits favorable combustion characteristics, potentially leading to lower emissions of harmful pollutants such as particulate matter and nitrogen oxides.

In summary, biodiesel produced from Aidan seeds presents a dual advantage, offering economic benefits while contributing to environmental sustainability. This section emphasizes the potential of Aidan seed-based biodiesel to create a positive impact on local economies, reduce waste, and play a role in mitigating climate change. The economic and environmental values of Aidan seed biodiesel align with the broader goal of transitioning toward cleaner, more sustainable energy sources.

Some critical issues, such as the search for cheap and ecologically friendly feedstock for biodiesel production, highly effective and reusable catalysts that are inexpensive, eco-friendly transesterification agents, and the utilization of nanoparticles in biodiesel synthesis have become very necessary in the quest for biodiesel production. With the rapid development of the automotive and chemical sectors in general, the worldwide catalyst market is expected to grow rapidly in the coming years. Vladimir Ipatieff accomplished some of the early mechanical scale responses, including the discovery and commercialization of oligomerization and the enhancement of hydrogenation catalysts [29]. A dramatic shifting market with risks and opportunities emerges from the catalyst sector as a result of environmental restrictions, increases in material prices, and denser feedstocks, which attract the attention of green chemistry as an incentive. Massive investments have been made or are being planned in emerging economies and industrialized regions to suit the market's expanding demands. Increased precious metal prices and a decrease in rare-earth supplies have prompted worldwide producers of leading catalysts to conduct extensive research and development of new techniques, technologies, and products [30]. It is critical to consider that the production of biodiesel does not interfere with the usage of farm areas for agriculture and the subsequent supply of food. The economic, social welfare, and environmental welfare should all be in balance. As a result, some governments have avoided using food crops to make biodiesel [31].

16.4.1 Challenges and Benefits of Utilizing Aidan Seeds as Feedstocks for Biodiesel Production

16.4.1.1 Challenges of Utilizing Aidan Seeds as Feedstocks

While Aidan seeds show promise as a raw material for biodiesel production, several challenges can be considered. These challenges encompass various aspects, ranging from cultivation to processing and economic feasibility:

1) *Cultivation and yield variability*: Aidan plants may exhibit variations in growth and yield based on factors such as soil conditions, climate, and agricultural practices. Inconsistent yields can pose challenges for establishing a reliable and scalable supply chain.

2) *Land use and competition*: As with any agricultural crop, the cultivation of Aidan seeds requires arable land. The competition for land use between food crops and biodiesel feedstocks can raise ethical concerns and impact global food security.
3) *Resource intensity*: Adequate resources, including water and nutrients, are essential for the successful cultivation of Aidan plants. The resource intensity of these crops may vary depending on the geographical location, potentially impacting the overall sustainability of biodiesel production.
4) *Processing technology*: The extraction of oil from Aidan seeds demands efficient and cost-effective processing technologies. The selection of appropriate extraction methods, such as solvent extraction or cold pressing, requires careful consideration to maximize oil yield and quality.
5) *Fatty acid composition*: The fatty acid composition of Aidan seed oil can influence the biodiesel's properties. High levels of unsaturated fatty acids may lead to instability and higher susceptibility to oxidation, requiring additional measures to enhance fuel stability.
6) *Economic viability*: The economic feasibility of Aidan seed biodiesel is crucial for its widespread adoption. Factors such as the cost of cultivation, extraction, and processing must be competitive with other biodiesel feedstocks and traditional diesel to ensure market competitiveness.
7) *Market acceptance*: The market acceptance of Aidan seed biodiesel may face challenges related to public perception, especially if it competes with food crops. Ensuring sustainable and ethical cultivation practices can play a role in overcoming potential resistance.
8) *Regulatory compliance*: Meeting regulatory standards and certifications for biodiesel derived from Aidan seeds is essential for market entry. Compliance with environmental, quality, and safety standards is crucial to gaining acceptance in the biodiesel market.
9) *Scaling up production*: Transitioning from small-scale production to large-scale commercial production poses logistical challenges. Scaling up cultivation and processing operations while maintaining quality standards requires strategic planning and investment.
10) *Global market dynamics*: Fluctuations in global markets, including the biodiesel market, can impact the economic viability of Aidan seed biodiesel. External factors such as changes in oil prices and competing technologies may influence the market dynamics. Addressing these challenges requires collaborative efforts from researchers, policymakers, and industry stakeholders to develop sustainable cultivation practices, efficient processing technologies, and market strategies for Aidan seed-based biodiesel.

16.4.1.2 Benefits of Utilizing Aidan Seeds as Feedstocks for Biodiesel Production

Utilizing Aidan seeds as feedstocks for biodiesel production offers several benefits, making them an attractive option in the quest for sustainable and renewable energy sources:

1) *Nonedible source*: Aidan seeds are nonedible, meaning they do not compete with food crops for agricultural resources. This characteristic address concerns related to food security and ethical land use, contributing to sustainable practices.
2) *Reduced environmental impact*: Aidan plants typically require minimal inputs in terms of water and nutrients for cultivation. Their ability to adapt to varying environmental conditions makes them a resource-efficient and environmentally friendly biodiesel feedstock.
3) *Carbon sequestration*: Aidan plants contribute to carbon sequestration during their growth phase, helping mitigate GHG emissions. The process of photosynthesis absorbs CO_2, offsetting the carbon footprint associated with biodiesel combustion.
4) *Biodiesel quality*: Aidan seed oil exhibits favorable fatty acid composition, providing good-quality biodiesel. The oil's characteristics contribute to biodiesel with desirable properties such as low viscosity, good cold flow properties, and potential for efficient combustion.
5) *Waste utilization*: By-products generated during Aidan seed biodiesel production, such as glycerol, can find applications in various industries. This adds value to the overall biodiesel production process by minimizing waste and utilizing by-products.
6) *Local economic development*: The cultivation and processing of Aidan seeds for biodiesel production can stimulate local economies by creating job opportunities and supporting the development of a regional biodiesel industry.
7) *Energy security*: Biodiesel produced from Aidan seeds contributes to diversifying the energy sources used in the transportation sector. This diversification enhances energy security by reducing reliance on traditional fossil fuels, which are subject to geopolitical and market fluctuations.
8) *Reduced dependence on imported fuels*: Cultivating and processing Aidan seeds locally for biodiesel production can reduce dependency on imported fossil fuels. This shift contributes to energy independence and mitigates the economic impact of fluctuating global oil prices.
9) *Biodiversity conservation*: Aidan seeds can be cultivated without encroaching on arable land used for food crops, contributing to biodiversity conservation. This reduces the pressure on ecosystems and promotes sustainable land use practices.
10) *Potential for rural development*: The cultivation of Aidan seeds for biodiesel production can be well-suited for small-scale and rural agricultural settings. This has the potential to uplift rural communities by providing alternative income sources and promoting sustainable agricultural practices.

11) *Compliance with sustainability goals*: Biodiesel from Aidan seeds aligns with broader sustainability goals, meeting criteria related to reduced environmental impact, social responsibility, and economic viability. This aligns with global initiatives aimed at transitioning to cleaner and more sustainable energy sources.

In summary, the utilization of Aidan seeds as feedstocks for biodiesel production presents a compelling case, offering a range of environmental, economic, and social benefits. These advantages contribute to the overall sustainability and viability of Aidan seed-based biodiesel in the renewable energy setting.

16.5 Conclusion

This study focused on the production of biodiesel using unique Aldin seeds as feedstock. It underscores the pivotal role that biofuels, particularly biodiesel, plays in alleviating the dependence on conventional fossil fuels and fostering environmental sustainability in industries, transportation, and other sectors. Biodiesel stands out as a compelling alternative to traditional diesel, given its cost-effectiveness, ease of production, and renewable nature. The unique attribute as a carbon-neutral fuel, wherein the CO_2 emitted during combustion is offset by plant absorption during growth, positions biodiesel as an environmentally beneficial solution. While significant strides have been made in biodiesel research, encompassing advancements in production technologies and its expanding utility across various industries, certain challenges persist. Notably, the high-production costs and the complex process of commercialization pose hurdles that have, to some extent, impeded widespread biodiesel adoption. Addressing these challenges requires concerted efforts from researchers, policymakers, and industry stakeholders to further enhance the economic viability and accessibility of biodiesel, ensuring its continued trajectory as a sustainable and transformative energy solution. Navigating the path toward a greener future, continued innovation and strategic initiatives are imperative to unlock the full potential of biodiesel and solidify its place as a cornerstone in the transition to cleaner and more sustainable transportation fuels.

References

1 Kumar, S., Singh, P., Shrivastava, V. et al. (2023). Potential non-edible oil seeds for biodiesel production: an Indian context. *Mater. Today Proc.*

2 Okonkwo, C.P., Ajiwe, V.I.E., Obiadi, M.C., and Okwu, M. (2021). A systematic review of biodiesel production with sustainable feedstock using assorted catalyst. *Am. J. Appl. Chem.* 9 (5): 154–163.

3 Oyetola, O. and Noor, A.A. (2019). A review of global current scenario of biodiesel adoption and combustion in vehicular diesel engines. *Energy Rep.* 5: 1560–1579.

4 Tayari, S., Abedi, R., and Rahi, A. (2020). Comparative assessment of engine performance and emissions fueled with three different biodiesel generations. *Renew. Energy* 147: 1058–1069.

5 Otori, A.A., Mann, A., Suleiman, M.A.T., and Egwim, E.C. (2019). Synthesis of heterogeneous catalyst from waste snail shells for biodiesel production using *Afzelia africana* seed oil. *Niger. J. Chem. Res.* 23: 35–51.

6 Giakoumis, E.G. and Sarakatsanis, C. (2019). A comparative assessment of biodiesel cetane number predictive correlations based on fatty acid composition. *Energies* 12: 422.

7 Shemelis, N.G. and Jorge, M.M. (2017). Review biodiesel production technologies: review. *AIMS Energy* 5: 425–457.

8 Sivanesan, S. (2009). Kinetic study of the esterification of free fatty acids in non-edible *Pongamia pinnata* oil using acid catalyst. *Indian J. Sci. Technol.* 2 (12): 20–24.

9 Akin-Idowu, P.E., Ibitoye, D.O., Ademoyegun, O.T., and Adeniyi, O.T. (2011). Chemical composition of the dry fruit of *Tetrapleura tetraptera* and its potential impact on human health. *J Herbs Spices Med Plants* 17 (1): 52–61.

10 Atawodi, S.E., Yakubu, O.E., Liman, M.L., and Iliemene, D.U. (2014). Effect of methanolic extract of *Tetrapleura tetraptera* (Schum and Thonn) Taub leaves on hyperglycemia and indices of diabetic complications in alloxaninduced diabetic rats. *Asian Pac. J. Trop. Biomed.* 4: 272–278.

11 Abii, T.A. and Amarachi, E. (2007). Investigation into the chemical composition of the dry fruit of *Tetrapleura tetraptera* (Ubukirihu). *J. Food Technol.* 5 (3): 229–232.

12 Berthiaume, D. and Tremblay, A. (2006): Study of the Rancimat test method in measuring the oxidation stability of biodiesel ester and blends NRCan project No.CO414CET-327O, LEOTEK Inc., Quebec, Canada. https://www.osti.gov/etdeweb/biblio/20977973.

13 Thomas, A. (2000). Fats and fatty oils. In: *Ullmann's Encyclopedia of Industrial Chemistry*. Weinheim: Wiley VCH. https://doi.org/10.1002/14356007.a10_173.

14 Istvan, B. and Ioan-Adrian, T. (2011). Biodiesel quality, standards and properties. In: *Biodiesel-Quality, Emissions and By-products* (ed. G. Monterro), 978-953-307-7840.

15 Vargončík, M. (2007). Waste vegetable oil as a diesel replacement. *Environ. Sci. Eng.* https://www.semanticscholar.org/paper/Waste-vegetable-oil-as-a-diesel-replacement-fuel Vargon%C4%8D%C3%ADk/9166083d42eff4035aebdfbef0d85f9dc26cad1e.

16 Ezekwe, C.C. and Ajiwe, V.I.E. (2011). The effects of biodiesel's composition on its viscosity. *Anachem* 5 (1): 1016–1028.

17 Knothe, G. and Steidley, K.R. (2005). Kinematic viscosity of biodiesel fuel components and related compounds. Influence of compound structure and comparison to petrol diesel fuel component. *Fuel* 84 (9): 1059–1065.

18 L. Cozzi, T. Gould, S. Bouckart, D. Crow, T.Y. Kim, C. Mcglade, P. Olejarnik, B. Wanner, D. Wetzel, World energy outlook 2020, IEA Paris, France (2020), 1–461.

19 Logesh, K., Karthick, M., Baskar, S., and Sharma, A. (2022). Performance and emissions characteristics of diesel engine run on *Citrullus colocynthis* biodiesel with zinc oxide additive. In: *Advanced Materials for Sustainable Energy and Engineering*, vol. II, 513–520. Springer.

20 Siddartha, G.N.V., Ramakrishna, C.S., Kujur, P.K. et al. (2022). Effect of fuel additives on internal combustion engine performance and emissions. *Mater. Today Proc.*

21 Aron, N.S.M., Chew, K.W., Ang, W.L. et al. (2022). Recovery of microalgae biodiesel using liquid biphasic flotation system. *Fuel* 317: 123368.

22 Sharma, A., Gupta, G., and Agrawal, A. (2020). Utilization of waste lubricating oil as a diesel engine fuel. *IOP Conf. Ser.: Mater. Sci. Eng.* 840: https://doi.org/10.1088/1757-899X/840/1/012015.

23 Sharma, A. and Murugan, S. (2017). Durability analysis of a single cylinder DI diesel engine operating with a non-petroleum fuel. *Fuel* 191: 393–402. https://doi.org/10.1016/j.fuel.2016.11.086.

24 Sharma, A. and Murugan, S. (2017). Effect of blending waste tyre derived fuel on oxidation stability of biodiesel and performance and emission studies of a diesel engine. *Appl. Therm. Eng.* 118: 365–374. https://doi.org/10.1016/j.applthermaleng.2017.03.008.

25 Sharma, A. and Murugan, S. (2015). Combustion, performance and emission characteristics of a di diesel engine fuelled with non-petroleum fuel: a study on the role of fuel injection timing. *J. Energy Inst.* 88: 364–375. https://doi.org/10.1016/j.joei.2014.11.006.

26 Sharma, A. and Murugan, S. (2014). Influence of fuel injection timing on the performance and emission characteristics of a diesel engine fueled with Jatropha methyl ester-tyre pyrolysis oil blend. *Appl. Mech. Mater.* 592–594: 1627–1631. https://doi.org/10.4028/www.scientific.net/AMM.592-594.1627.

27 Kumar, S.S., Rajan, K., Mohanavel, V. et al. (2021). Combustion, performance, and emission behaviors of biodiesel fueled diesel engine with the impact of alumina nanoparticle as an additive. *Sustainability* 13 (21): 12103.

28 Devaraj, A., Nagappan, M., Yogaraj, D. et al. (2022). Influence of nano-additives on engine behaviour using diesel-biodiesel blend. *Mater. Today Proc.*

29 Khatri, D., Goyal, R., and Sharma, A. (2021). Effects of silicon dioxide nanoparticles on the combustion features of diesel engine using water diesel emulsified fuel. In: *Energy Systems and Nanotechnology*, 119–130. Springer.

30 Nicholas, C.P. (2018). Dehydration, dienes, high octane, and high pressures: contributions from Vladimir Nikolaevich Ipatieff, a father of catalysis. *ACS Catal.* 8: 8531–8539.

31 Knothe, G., Krahl, J., and Van, G. (2015). *The Biodiesel Handbook*. Salt Lake City: Elsevier.

17

Integrating Response Surface Methodology (RSM) and Grey Wolf Optimization (GWO) for Sustainable Biodiesel Production from Tobacco Seeds

17.1 Introduction

In the quest for sustainable and renewable energy sources, the utilization of nonfood crops for biodiesel production has gained significant attention. One such promising feedstock is tobacco seed oil, derived from the seeds of the tobacco plant (*Nicotiana tabacum*). Historically cultivated for tobacco production, the versatile nature of the tobacco plant extends beyond its traditional use. This introduction explores the potential of tobacco seed oil as a viable feedstock for biodiesel production, considering its composition, cultivation characteristics, and the imperative need to diversify feedstock options in the bioenergy sector. The conventional biodiesel industry primarily relies on feedstocks such as soybean, rapeseed, and palm oil. However, the increasing global demand for biodiesel, coupled with concerns about food crop-based feedstocks, necessitates exploration into alternative sources. Tobacco, a nonfood crop, emerges as a prospective candidate due to its high oil content in its seeds and the possibility of repurposing tobacco cultivation for sustainable energy production.

Environmentally friendly biodiesel took the place of imported oils, and depending on crop and oil prices, it may eventually replace diesel [1]. With its leaves used to make cigarettes, tobacco is the most profitable and important nonfood agricultural crop in the world. Thousands of families rely on the crop for their financial well-being, even though it has been demonstrated that cigarettes are harmful to human health [2]. As an economical option to petroleum-based diesel fuel, biodiesel derived from vegetable oil is available. It is biodegradable, nontoxic, sulfur-free, and least polluting. It also has an environmentally beneficial profile. Gasoline that has been mixed with biodiesel has a higher cetane rating and is more lubricious, which may result in easier starting and less waiting for the spark. It can be made with vegetable oil or animal fat, both of which can be recycled from food waste products from

Feedstocks for Sustainable Biodiesel Production: Characterization, Selection, and Optimization, First Edition. Chinwe P. Okonkwo and Modestus O. Okwu.

processing facilities that handle hogs, cattle, and poultry because the new technologies have the potential to increase the fuel efficiency of CI engines and reduce emissions [3–5]. Since B100 biodiesel can provide your engine with 100–200 more horsepower than diesel, the National Biodiesel Board (NBB) promotes the use of biodiesel as a fuel source on behalf of the organizations it represents.

Currently, the biodiesel industry in the United States may produce up to two billion gallons of biodiesel annually. However, the fact that a straight B100 tends to be dirty when it is idling is frequently not a problem because spectators at pulling competitions and drag races enjoy seeing a lot of black smoke [6–8]. Environmental concerns have led to an increase in the cost of conventional fossil fuels, which has prompted research into alternative fuel sources for automobiles. It has been demonstrated that taking preventative measures to reduce the usage of fossil fuels can reduce pollution and save developing countries money [8]. Tobacco seeds are a valuable oil resource, using inedible oils for biodiesel production [9–12]. In general, tobacco, or *N. tabacum*, is an annual plant in the Solanaceae family whose leaves are plucked and used to make cigarettes. It is widely grown in China, North America, and South America. China alone contributes more than 40% of the world's tobacco production, which is grown on almost 3.4 million ha of field space across about 130 nations [13–15]. Tobacco seeds typically have an oil content of 30–43 wt. %, with unsaturated fatty acids making up the majority of the oil [16]. The main fatty acids found in tobacco seed oil are, in order, linoleic acid (64.38%–79.0%), oleic acid (9.97%–13.49%), palmitic acid (8.72%–12.3%), and stearic acid (2.1%–3.3%) [12].

17.1.1 Oil Composition of Tobacco Seeds

Tobacco seeds, renowned for their significant oil content, typically range between 30% and 40% by weight [17]. The composition of tobacco seed oil is characterized by a notable percentage of fatty acids, making it a valuable and versatile source for various applications. Among the predominant fatty acids found in tobacco seed oil, oleic acid, and linoleic acid stand out. Oleic acid, a monounsaturated omega-9 fatty acid, constitutes a substantial portion of the fatty acid profile in tobacco seed oil. This acid is recognized for its stability and resistance to oxidation, contributing to the overall quality of the oil. In addition, oleic acid's favorable properties, such as its low freezing point and high boiling point, enhance the suitability of tobacco seed oil as a feedstock for biodiesel production.

Linoleic acid, on the other hand, is a polyunsaturated omega-6 fatty acid that plays a crucial role in the composition of tobacco seed oil. Its presence contributes to the fluidity and stability of the oil, especially under varying temperature conditions. Linoleic acid is particularly valuable in the context of biodiesel production, as it enhances the cold-flow properties of the fuel, preventing crystallization and ensuring optimal performance in diverse climates. The combination of these fatty acids in tobacco seed oil creates a favorable fatty acid profile, positioning it as an excellent raw material for biodiesel production. The desirable properties of oleic acid and linoleic acid, including their stability, resistance to oxidation, and impact on the cold-flow characteristics of the resulting biodiesel, underscore the potential of tobacco seed oil in the sustainable production of biofuels. As the demand for renewable and environmentally friendly energy sources continues to grow, the composition of tobacco seed oil makes it a promising candidate for contributing to the advancement of biodiesel technology and the broader field of sustainable energy.

17.1.2 Cultivation Characteristics

Tobacco, historically cultivated in various regions across the globe primarily for the production of tobacco leaves, exhibits remarkable adaptability to diverse climates and soil conditions. This inherent adaptability positions tobacco as a versatile and resilient crop, capable of thriving in a wide range of environmental settings. Leveraging the well-established infrastructure already in place for tobacco cultivation presents a unique opportunity for the extraction of oil from tobacco seeds. This dual-purpose approach, where tobacco serves both as a source of leaves for traditional use and as a feedstock for biodiesel production, holds the potential to create an economically viable and sustainable pathway for biofuel production.

The existing tobacco cultivation infrastructure, including farms, processing facilities, and distribution networks, can be repurposed to incorporate oil extraction from tobacco seeds seamlessly. This repurposing not only optimizes the use of available resources but also minimizes the need for significant changes in agricultural practices, thereby reducing the overall environmental impact. By tapping into the established knowledge base and practices associated with tobacco cultivation, this integrated approach offers a pragmatic solution for diversifying the agricultural output associated with tobacco farming. The strategic alignment of this dual-purpose cultivation approach with the principles of sustainable agriculture is particularly noteworthy. Sustainable agriculture emphasizes the harmonious coexistence of economic viability, environmental responsibility, and social equity. Repurposing tobacco cultivation for oil extraction not only addresses the economic aspect by providing an additional revenue stream for farmers but also aligns with environmental sustainability goals by promoting the utilization of a by-product for biofuel production.

Furthermore, this innovative approach holds the potential to contribute significantly to the revitalization of economies that are historically dependent on tobacco cultivation. By diversifying the use of tobacco crops, communities and regions that have traditionally relied on tobacco production can transition to a more sustainable and economically viable model. This diversification is in line with broader efforts to promote resilient and adaptable agricultural practices that can withstand changing market dynamics and global economic shifts.

In summary, the cultivation characteristics of tobacco, coupled with the strategic repurposing of existing infrastructure, present an opportunity to establish an economically viable and sustainable pathway for biodiesel production. This integrated approach not only leverages the adaptability of tobacco to diverse climates but also aligns with the principles of sustainable agriculture, offering a potential solution for the revitalization of tobacco-dependent economies.

17.1.3 Rationale for Exploration

The exploration of tobacco seed oil for biodiesel production addresses multiple critical issues. First, it offers an alternative to conventional feedstocks, reducing the competition for land and resources with food production. Second, repurposing tobacco cultivation for biodiesel production aligns with the global movement toward sustainable practices, facilitating a transition from traditional tobacco farming. Third, biodiesel derived from tobacco seed oil has the potential to meet stringent environmental standards, contributing to the reduction of greenhouse gas emissions and fostering a more sustainable energy landscape.

17.1.4 Objectives of the Study

This study aims to comprehensively investigate the feasibility of utilizing tobacco seed oil as a biodiesel feedstock. Key objectives include assessing the oil extraction process, analyzing the fatty acid profile, evaluating the biodiesel production potential and performing parametric studies using response surface methodology (RSM) and grey wolf optimization (GWO), and finally, considering the economic and environmental implications. By addressing these objectives, this research seeks to contribute valuable insights to the growing discourse on sustainable bioenergy alternatives and broaden the scope of feedstock options in the biodiesel industry. The synthesis of biodiesel derived from tobacco is a highly promising form of second-generation biodiesel. The current study aims to clarify the process of producing biodiesel from tobacco seed oil by analyzing and contrasting the responses surface methodology and GWO.

17.2 Review of Relevant Literature

17.2.1 Features of Tobacco Plant

The tobacco plant's leaves and seeds hold commercial significance. While the tobacco industry primarily utilizes leaves for cigarette manufacturing, there is a growing emphasis on alternative uses of tobacco by-products [18]. Researchers suggest exploring other applications of tobacco, such as biofuel, possibly driven by increased awareness of the health risks associated with cigarette smoking [19]. Notably, tobacco seeds, a by-product devoid of nicotine, can be employed in the production of tobacco-based biodiesel. This alternative has the potential to mitigate health concerns related to cigarette smoking and contribute to socio-economic equilibrium in Africa and globally [20] (Figures 17.1 and 17.2).

Worldwide cultivation of the tobacco plant (*Nicotiana tabacum* L.) is a significant economic crop [21]. Typically, its leaves are processed to make a variety of tobacco products, including gutka, bidis, snuffs, cigars, and hookah. However, tobacco seeds are discarded debris from the growing of commercial leaves. Many healthy compounds, including triacylglycerols, phospholipids,

Figure 17.1 Tobacco leaves (*Sources:* Derek Ramsey/Wikimedia Commons/CC BY-SA 4.0 and Kgroten/Wikimedia Commons/CC BY-SA 3.0).

Figure 17.2 Leaves and seeds of tobacco seed oil (*Source:* Dmytro Synelnychenko/Adobe Stock and Courtesy of NWT Seeds).

tocopherols, sterols, and unsaturated fatty acids, are found in tobacco seed oil [22]. Furthermore, linoleic acid, oleic acid, palmitic acid, and stearic acid are the main unsaturated fatty acids found in tobacco seed oils [23]. It may be possible to treat heart and brain disorders with unsaturated fatty acids [24]. A valuable source of dietary oil, tobacco seeds are also nutrient-rich, particularly in polyunsaturated fatty acids. Therefore, more research on the chemical makeup and biological activity of tobacco seed oils for commercial application is desperately needed [8].

17.2.2 Assessing the Oil Extraction Process

The tobacco seed oil extraction process includes the following: Seed preparation, oil Extraction, oil yield measurement, quality analysis, and process optimization.

1) *Seed preparation*: The process involved in the seed preparation includes cleaning and sorting; drying. Cleaning and sorting involve removing impurities and sorting seeds to ensure uniformity. The drying process is a way of reducing moisture content.
2) *Oil extraction methods*: The methods involved in oil extraction include solvent extraction, cold pressing, and supercritical fluid extraction. Solvent extraction involves using organic solvents to dissolve oil from seeds. Cold pressing has to do with the application of mechanical pressure to extract oil without heat and the supercritical fluid extraction has to do with the use of supercritical carbon dioxide for high efficiency.
3) *Oil yield measurement*: To measure tobacco seed oil yield requires the determination of the quantity of oil extracted per unit of seeds and the calculation of oil yield percentage to assess process efficiency.
4) *Quality analysis*: To measure tobacco seed oil quality requires the evaluation of oil quality for factors like color, odor, and impurities. Then, the performance tests such as acid value, peroxide value, and moisture content.
5) *Process optimization*: The optimization process of Tobacco seed oil include: Adjusting parameters like temperature, pressure, and extraction time for optimal yield and consideration of economic and environmental factors in process design.

17.2.3 Analyzing the Fatty Acid Profile for Tobacco Seed Oil

Analyzing the fatty acid profile in biodiesel production from tobacco seed oil involves several steps to determine the composition of fatty acids, which is crucial for understanding the properties of the resulting biodiesel. Below is a detailed explanation of each process involved: Extraction of lipids, fatty acid methyl ester (FAME) preparation, gas chromatography (GC) analysis, fatty acid identification and quantification, quality parameter analysis. The first step which involves lipid extraction require from tobacco seeds involve using a suitable solvent such as hexane, complete extraction is ensured to obtain a representative sample for fatty acid analysis. Then, the FAMEpreparation involves transesterification reaction, which has to do with transesterifying the extracted lipids to convert triglycerides into FAME, this can be achieved by using methanol or ethanol in the presence of a catalyst such as sodium hydroxide or potassium hydroxide. It is equally important to note the reaction conditions focusing on control reaction temperature and time to ensure complete transesterification then the resulting FAME is neutralized and washed to remove impurities and catalyst residues. The GCanalysis involves the instrumentation process such as utilizing GC, equipped with a flame ionization detector (FID) or mass spectrometry (MS), proper calibration is ensured using standard FAME mixtures. The next process involves sample injection by injecting a prepared FAME sample into the GC column using a split or splitless injection method. After this, FAMEs are separated based on their boiling points and molecular weights as they travel through the GC column. Individual FAMEs are then detected and quantified as they exit the column, and a chromatogram representing the fatty acid profile is generated. To identify and quantify fatty acid, retention time analysis is required which can be achieved by identifying FAMEs based on their retention times, and comparing the result obtained to known standards. Internal and external standards are used for accurate identification. Quantifying each fatty acid as a percentage of the total FAME peak area becomes important at this point, this is possible by calculating the composition of individual fatty acids in the sample. Then, the quality parameter analysis is performed based on *cetane number*, which indicates the ignition quality of biodiesel; *kinematic viscosity* to assess the fluidity of the biodiesel, *oxidative stability evaluation* which indicates resistance to oxidation during storage. The next process is the data interpretation and reporting. The interpretation aspect involves analyzing the fatty acid profile to understand the composition and properties of the biodiesel and the assessment of the balance of saturated and unsaturated fatty acids. Then, the reporting has to do with presenting the results based on a comprehensive statement, detailing the composition and quality parameters of the biodiesel. *The results are* compared with biodiesel standards. The production process is optimized based on the analyzed fatty acid profile, and quality control measures are implemented to ensure consistency and compliance with biodiesel standards.By following the aforementioned steps, detailed information about the fatty acid profile of biodiesel produced from tobacco seed oil can be obtained, enabling informed decisions about the optimization of the production process and ensuring the quality of the final product.

17.2.4 Grey Wolf Optimization Algorithm

A meta-heuristic optimization technique called the GWO was developed after studying the ways in which GW hunt. The construction optimization assignment was completed by coupling and coding the optimization algorithm. After investigating how algorithm settings affected GWO's optimization performance, it was discovered that GWO might offer the greatest results. In the future, biodiesel will be a key component of energy investments and will have a significant impact on the growth of the industrial and local auto industries.

A brand-new global optimization meta-heuristic algorithm, GWO is designed to mimic the natural hunting strategy and leadership structure of GW. Alpha, beta, delta, and omega are the four varieties of GW that are used to model the leadership hierarchy. Furthermore, to carry out optimization, three primary phases – hunting, seeking prey, encircling prey, and attacking prey are used [25]. A number of optimization algorithms that were previously used in BOPs have recently been proposed in the literature and have demonstrated better performance than the state-of-the-art techniques. But since these cutting-edge optimization algorithms have never been used in BOPs, it is unknown how well they perform. Therefore, it is necessary to look into how new optimization techniques can be used for BOPs. The foraging social behavior of GW served as the model for the new meta-heuristic optimization technique known as the GWO algorithm.

17.2.4.1 Exploration of GWO Algorithm

Mirjalili et al. [26] first proposed the GWO in 2014. Faris et al. [27] recently examined GWO scientific uses. GWO had showed promising outcomes in a wide range of optimization challenges, they claimed. The outstanding properties of GWO above other swarm intelligence approaches may account for its high degree of effectiveness in dealing with optimization challenges in the literature. This review emphasizes that GWO requires no derivation information on the search space and has only a few parameters. Furthermore, GWO is scalable, adaptable, simple to use, and straightforward. Through the search process, the algorithm benefits from a mix of exploration and exploitation, resulting in excellent convergence. As a result, the GWO has piqued the interest of scientists and engineers from a variety of disciplines. Since 2014, the GWO algorithm has been widely employed in the optimization of a wide range of scientific fields. Mirjalili et al. [28] have also evaluated the use of GWO in computational fluid dynamics experiments. The GWO and its modified versions have been used in a variety of engineering applications, including feature selection methods for classification, path planning, groundwater remediation, improving biodiesel production, machining of multiwall carbon nanotube/polymer nanocomposites, wireless sensor network coverage optimization [29–36], optimization of construction duration and schedule robustness [37], prediction of soil electrical conductivity [38], and a favorable reservoir of oil. As can be seen from the literature review, GWO has had great success in a variety of engineering domains.

17.2.5 Response Surface Methodology

RSM is a statistical and mathematical technique used for designing experiments, analyzing data, and optimizing processes. Developed primarily for applications in the field of experimental design and response optimization, RSM plays a crucial role in understanding the relationship between multiple input variables and the corresponding output responses. The fundamental objective of RSM is to model the complex relationships between input variables and responses, allowing for the prediction and optimization of these responses within a specified range of input values. This methodology is particularly valuable in situations where experiments involve multiple factors, and the aim is to identify the optimal combination of these factors to achieve desired outcomes.

17.2.5.1 Key Components of Response Surface Methodology

Experimental design: RSM typically involves the design of carefully planned experiments where various factors are manipulated systematically to observe their impact on the response variable of interest. These experiments are designed to efficiently gather relevant data for constructing response surface models.

Mathematical modeling: Response surfaces are mathematical models that represent the relationship between input variables and the response. These models are often polynomial equations that capture the curvature and interactions among factors.

Analysis of variance (ANOVA): ANOVA is employed to analyze the statistical significance of the model and the individual factors. It helps identify which factors significantly influence the response and which can be considered negligible.

Optimization: Once the response surface model is validated, optimization techniques are applied to find the optimal set of input variables that yield the desired response. This enables researchers and practitioners to identify the most efficient and effective conditions for a given process.

Graphical representation: Visualization tools, such as contour plots and three-dimensional response surface plots, are commonly used in RSM to provide a graphical representation of the relationships between variables, aiding in the interpretation and communication of results.

The purpose of this research is to understudy the tobacco plant as a feedstock and perform experimental analysis to predict the effectiveness of GWO and RSM for biodiesel production using dataset obtained from tobacco seed.

17.3 Materials and Method

17.3.1 Experimental Material and Equipment

In this study, 20 experimental runs were obtained from a study conducted elsewhere [17] and used as the experimental runs matrix. Each experimental run comprising the reaction time (min), methanol to oil mole ratio and catalyst (%) was used as input parameters, while the tobacco seed oil methyl esther (TSOME) Yield (%) as the output parameters respectively as presented in Table 17.2. Twenty experimental runs were performed to obtain the experimental values presented in Table 17.2. Prediction was performed based on the experimental value in the form of TSOME Yield (%) using the GW meta heuristic algorithm and RSM. The input variables include the reaction time (min), methanol to oil mole ratio, and catalyst (%).

17.3.2 GWO Process Modeling

GWO is one of the innovative algorithms which is also a population-based swarm optimization technique. The biomimicry approach was used for the formulation of the algorithm, with the motivation from the hunting pattern of GW. It is also imperative to take note of the GW's rock-hard pyramid architecture as shown in Figure 17.3, this has been implemented in previous studies [26]. In GW system mimicking, it is worth noting that the Deltas equally known as the wolves' park are very subservient and devoted to the parks of the foremost dual wolves',"Alphas and Betas," the omegas tend to be dominated. They, typically work as a watchman, detective, curator, predator, etc. [40].

Figure 17.3 is a flowchart showing the GWO technique. The first phase in the optimization process, as depicted in the flowchart, represents the GW random initialization. A choice is made based on the parameter of the stochastic input GWO, and the membership values of the stochastic input variables are literally mapped using the relevant algorithm for the disordered number and process variable at the initialization stage [41]. The fitness of all GW adjusted is calculated using standard index functions which is sorted by fitness. After the arrangement, the first wolf is obtained which is assumed to be Alpha, while another wolf which is the second is assumed to be assumed to be Beta and the third wolves is the Delta. Following the progression, the process of evaluation fitness is carried out to determine the wolves which appear as the top three. There is the repetition of the sequence, with an update on the disarrayed map and all the GW positions. Having completed the process, a fitness assessment follows, after the assessment is the checking best and worst fitting gray, the worst is replaced with the best. Then, the subsequent stage is to make a decision on the possible criteria that must be met so as to make a decision on the output solution that appears to be the best.

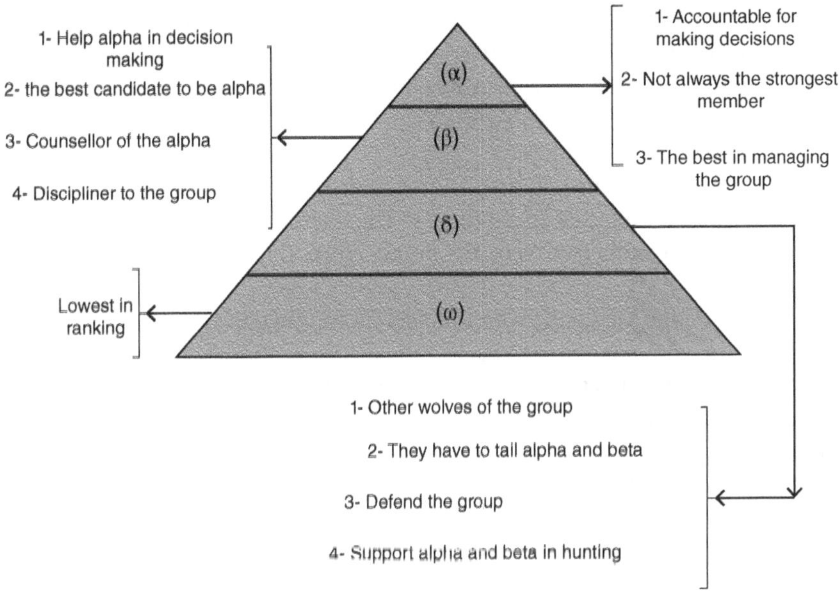

Figure 17.3 Grey wolf hierarchy (*Source:* Adapted from Yk and Llsb [39]).

17.3.2.1 Essential Concept of Gray Wolf Optimizer (GWO)

The inspiration for the GWO method is the GW leadership and hunting mechanisms. The main steps in GW hunting are as follows: Hunting for prey; trailing, hurtling, and approaching the prey; pursuing and encircling the prey until the GW gets full hold of it; finally, is attacking the prey [26].

In summary, the steps are as follows: search, pursue, encircle, and attack. The fitness solution Alpha is translated into a mathematical model. The second and third fittest solutions are beta and delta wolves, respectively. Omega represents the trail of the three wolves. The following mathematical equation is used to model encircling the prey:

$$\vec{D} = \left|\vec{C} . \vec{X}_p - \vec{X}(t)\right| \tag{17.1}$$

$$\vec{X}(t+1) = \left|\vec{X}_p(t) - \vec{A} . \vec{D}\right| \tag{17.2}$$

$\vec{A}$, $\vec{C}$ which represent the coefficient of vectors are calculated as:

$$\vec{A} = 2 . \vec{a} . \vec{r_1} - \vec{a}; \tag{17.3}$$

$$\vec{C} = 2 . \vec{r_2}; \tag{17.4}$$

$\vec{r}_1, \vec{r}_2$ = random vectors range [0,1]; $\vec{X}$ represent the position of the GW; *t* represents the existing iterations; $\overrightarrow{X_p}$ is theprey location; $\vec{A}$, $\vec{C}$ represent the coefficient vectors.

It is important to note that $\vec{a}$ decreases linearly from 2 to 0 during the course of the iterations. The formulated equations can be used to update the position of the GW based on the prey position.

By adjusting the values of $\vec{A}$ and $\vec{C}$, around the best search agents, several positions can be obtained in relation to the current position. The wolf can easily move between two specific points with the help of random vectors $\vec{r}_1$ and $\vec{r}_2$.

To further define the mathematical model for hunting, the process is described as follows:

Generally, Alpha directs the hunting process. This procedure is such that Delta, Beta, and Alpha are positioned for prompt location of the prey in terms of optimal solution. The position of other wolves can be adjusted based on the location of Delta, Beta, and Alpha [26].

$$\overrightarrow{D_\alpha} = \left|\overrightarrow{C_1}.\overrightarrow{X_\alpha} - \vec{X}(t)\right| \tag{17.5}$$

$$\overrightarrow{D_\beta} = \left|\overrightarrow{C_2}.\overrightarrow{X_\beta} - \vec{X}(t)\right| \tag{17.6}$$

$$\overrightarrow{D_\delta} = \left|\overrightarrow{C_3}.\overrightarrow{X_\delta} - \vec{X}(t)\right| \tag{17.7}$$

$$\overrightarrow{X_1} = \left|\overrightarrow{X_\alpha} - \overrightarrow{A_1}.\overrightarrow{D_\alpha}\right| \tag{17.8}$$

$$\overrightarrow{X_2} = \left|\overrightarrow{X_\beta} - \overrightarrow{A_2}.\overrightarrow{D_\beta}\right| \tag{17.9}$$

$$\overrightarrow{X_3} = \left|\overrightarrow{X_\delta} - \overrightarrow{A_3}.\overrightarrow{D_\delta}\right| \tag{17.10}$$

Updating the position of the GW will be possible using the equation presented in Eq. (17.11):

$$\vec{X}(t+1) = \frac{\overrightarrow{X_1} + \overrightarrow{X_2} + \overrightarrow{X_3}}{3} \tag{17.11}$$

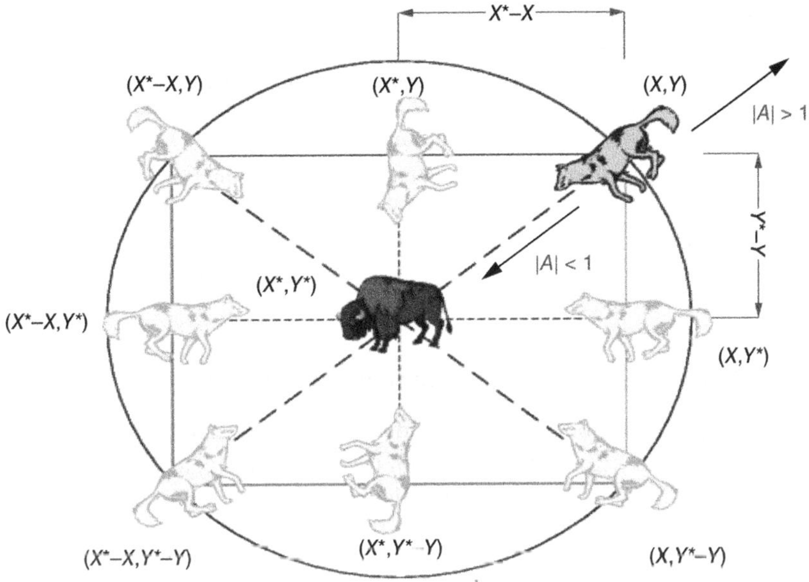

Figure 17.4 Grey wolf positioning for search agent (*Source:* Mirjalili et al. [26]/with permission of Elsevier).

$\overrightarrow{X_{\propto}}$ is the position vector of $\propto, \overrightarrow{X_{\beta}}$ is the position vector of $\beta, \overrightarrow{X_{\delta}}$ is the position vector of δ.

The coefficient vectors are presented as $\overrightarrow{A_1}, \overrightarrow{A_2}$, and $\overrightarrow{A_3}$. To formulate the mathematical model that enables the GW attack they prey, the description is as follows: Wolf attack the prey when it stops moving so as to complete the hunting process. This is represented mathematically by the reduction of $\vec{a}$ from 2 to 0 in the process of iterations. $\vec{A}$ decreases as a result of decrease in $\vec{a}$. The wolf is forced to launch an attach targeted toward the prey when the value of $A < 1$. Conversely, if $|A| > 1$, the wolf will seek out better prey. This is presented in Figure 17.3 [26]. A close observation of the $\vec{C}$ vectors shows arbitrary values in the range [0, 2]. The $\vec{C}$ vectors improve the weighting strength of the prey, making it quite difficult for the wolves to trace.

It is also important to note that when $C > 1$, there is high priority or relevance given to the prey increases, while when $C = 1$, the priority given to the prey decreases. A chat representing the positioning of the search agent is presented in Figure 17.4.

The terms used in GWO are significant in biodiesel parametric studies. The equivalent significant terminologies are further presented in Table 17.1.

Table 17.1 Grey wolf optimization terminology for biodiesel prediction.

GWO terms	TiO_2 nano-lubricant machining
Wolves position	Input parameters of biodiesel
Wolves searching space for prey.	Minimum and maximum limits in terms of input values required to select the optimal results.
Position of prey's	Final optimal sets of predicted value
Alpha wolf	The best set of input parameters

17.3.3 Response Surface Methodology Process Modeling

By following the methodology, this study aims to systematically investigate the influence of reaction time, methanol to oil mole ratio, and catalyst concentration on TSOME Yield in the biodiesel production process. The application of RSM will facilitate the optimization of these parameters to achieve enhanced biodiesel yield.

1) *Experimental design*: AnRSM to optimize the biodiesel production process utilized a central composite design (CCD) with three factors at three levels for each input parameter.
2) *Input parameters*:

   Factor 1: Reaction Time (min): The reaction time was varied within a specified range to observe its impact on TSOME Yield using the available dataset and considering low, central, and high levels chosen for comprehensive exploration.

   Factor 2: Methanol to oil mole ratio: The influence of different mole ratios of methanol to oil on biodiesel yield was examined. Three levels were chosen for thorough analysis.

   Factor 3: Catalyst concentration (%): The effect of varying catalyst concentrations on the TSOME Yield was investigated. For a comprehensive study, low, central, and high levels are selected.
3) *Experimental runs*: A total of 20 experimental runs based on the CCD design were conducted. Each run represents a unique combination of the three input parameters.
4) *CCDmatrix*: A CCD matrix with coded and uncoded values for each factor was generated. Randomization of experimental runs to minimize systematic errors was ensured.
5) *Biodiesel production process*: The transesterification process was executed for each experimental run. Reaction conditions as per the CCDwere monitored.

6) *Response variable*: TSOME Yield (%): TSOME Yield as the response variable to be optimized was defined for all 20 experimental runs, ensuring precision and accuracy in measurement and recording.
7) *Model validation*: Assessed the adequacy of the model using statistical tools
8) *Optimization*: A validated model was utilized to identify optimal conditions for maximizing TSOME Yield. Response surface plots were employed for visualizing the impact of factors.
9) *Verification runs*: Verification runs were conducted based on the optimal conditions predicted by the model. The accuracy of the model predictions was verified against experimental outcomes.
10) *Statistical analysis*: Statistical techniques were applied to evaluate the significance of each input parameter.

17.4 Result and Discussions

17.4.1 The Operational Data

In this study, 20 experimental runs were obtained from a study conducted elsewhere [17], and used as the Experimental runs matrix. Each experimental run comprising the reaction time (min), methanol to oil mole ratio and catalyst (%) was used as input parameters, while the TSOME Yield (%) as the output parameters, respectively, as presented in Table 17.2.

17.4.2 Modeling and Prediction Using the Response Surface Methodology

In this study, an attempt is made to develop a second order mathematical relationship between selected input variables, namely; Reaction time (min), methanol to oil mole ratio and catalyst (%) coupled with one response variables, namely; TSOME Yield (%) using the RSM.

17.4.3 Modeling and Prediction of the TSOME Yield (%)

To ascertain the appropriateness and efficacy of the quadratic model for comprehensively analyzing the experimental data, a rigorous validation process was undertaken. This involved the calculation of the sequential model sum of squares, a critical statistical metric used to assess the adequacy of the model in capturing the variation in the TSOME Yield (%), as presented in Table 17.2. This analysis serves as a pivotal step in ensuring the robustness and reliability of the quadratic model in its capacity to effectively model and predict the TSOME Yield (%) response. The sequential model sum of squares not only aids in validating the

Table 17.2 Experimental runs matrix.

Run	Reaction time (min)	Methanol to oil mole ratio	Catalyst (%)	TSOME Yield (%)
1	40	4	0.5	48.06
2	80	4	0.5	55.63
3	40	8	0.5	56.52
4	80	8	0.5	56.04
5	40	4	1.5	62.97
6	80	4	1.5	76.01
7	40	8	1.5	65.13
8	80	8	1.5	73.45
9	40	6	1	79.38
10	80	6	1	89.75
11	60	4	1	81.15
12	60	8	1	78.18
13	60	6	0.5	59.71
14	60	6	1.5	74.71
15	60	6	1	90.07
16	60	6	1	86.04
17	60	6	1	86.04
18	60	6	1	90.07
19	60	6	1	85.13
20	60	6	1	85.13

model's fit but also provides valuable insights into the contribution of each term, thereby enhancing the precision and confidence in the model's predictive capabilities.

Upon careful examination of the findings presented in Table 17.3, a notable observation emerged – namely, the cubic polynomial was identified as being aliased. This inherent aliasing implies that the cubic polynomial, due to its intricate and potentially confounding nature, cannot be effectively employed to contribute to the final model. In light of this revelation, a discerning choice was made, steering the analysis toward the quadratic and 2FI (two-factor interaction) models as the most viable and fitting alternatives. This judicious selection not only streamlines the complexity associated with the modeling process but also substantiates

Table 17.3 Sequential model sum of squares.

Source	Sum of squares	*df*	Mean square	*F*-value	*p*-Value	
Mean vs.total	1.094E+05	1	1.094E+05			
Linear vs.mean	736.05	3	245.35	1.45	0.2661	
2FI vs.linear	56.58	3	18.86	0.0923	0.9629	
Quadratic vs. 2FI	**2602.65**	**3**	**867.55**	**165.87**	**< 0.0001**	**Suggested**
Cubic vs.quadratic	16.03	4	4.01	0.6627	0.6403	Aliased
Residual	36.27	6	6.05			
Total	1.128E+05	20	5642.24			

the utilization of the quadratic polynomial as the preferred analytical tool in this comprehensive investigation.

To evaluate the quadratic model's capability in elucidating the inherent variation within the experimental data, a crucial diagnostic step involved estimating the lack of fit for the response variable. The lack of fit test serves as a robust metric, gauging how well the quadratic model accounts for the residual discrepancies between the predicted and observed values. In Table 17.3, the results of the computed lack of fit for the TSOME Yield (%) are meticulously presented. These results provide valuable insights into the adequacy and reliability of the quadratic model, shedding light on its capacity to effectively capture and explain the intricacies inherent in the experimental data. The lack of fit analysis contributes significantly to the overall validation of the chosen quadratic model, ensuring a comprehensive understanding of its predictive capabilities and appropriateness for the analytical task at hand.

Analysis of the results presented in Table 17.4 revealed a critical criterion for model selection – the model should exhibit an insignificant lack of fit to be considered suitable for predictive purposes. In light of this criterion, any model showcasing a significant lack of fit becomes unsuitable for reliable predictions. In our scrutiny, it was reaffirmed that the quadratic polynomial exhibited a nonsignificant lack of fit, rendering it well-suited for model analysis and subsequent predictions. Conversely, the cubic polynomial, with its noteworthy lack of fit, was deemed significant and consequently aliased, thereby precluding its utilization in the predictive modeling process.

To further substantiate the appropriateness of the selected model, comprehensive model statistics were computed for the response variable of centrifugal force. These statistics, derived from the model sources, are meticulously presented in Table 17.5. This table serves as a repository of critical information, offering

Table 17.4 Lack of fit tests.

Source	Sum of squares	*df*	Mean square	*F*-value	*p*-Value	
Linear	2683.88	11	243.99	44.12	0.0003	
2FI	2627.30	8	328.41	59.39	0.0002	
Quadratic	**24.65**	**5**	**4.93**	**0.8917**	**0.5485**	**Suggested**
Cubic	8.63	1	8.63	1.56	0.2669	Aliased
Pure error	27.65	5	5.53			

Table 17.5 Model summary statistics.

Source	Std. Dev.	R^2	Adjusted R^2	Predicted R^2	PRESS	
Linear	13.02	0.2135	0.0660	−0.3622	4696.44	
2FI	14.29	0.2299	−0.1255	−4.1064	17604.57	
Quadratic	**2.29**	**0.9848**	**0.9712**	**0.9369**	**217.62**	**Suggested**
Cubic	2.46	0.9895	0.9667	−2.0848	10635.04	Aliased

insights into the performance and reliability of the chosen model in explaining the variation in the centrifugal force response. The model statistics, including regression coefficients and associated *p*-values, contribute to the overall understanding of the model's effectiveness in representing the complex relationships inherent in the experimental data. The significance of each model term is judiciously examined, allowing for a nuanced evaluation of the model's predictive capacity and its potential application in practical scenarios.

Based on the information gleaned from Table 17.5, the quadratic polynomial model was recommended for consideration, while the cubic polynomial model was deemed aliased. Consequently, the quadratic polynomial model emerged as the preferred choice for this analytical investigation. To gauge the effectiveness of the quadratic model in minimizing the TSOME Yield (%), a one-way ANOVAtable was constructed and is delineated in Table 17.5. This table serves as a comprehensive tool for evaluating the statistical significance of the quadratic model's terms and provides valuable insights into the model's efficacy in explaining variations in the TSOME Yield (%). The ANOVA results contribute to the robustness and reliability assessment of the chosen quadratic model, substantiating its suitability for the specific analytical objectives at hand.

Examining the contents of Table 17.6, it becomes evident that an essential step in validating the model's significance and comprehensively assessing the

Table 17.6 ANOVA for quadratic model.

Source	Sum of squares	*df*	Mean square	*F*-value	*p*-Value	
Model	3395.27	9	377.25	72.13	<0.0001	Significant
A: Reaction time	150.70	1	150.70	28.81	0.0003	
B: Methanol to oil mole ratio	3.03	1	3.03	0.5784	0.4645	
C: Catalyst	582.32	1	582.32	111.34	< 0.0001	
AB	20.38	1	20.38	3.90	0.0766	
AC	25.45	1	25.45	4.87	0.0519	
BC	10.74	1	10.74	2.05	0.1823	
A^2	2.22	1	2.22	0.4242	0.5295	
B^2	92.45	1	92.45	17.68	0.0018	
C^2	916.24	1	916.24	175.18	< 0.0001	
Residual	52.30	10	5.23			
Lack of fit	24.65	5	4.93	0.8917	0.5485	Not significant
Pure error	27.65	5	5.53			
Cor Total	3447.58	19				

substantial contributions of individual variables, as well as their combined and quadratic effects on each response, involved the application of ANOVA. This statistical method serves the dual purpose of determining the overall significance of the model and delving into the specific influences exerted by each variable and their interactions.

The Model *F*-value, a pivotal metric derived from the ANOVA, stands at an impressive 72.13. This signifies a noteworthy level of significance in the model. The magnitude of the *F*-value is such that there exists only a 0.01% probability that a value of this magnitude could arise purely due to random noise in the data. This statistical insight emphasizes the robustness of the model, highlighting that the observed effects are highly unlikely to be a result of chance or random variability. The Model *F*-value, therefore, provides strong evidence in support of the model's significance, bolstering confidence in its ability to effectively capture and explain the variations in the responses under consideration.

P-values below 0.0500 serve as an indicator of the significance of model terms. In this particular instance, model terms A, C, B^2, and C^2 exhibit significance, as

Table 17.7 Fit statistics.

Std. Dev.	2.29	R^2	0.9848
Mean	73.96	**Adjusted** R^2	0.9712
C.V. %	3.09	**Predicted** R^2	0.9369
		Adeq precision	25.3337

their *p*-values fall below this threshold. Conversely, values exceeding 0.1000 suggest that the corresponding model terms lack significance. If numerous model terms are deemed insignificant, excluding those essential for maintaining hierarchy, a reduction in the model may enhance its effectiveness. Moving to the Lack of Fit *F*-value, which stands at 0.89, its nonsignificance relative to the pure error is noteworthy. The implication is that there is a 54.85% probability that a Lack of Fit *F*-value of this magnitude could arise due to random noise. A lack of significance in the Lack of Fit is advantageous for the model's fitting, signifying that the observed discrepancies are likely not due to unaccounted factors or noise.

To ascertain the quadratic model's efficacy in minimizing centrifugal force, a thorough examination of fit statistics is presented in Table 17.7. This analysis serves as a crucial step in validating the adequacy of the quadratic model and its ability to effectively capture and minimize variations in centrifugal force.

The Predicted R^2, registering at 0.9369, exhibits a commendable level of concordance with the Adjusted R^2, which stands at 0.9712. Notably, the difference between these two values is less than 0.2, signifying a robust agreement. This alignment underscores the reliability of the model in predicting responses, emphasizing that the model's predictive capabilities are well-maintained even when accounting for variations in the number of terms.

An essential metric, Adeq Precision, provides insights into the signal-to-noise ratio. A ratio exceeding 4 is considered desirable, and in this context, your ratio of 25.334 signifies a highly adequate signal. This suggests that the model possesses a substantial signal component relative to the noise, reinforcing the model's reliability for navigation within the design space. The high Adeq Precision ratio lends credibility to the model's predictive capabilities and supports its practical utility.

In pursuit of the optimal solution, meticulous consideration is given to the coefficient statistics and their corresponding standard errors. These statistics, essential for determining the significance and reliability of each coefficient, are pivotal in discerning the impact of each variable on the response. In Table 17.8,

Table 17.8 Coefficients in terms of coded factors.

Factor	Coefficient estimate	*df*	Standard error	95% CI low	95% CI high	VIF
Intercept	86.43	1	0.7862	84.68	88.19	
A: Reaction time	3.88	1	0.7232	2.27	5.49	1.0000
B: Methanol to oil mole ratio	0.5500	1	0.7232	−1.06	2.16	1.0000
C: Catalyst	7.63	1	0.7232	6.02	9.24	1.0000
AB	−1.60	1	0.8086	−3.40	0.2053	1.0000
AC	1.78	1	0.8086	−0.0178	3.59	1.0000
BC	−1.16	1	0.8086	−2.96	0.6428	1.0000
A^2	−0.8982	1	1.38	−3.97	2.17	1.82
B^2	−5.80	1	1.38	−8.87	−2.73	1.82
C^2	−18.25	1	1.38	−21.33	−15.18	1.82

the coefficient statistics for the TSOME Yield (%) response variable are meticulously presented. This comprehensive analysis aids in refining the model and fine-tuning the coefficients to attain an optimized and reliable solution within the experimental design.

Extracting insights from Table 17.8, it is crucial to understand that the coefficient estimate holds significance as it denotes the anticipated alteration in the response for every unit shift in factor value, with the condition that all other factors remain constant. In the context of an orthogonal design, the intercept assumes the role of the overall average response across all experimental runs. The coefficients, then, act as adjustments around this average, contingent upon the specific factor settings.

In scenarios where factors are orthogonal, variance inflation factors (VIFs) assume a value of 1, indicating an absence of multicollinearity. VIFs exceeding 1, on the other hand, signal the presence of multicollinearity, and the severity of correlation among factors intensifies with higher VIF values. As a general guideline, VIFs below 10 are considered tolerable, implying a manageable degree of multicollinearity. This metric proves valuable in gauging the degree of interdependence among factors, thereby influencing the reliability and stability of the model.

The optimal equation which shows the individual effects and combine interactions of the selected input variables against the response is presented based on the coded variables in Eq. (17.12).

Final equation in terms of coded factors (17.12)

TSOME Yield (%)	=
+86.43	
+3.88	A
+0.5500	B
+7.63	C
−1.60	AB
+1.78	AC
−1.16	BC
−0.8982	A^2
−5.80	B^2
−18.25	C^2

The equation in terms of coded factors can be used to make predictions about the response for given levels of each factor. By default, the high levels of the factors are coded as +1 and the low levels are coded as −1. The coded equation is useful for identifying the relative impact of the factors by comparing the factor coefficients.

Final equation in terms of actual factors (17.13)

TSOME Yield (%)	=
−86.02098	
+0.524617	Reaction time
+21.22267	Methanol to oil mole ratio
+157.53745	Catalyst
−0.039906	Reaction time × methanol to oil mole ratio
+0.178375	Reaction time × catalyst
−1.15875	Methanol to oil mole ratio × catalyst
−0.002245	Reaction time^2
−1.44955	Methanol to oil mole ratio^2
−73.01273	Catalyst^2

The equation in terms of actual factors can be used to make predictions about the response for given levels of each factor. Here, the levels should be specified in the original units for each factor. This equation should not be used to determine the relative impact of each factor because the coefficients are scaled to accommodate the units of each factor and the intercept is not at the center of the design space.

The diagnostics case statistics which show the observed values of centrifugal force against their predicted values is presented in Table 17.9. The diagnostic case statistics actually give insight into the model strength and the adequacy of the optimal second order polynomial equation.

To accept any model, its satisfactoriness must first be checked by an appropriate statistical analysis output. To diagnose the statistical properties of the response surface model, the normal probability plot of residual for the TSOME Yield (%)is represented in Figure 17.5.

The normal probability plot of residuals is the most significant assumption for checking the sufficiency of a statistical model. The result revealed that the computed residuals are approximately normally distributed an indication that the model developed is satisfactory.

To detect for the presence of mega patterns or expanding variance a plot of residuals and the predicted was produced for the TSOME Yield (%) which is shown in Figure 17.6.

Figure 17.6 provides a visual representation in the form of a graph, depicting the residuals plotted against ascending response values. This visualization serves the purpose of evaluating the assumption of constant variance within the model. The aim is to scrutinize the distribution of residuals across the range of response values and ascertain whether the variability remains relatively constant.

In a complementary effort to unveil any values or groups of values that might escape easy detection by the model, Figure 17.7 is presented. This figure features a plot showcasing the relationship between predicted values and actual values specifically for TSOME Yield (%). By comparing the predicted and actual values on this plot, we gain insights into how well the model aligns with the observed data, facilitating the identification of any discrepancies or outliers that could impact the overall model performance.

As upon close examination of the graph, it is evident that the data points closely align with the fitted line, indicating a strong correlation between the predicted values and the actual observations. The model exhibits a noteworthy capability to accurately predict the majority of the data points, showcasing its effectiveness in capturing the underlying patterns in the experimental data.

To further scrutinize the dataset for potential outliers or influential observations, a Cook's distance plot was generated for the responses. Cook's distance is a statistical measure that assesses the impact of individual data points on the regression model. In this context, Figure 17.8 displays the Cook's distance specifically for the centrifugal force. This plot provides valuable insights into the influence of each data point on the model, helping to identify any observations that might significantly affect the model's performance or assumptions.

Table 17.9 Diagnostics case statistics report of observed vs. predicted centrifugal force.

Run order	Actual value	Predicted value	Residual	Leverage	Internally studentized residuals	Externally studentized residuals	Cook's distance	Influence on fitted value DFFITS	Standard order
1	48.06	48.45	−0.3895	0.793	−0.374	−0.358	0.054	−0.701	1
2	55.63	55.84	−0.2085	0.793	−0.200	−0.191	0.015	−0.373	2
3	56.52	55.06	1.46	0.793	1.404	1.487	0.756	2.912	3
4	56.04	56.06	−0.0235	0.793	−0.023	−0.021	0.000	−0.042	4
5	62.97	62.46	0.5085	0.793	0.489	0.469	0.092	0.919	5
6	76.01	76.99	−0.9755	0.793	−0.938	−0.932	0.337	−1.825	6
7	65.13	64.44	0.6935	0.793	0.667	0.647	0.171	1.267	7
8	73.45	72.58	0.8745	0.793	0.841	0.827	0.271	1.621	8
9	79.38	81.65	−2.27	0.491	−1.393	−1.472	0.187	−1.446	9
10	89.75	89.42	0.3329	0.491	0.204	0.194	0.004	0.190	10
11	81.15	80.09	1.06	0.491	0.653	0.633	0.041	0.621	11
12	78.18	81.19	−3.01	0.491	−1.842	−2.149	0.327	−2.110	12
13	59.71	60.55	−0.8391	0.491	−0.514	−0.494	0.025	−0.486	13
14	74.71	75.81	−1.10	0.491	−0.675	−0.655	0.044	−0.643	14
15	90.07	86.43	3.64	0.118	1.693	1.902	0.038	0.696	15
16	86.04	86.43	−0.3933	0.118	−0.183	−0.174	0.000	−0.064	16
17	86.04	86.43	−0.3933	0.118	−0.183	−0.174	0.000	−0.064	17
18	90.07	86.43	3.64	0.118	1.693	1.902	0.038	0.696	18
19	85.13	86.43	−1.30	0.118	−0.607	−0.587	0.005	−0.215	19
20	85.13	86.43	−1.30	0.118	−0.607	−0.587	0.005	−0.215	20

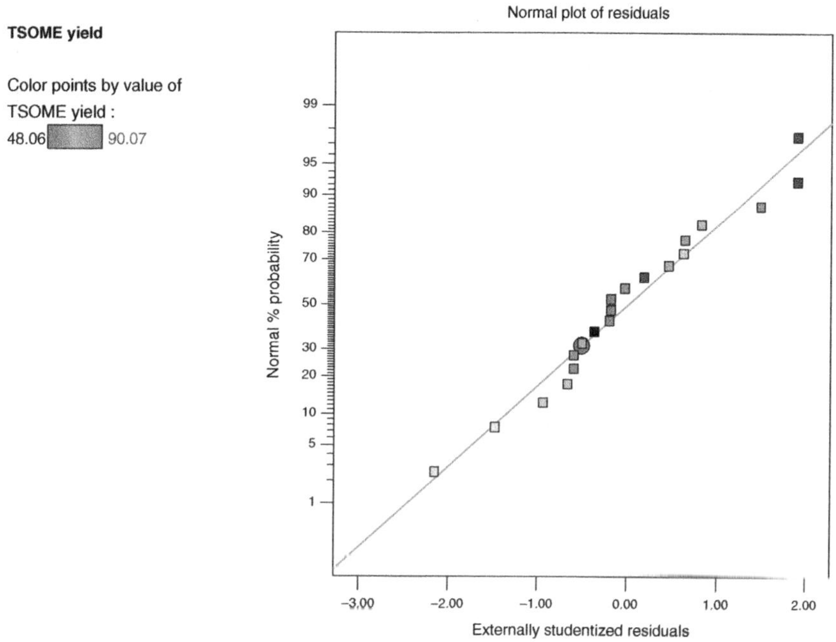

Figure 17.5 Normal probability plot of studentized residuals for TSOME Yield (%).

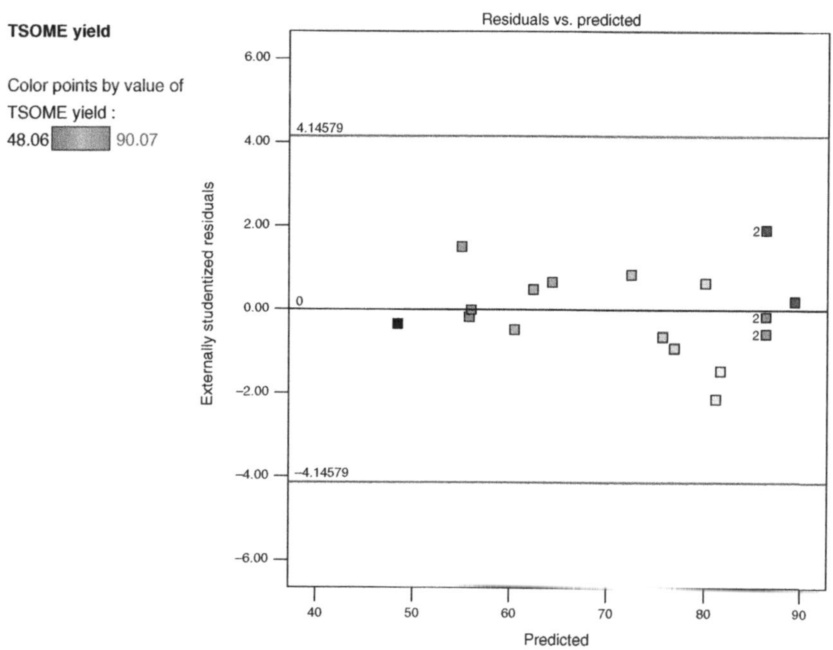

Figure 17.6 Plot of residual vs predicted for TSOME Yield (%).

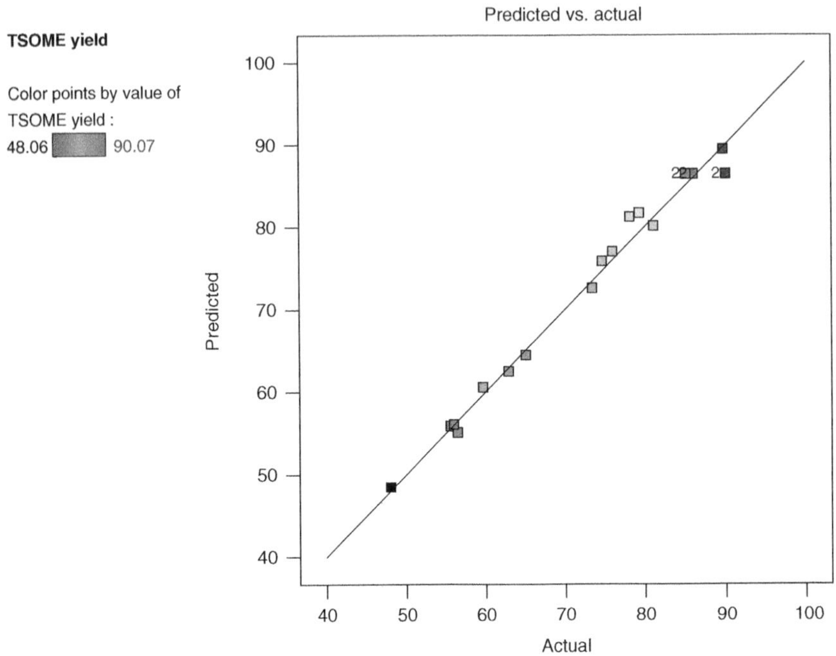

Figure 17.7 Plot of predicted vs actual for TSOME Yield (%).

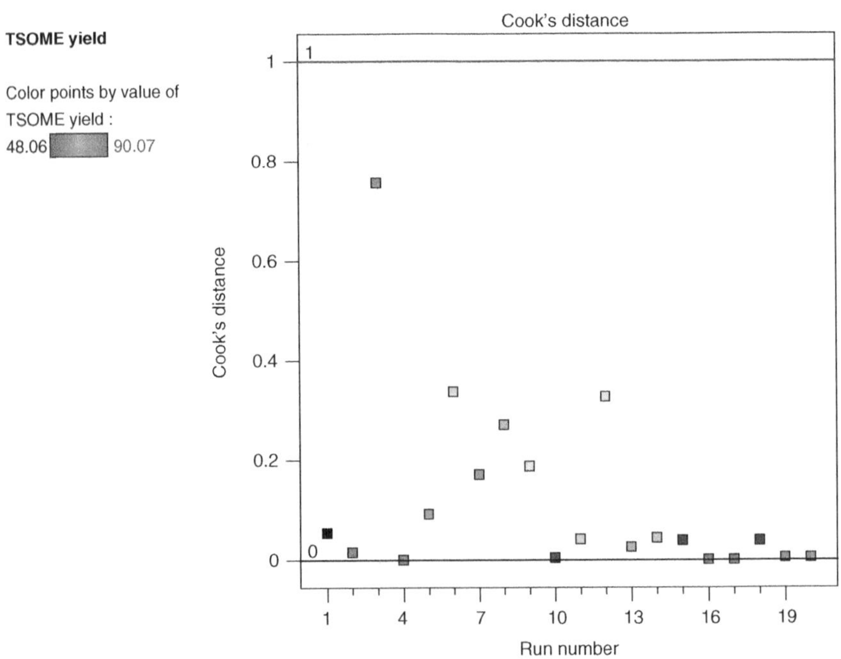

Figure 17.8 Generated cook's distance for TSOME Yield (%).

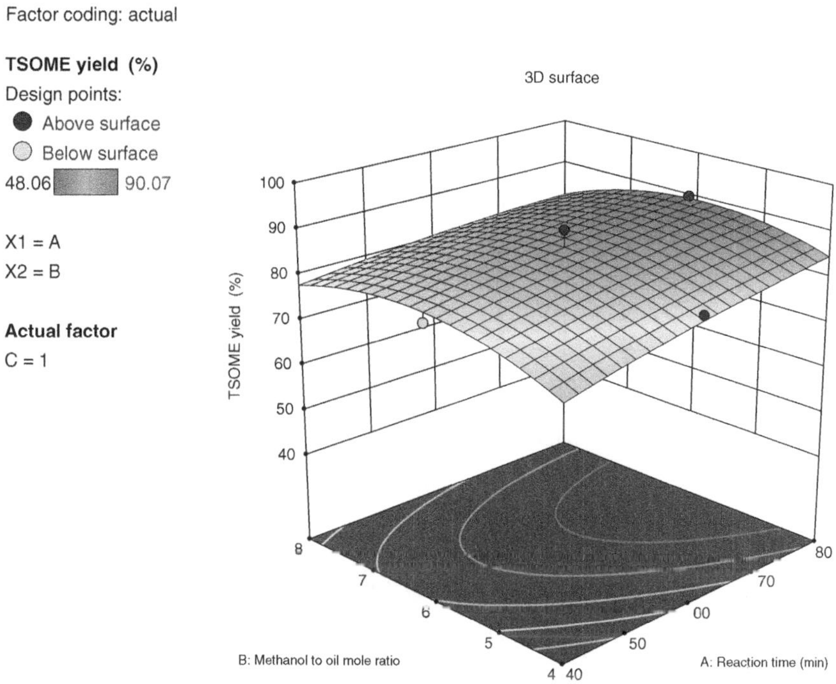

Figure 17.9 3D surface plot of TSOME Yield, methanol to oil mole ratio, and reaction time.

The Cook's distance plot is characterized by an upper limit of 1.00 and a lower limit of 0.00. In this context, experimental values falling below the lower bound or exceeding the upper bound are deemed as potential outliers, warranting thorough investigation and scrutiny. Identifying and understanding such outliers is crucial for ensuring the robustness and reliability of the analysis.

To comprehensively examine the influence of the combined input variables on TSOME Yield (%), a series of 3D surface plots is presented in Figure 17.9. These plots offer a visual representation of the relationships and interactions between multiple input variables and their impact on the TSOME Yield. By exploring these surface plots, one can gain valuable insights into how variations in the input variables collectively contribute to the observed changes in the TSOME Yield, aiding in the interpretation of the overall system behavior.

To study the effects of combine input variables, catalyst, and reaction timeon the TSOME Yield, the 3D surface plots presented in Figure 17.10 was generated as follows.

In order to comprehensively investigate the impact of the combined input variables, specifically catalyst and methanol to oil mole ratio, on the TSOME Yield, a set of informative 3D surface plots was generated, as illustrated in Figure 17.8.

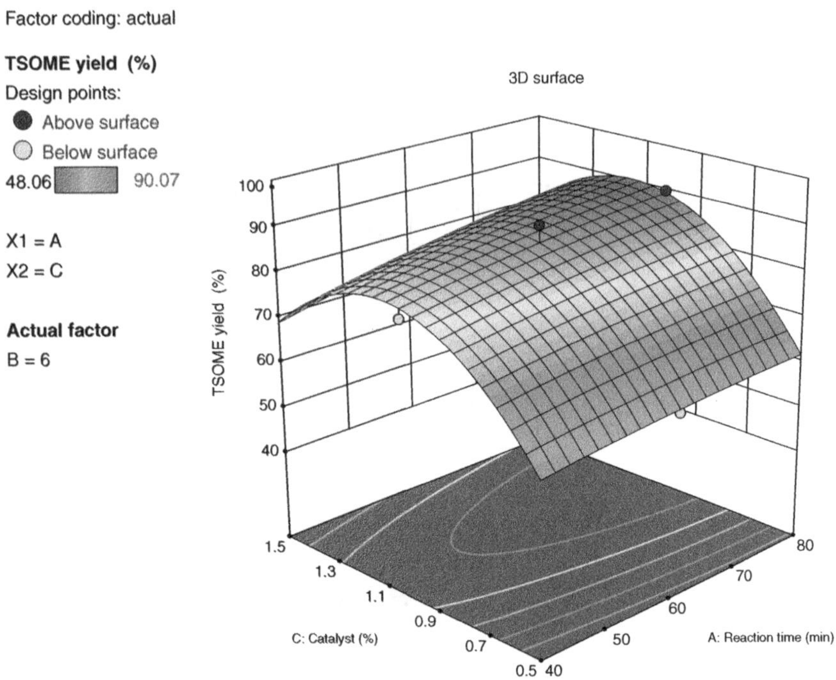

Figure 17.10 3D surface plot of TSOME Yield, catalyst, and reaction time.

These plots serve as a visual representation to elucidate the intricate relationships and interactions between the specified input variables and their collective influence on the TSOME Yield.

The process involved in creating these 3D surface plots entails a systematic exploration of various combinations of catalyst and methanol to oil mole ratio, generating corresponding TSOME Yield values. By carefully mapping this multidimensional dataset onto a 3D surface, the plots provide an intuitive visualization of how alterations in the levels of catalyst and methanol to oil mole ratio jointly contribute to the fluctuations in TSOME Yield. This approach facilitates a nuanced understanding of the interdependencies between the input variables and the resultant TSOME Yield, aiding in the interpretation of the complex dynamics governing the system (Figure 17.11).

The contour plots, depicted in Figure 17.12, vividly showcase the variation in the TSOME Yield response variable concerning changes in both catalyst and methanol to oil mole ratio. These plots provide a visual representation of the intricate relationships and interactions between the two input variables, offering valuable insights into the combined effects on the TSOME Yield.

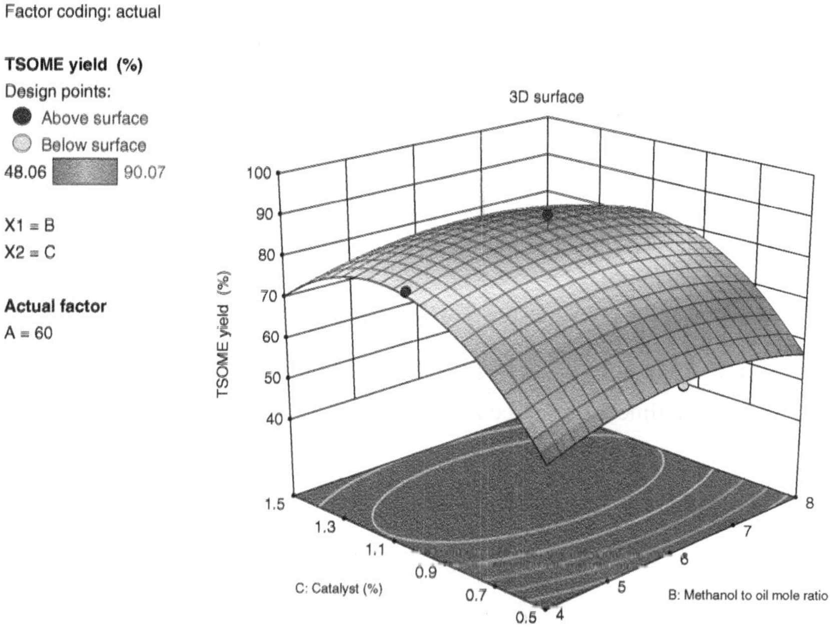

Figure 17.11 3D surface plot of TSOME Yield, catalyst, and methanol to oil mole ratio.

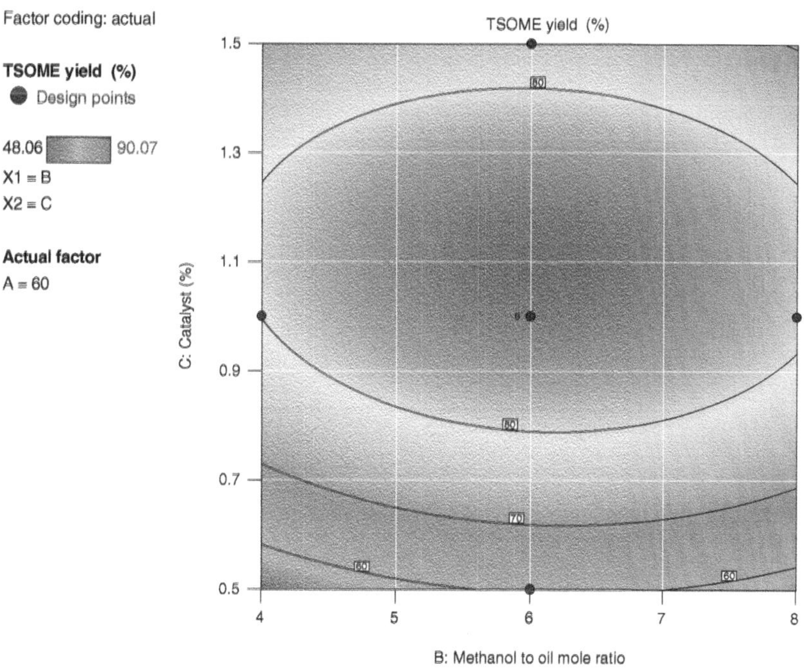

Figure 17.12 The contour plots of TSOME Yield, catalyst, and methanol to oil mole ratio.

By presenting contours of constant TSOME Yield levels on the catalyst–methanol to oil mole ratio plane, the plots facilitate an intuitive understanding of how alterations in these input variables collectively influence the response variable. The contour lines connect points with equal TSOME Yield values, enabling the identification of regions where the response is optimized or minimized based on different combinations of catalyst and methanol to oil mole ratio. This graphical representation aids in discerning the optimal conditions for achieving desired TSOME Yield levels, contributing to a more comprehensive interpretation of the experimental outcomes.

The contour plots showing the TSOME Yield response variable against the methanol to oil mole ratio and reaction time are presented in Figure 17.13.

The contour plots showing the TSOME Yield response variable against the catalyst and reaction time is presented in Figure 17.14.

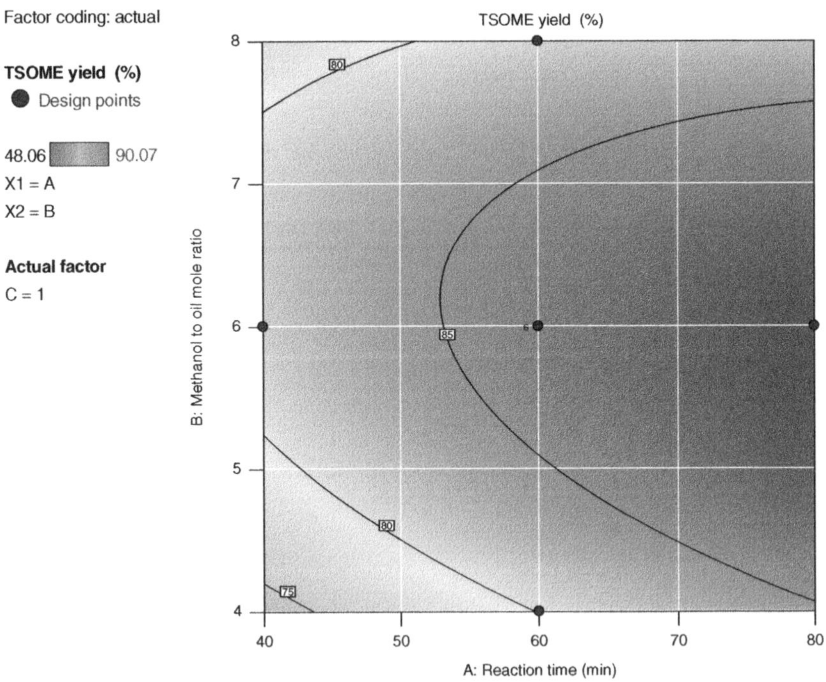

Figure 17.13 The contour plots of TSOME Yield against methanol to oil mole ratio and reaction time.

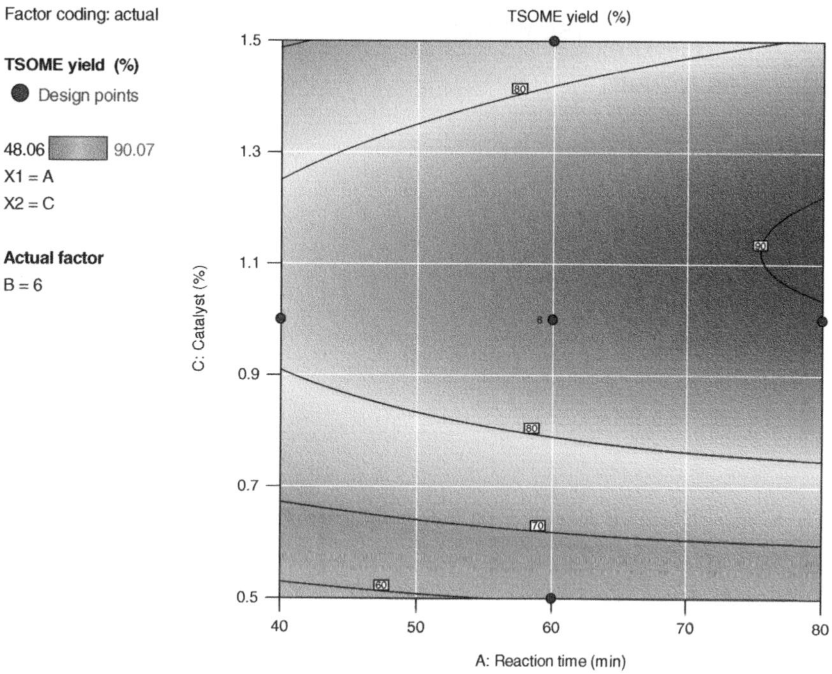

Figure 17.14 The contour plots of TSOME Yield against the catalyst and reaction time.

17.4.4 Optimization Using the Response Surface Methodology

The target of the optimization model was to maximize the TSOME Yield as shown in Figure 17.15.

From Figure 17.15, the maximum TSOME Yield exists between 48.06 and 90.07.

The constraints of the optimization are shown in Table 17.10.

The numerical optimization produces about 15 optimal solutions which a represented in Table 17.11.

The regression equation obtained for predicting TSOME (tobacco seed oil methyl ester) Yield based on the response surface methodology (RSM) is as follows (Figure 17.16):

$$\text{TSOME Yield} = -0.003 + 1.000\text{RSM TSOME Yield}$$

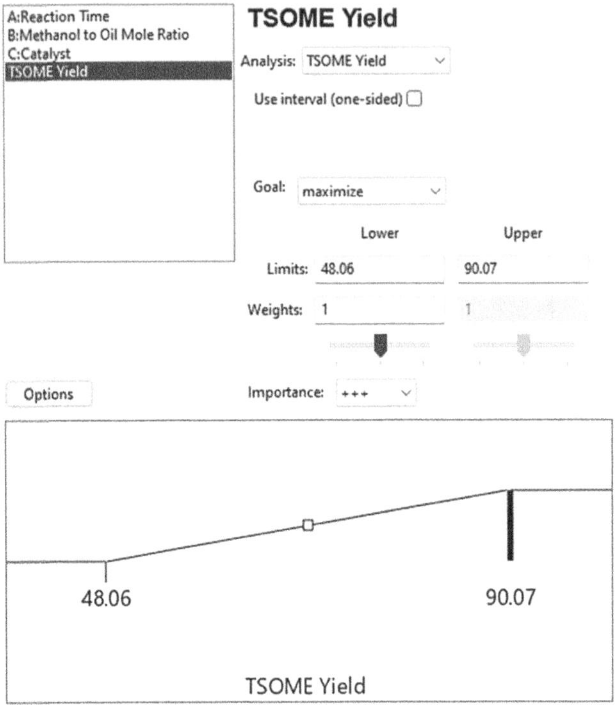

Figure 17.15 Maximum TSOME limits.

Table 17.10 Constraints.

Name	Goal	Lower limit	Upper limit	Lower weight	Upper weight	Importance
A:Reaction time	Is in range	40	80	1	1	3
B:Methanol to oil mole ratio	Is in range	4	8	1	1	3
C:Catalyst	Is in range	0.5	1.5	1	1	3
TSOME Yield	Maximize	48.06	90.07	1	1	3

This equation below provides a quantitative relationship between the independent variable (RSM TSOME Yield) and the dependent variable (TSOME Yield). The coefficients indicate the magnitude and direction of the impact of the RSM TSOME Yield on the predicted TSOME Yield.

Table 17.11 The optimal solutions.

Number	Reaction time	Methanol to oil mole ratio	Catalyst	TSOME yield	Desirability	
1	**77.918**	**5.904**	**1.125**	**90.399**	**1.000**	**Selected**
2	79.272	5.916	1.108	90.548	1.000	
3	76.436	5.827	1.136	90.191	1.000	
4	78.147	6.087	1.114	90.315	1.000	
5	78.254	5.978	1.061	90.100	1.000	
6	78.668	5.928	1.142	90.480	1.000	
7	76.185	5.972	1.147	90.091	1.000	
8	77.778	5.807	1.130	90.396	1.000	
9	79.613	6.187	1.166	90.299	1.000	
10	76.914	5.754	1.162	90.173	1.000	

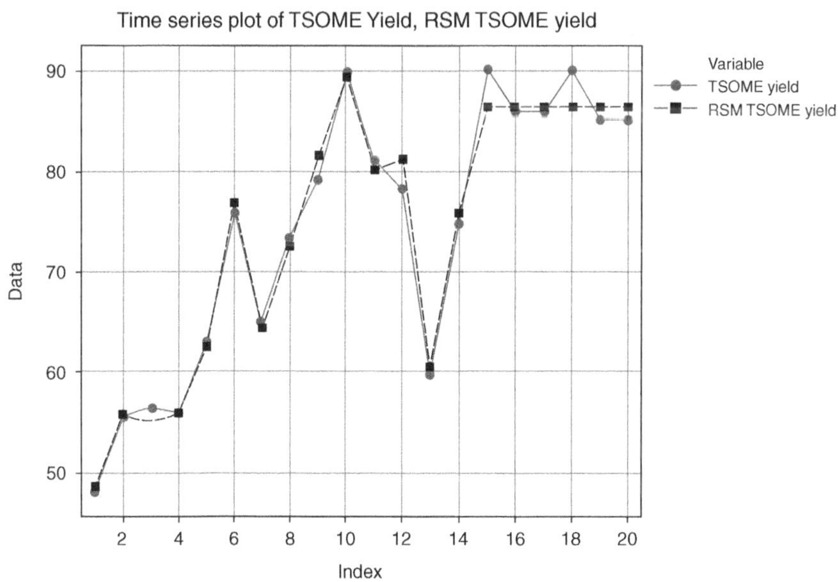

Figure 17.16 Time series plot of TSOME Yield, RSM TSOME Yield.

Model summary:

S	*R*-sq	*R*-sq(adj)
1.70500	98.48%	98.40%

The regression equation and model summary collectively suggest that the developed model is highly effective in predicting TSOME Yield based on the input variable RSM TSOME Yield.
Model summary:

Standard error (S): The standard error (S) is a measure of the variability of the actual TSOME Yield values around the predicted values. In this model, the standard error is1.70500, 1.70500.

R-squared (R-sq): *R*-squared (*R*-sq) represents the proportion of the variance in the TSOME Yield that is predictable from the RSM TSOME Yield. The obtained *R*-squared value is 98.48, indicating that the model explains 98.48% of the variability in TSOME Yield.

Adjusted R-squared (R-sq(adj)): Adjusted *R*-squared (*R*-sq(adj)) accounts for the number of predictors in the model, providing a more reliable measure of model fit. The adjusted *R*-squared value is 98.40, indicating the proportion of variance adjusted for the number of predictors.

Interpretation: The high *R*-squared values (98.48) indicate that the regression equation is an excellent fit for predicting TSOME Yield based on RSM TSOME Yield. The adjusted *R*-squared value (98.40) reinforces the model's reliability, considering the number of predictors. The standard error (1.70500) provides an estimate of the variability in TSOME Yield predictions, with lower values indicating a more precise model.

17.4.5 Analysis of Variance

The summary of the data suggests an RSM analysis commonly employed in experimental design to optimize and understand the relationship between multiple variables. The ANOVA table is a crucial component of RSM, aiding in the interpretation of the model's statistical significance. From the analysis of the variance table:

Regression: The significant F-statistic (1167.95) and extremely low *p*-value (0.000) for the regression indicate that the model is statistically significant. The regression model explains a significant amount of variability in the response variable.

Error: The error term represents the unexplained variability in the data. The low mean square error (MSE) suggests that the residual variability within the model is relatively small compared to the explained variability.

Total: The total sum of squares represents the overall variability in the data. The model has successfully accounted for a substantial portion of this variability.

Conclusion: The highly significant regression results suggest that the model fits the data well. The explanatory variables in the model contribute significantly to the observed variation in the response variable. In summary, the RSM analysis, as indicated by the ANOVA table, provides strong evidence that the regression model is a good fit for the data. This model can be further used for prediction and optimization based on the relationships between the variables under investigation.

17.4.6 Evaluation Metrics for Model Performance Using Statistical Parameters

The effectiveness of the constructed model was assessed through various performance evaluation metrics, including MSE, coefficient of determination (R^2), root mean square error (RMSE), and mean absolute percentage error (MAPE). The MAPE quantifies the mean absolute percentage difference between observed and expected values, while the RMSE measures the variance between predicted and actual data to evaluate the accuracy of the model. It is noteworthy that RMSE always maintains a positive value. The R^2 serves as a metric indicating the strength of the linear relationship between predicted and measured values. Lower values of MAPE and RMSE signify greater precision in predicting of TSOME, with the ideal scenario being both metrics approaching zero. The R^2 is a numerical value ranging from 0 to 1, where an R^2 value near +1 signifies a perfect linear relationship. The equations expressing, R^2, MSE, RMSE, and MAPE (%) are represented as follows:

$$R = \left(\frac{\sum_{m=1}^{n} \left(Y_{\text{Pred}.m} - y_{\text{Pred}}\right)\left(Y_{\text{exp}.m} - y_{\text{exp}}\right)}{\sqrt{\sum_{m=1}^{n} \left(Y_{\text{Pred}.m} - y_{\text{Pred}}\right)^2 \sum_{m=1}^{n} \left(Y_{\text{exp}.m} - y_{\text{exp}}\right)^2}} \right) \quad (17.14)$$

$$R^2 = 1 - \frac{\sum_{i=1}^{n} \left(Y_{i,\text{P}} - Y_{1,\text{e}}\right)^2}{\sum_{i=1}^{n} \left(Y_{i,\text{P}} - Y_{\text{e,ave}}\right)^2} \quad (17.15)$$

$$\text{RMSE} = \sqrt{\frac{\sum_{i=1}^{n} \left(Y_{i,\text{e}} - Y_{i,\text{p}}\right)^2}{n}} \quad (17.16)$$

$$\text{MAE} = \sum_{i=1}^{n} \frac{\left(Y_{i,\text{e}} - Y_{i,\text{p}}\right)}{n} \quad (17.17)$$

$$\text{SEP} = \frac{\text{RMSE}}{Y_{\text{e,ave}}} \quad (17.18)$$

$$\text{AAD} = \frac{100}{n} \sum_{i=1}^{n} \frac{\left(Y_{i,\text{e}} - Y_{i,\text{p}}\right)}{Y_{i,\text{e}}} \quad (17.19)$$

In determining the dominance and predictive capacity of the model techniques, statistical variables such as correlation coefficient (R), regression coefficient (R^2), RMSE and others are applied. The equation for computing the above-mentioned statistical variables is presented in systems of Eqs. (17.12–17.17). Analysis performed based on Table 17.12 shows results obtained using classical RSM and computational GW Algorithm. Table 17.13 presents the statistical results based on R, R^2 present. The plot showing experimental vs RSM and experimental result vs GWO is presented in Figures 17.17 and 17.18.

The GWO algorithm was employed as a predictive model to estimate the yield of waste Tobacco Seed Oil Methyl Esther (TSOME). In this context, three parameters, namely reaction time, methanol/oil molar ratio, and catalyst amount, were

Table 17.12 The experimental runs matrix with predicted TSOME yield.

Run	Reaction time (min)	Methanol to oil mole ratio	Catalyst (%)	TSOME Yield(%)	RSM TSOME Yield (%)	GWO TSOME Yield (%)
1	40	4	0.5	48.06	48.45	48.12
2	80	4	0.5	55.63	55.84	55.65
3	40	8	0.5	56.52	55.06	55.98
4	80	8	0.5	56.04	56.06	56.05
5	40	4	1.5	62.97	62.46	62.68
6	80	4	1.5	76.01	76.99	76.54
7	40	8	1.5	65.13	64.44	64.98
8	80	8	1.5	73.45	72.58	73.10
9	40	6	1	79.38	81.65	80.21
10	80	6	1	89.75	89.42	89.56
11	60	4	1	81.15	80.09	80.96
12	60	8	1	78.18	81.19	79.24
13	60	6	0.5	59.71	60.55	59.98
14	60	6	1.5	74.71	75.81	75.26
15	60	6	1	90.07	86.43	88.06
16	60	6	1	86.04	86.43	86.25
17	60	6	1	86.04	86.43	86.18
18	60	6	1	90.07	86.43	88.56
19	60	6	1	85.13	86.43	85.98
20	60	6	1	85.13	86.43	85.92

Table 17.13 Table showing statistical results obtained for GWO and RSM.

Metric	GWO TSOME yield	RSM TSOME yield
R	0.998437	0.992382
R^2	0.996868	0.984822
RMSE	0.734721	1.617501
SEP	0.734721	1.617501
MAE	0.5275	1.22
AAD	0.5275	1.22

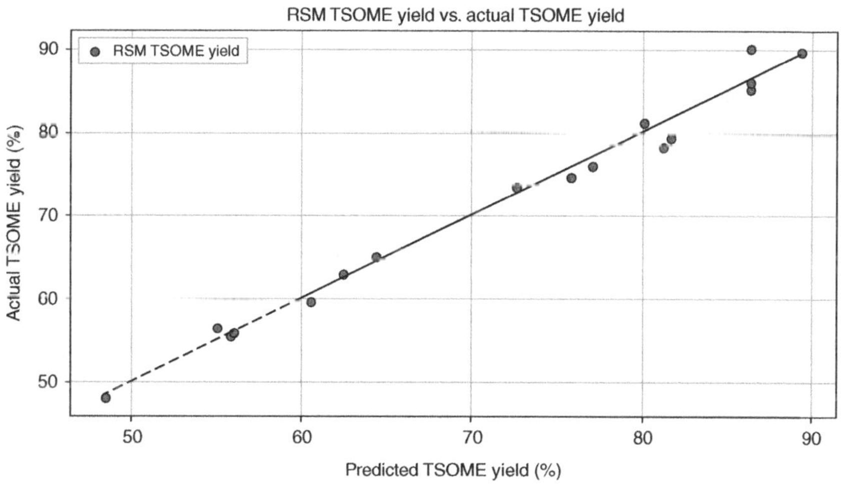

Figure 17.17 Predicted yield by RSM model (%).

selected as the model inputs. Concurrently, the TSOME yield was identified as the response variable. To optimize the parameters of the power equation, the boundary conditions established for the RSM model were applied to both the inputs and outputs variables for GWO. The analyses were executed using MATLAB R2023 software. Table 17.12 outlines the optimal values determined by the GWO for achieving the highest TSOME yield (Figure 17.18).

To assess the superiority and predictive capacity of the model techniques, statistical variables including the correlation coefficient (R), regression coefficient (R^2), root-mean-square error (RMSE), mean average error (MAE), standard prediction error (SEP), and absolute average deviation (AAD) were utilized, as outlined in

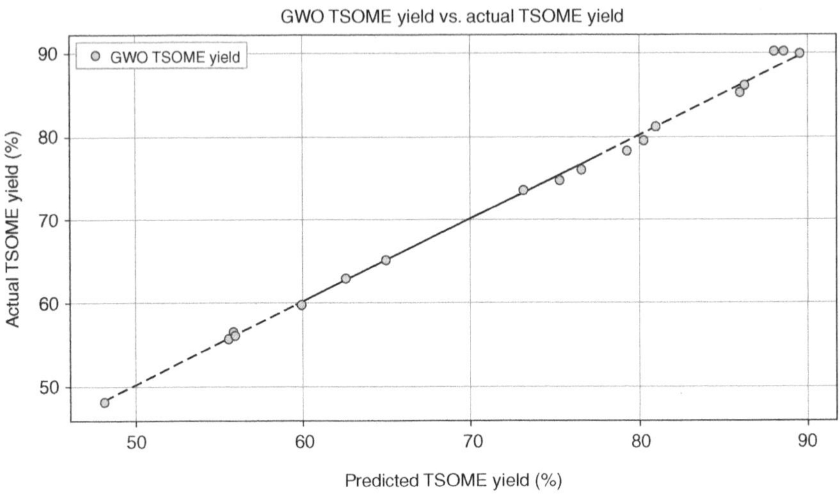

Figure 17.18 Predicted yield by GWO model (%).

Eqs. (17.12–17.17). Here, Y_{pred} represents the predicted value of samples, Y_{exp} is the experimental value of samples, Y_i is the observed value of samples, Y_e is the estimated value of samples, and N represents the number of samples. The variables A, B, and C denote independent variables, while β_{ij} signifies the coefficient of linear terms. The predicted indices for the models are highlighted in Table 17.4. As realized in Table 17.3, ANN possesses a higher regression coefficient and lower errors compared to the RSM values. Also obtained are values of R, R^2, RMSE, SEP, MAE, and AAD respectively, which clearly showed the accuracy of the ANN model and its superiority to that of the RSM model.

Table 17.12 illustrates the actual yields alongside the predictions from the RSM and GWO models. Notably, the biodiesel yield model generated by GWO demonstrates closer proximity to the experimental yield in comparison to the RSM predictions across all runs. This observation aligns with findings reported by researchers in other studies [34, 42].

Figures 17.17 and 17.18 provide visual representations of the experimental yield vsRSM-predicted yield and experimental yield vsGWO-predicted yield, respectively. The values obtained are deemed suitable for capturing the variations in experimental and model-based yields, attributed to high regression coefficients (R^2) of 0.984822 and 0.996868 for RSM and GWO, respectively. An R^2 of 0.9848 in the RSM model and 0.997 in the GWO model indicates that over 98.5% and 99.7% of the data fit into the respective models. Consequently, the GWO model exhibits a higher accuracy in predicting yield compared to the RSM model. This corroborates similar findings reported by researchers elsewhere [34, 42].

In Figure 17.19, estimated statistical indices, including correlation coefficient (R), regression coefficient (R^2), RMSE, standard error of prediction (SEP), mean average error (MAE), and AAD are presented. The computed values for the RSM model, such as R of 0.992382, R^2 of 0.984822, RMSE of 1.617501, SEP of 1.617501, MAE of 1.22, and AAD of 1.22, are compared to those of the GWO model: R (0.998437), R^2 (0.996868), RMSE (0.734721), SEP (0.734721), MAE (0.5275), and AAD (0.5275). The high R^2 values and low standard error indicate the model's robustness and accuracy in capturing the variability in biodiesel yield. These analytical results affirm the efficacy of the developed RSM model for biodiesel production from tobacco seed oil, providing valuable insights into the relationship between process parameters and TSOME Yield [43]. However, these results underscore the superior performance of the GWO model over the RSM model, as evidenced by higher R and R^2 values and lower values for other statistical indices.

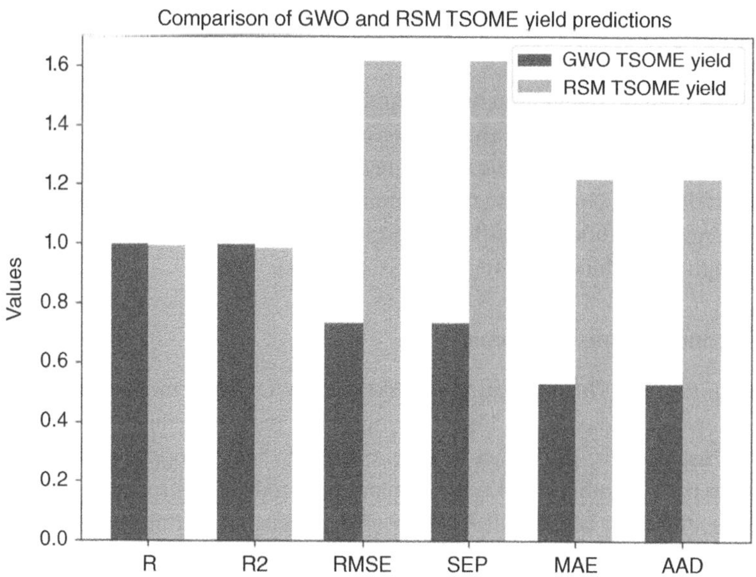

Figure 17.19 Comparison of GWO and RSM TSOME yield predictions.

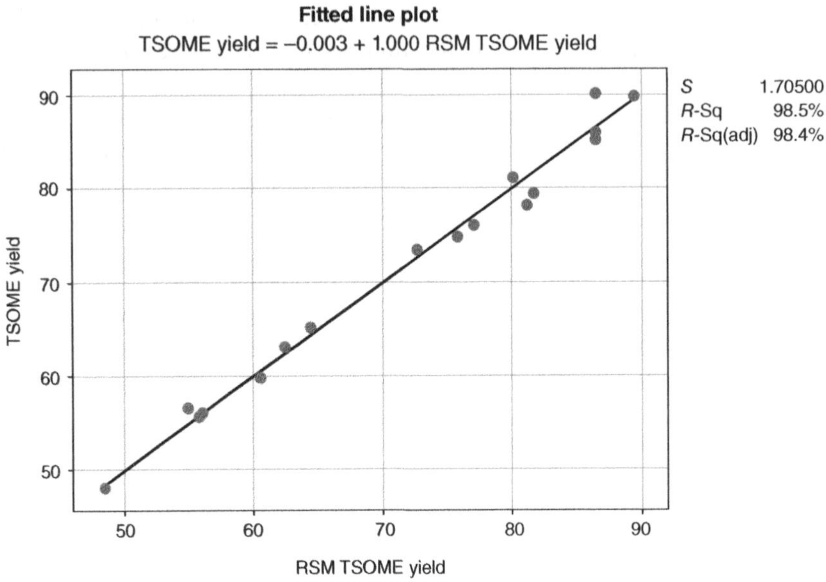

Figure 17.20 Fitted line plot.

17.5 Economic and Environmental Implications of the Study

This study focused on integration of RSM and GWO for modeling biodiesel using tobacco seeds as feedstock, not only offers economic benefits through cost optimization but also contributes significantly to reducing the environmental impact of biodiesel production. This sustainable approach aligns with global initiatives for cleaner energy and underscores the importance of advanced methodologies in shaping the future of biodiesel production from tobacco seeds.

17.5.1 Economic Considerations

1) *Cost optimization*: The integration of RSM and GWO allows for the identification of optimal process conditions, minimizing resource consumption and waste. Cost-effective parameter settings obtained through optimization contribute to the economic viability of the biodiesel production process.
2) *Resource efficiency*: Efficient utilization of resources, including catalysts and reactants, reduces production costs. Optimization leads to a more streamlined and economical process, ensuring competitiveness in the biodiesel market.

3) *Process scale-up*: The developed optimization model facilitates scaling up the biodiesel production process. Successful implementation on a larger scale enhances production efficiency and overall economic benefits.

17.5.2 Environmental Impacts

1) *Reduced environmental footprint*: Optimization of reaction conditions and resource utilization reduces waste and minimizes the overall environmental footprint of biodiesel production. Improved efficiency contributes to sustainable practices and aligns with environmental regulations.
2) *Energy efficiency*: The integration of RSM and GWO ensures the efficient use of energy resources in the biodiesel production process.
   Reduced energy consumption directly translates into lower greenhouse gas emissions and environmental impact.
3) *Raw material sustainability*: Biodiesel production from tobacco seeds, when optimized using advanced methodologies, contributes to the sustainable use of raw materials. The environmental impact associated with the cultivation and utilization of tobacco seeds is mitigated through efficient processes.
4) *Green chemistry practices*: The application of optimization techniques aligns with green chemistry principles. Reduced waste, minimal use of hazardous materials, and efficient processes contribute to environmentally friendly biodiesel production.

17.5.3 Sustainability and Future Prospects

1) *Long-term viability*: Implementing sustainable practices in biodiesel production enhances the long-term viability of the industry. Improved economic and environmental performance positions biodiesel as a renewable and competitive energy source.
2) *Technological innovation*: The integration of RSM and GWO reflects technological innovation in biodiesel production. Continued research and development in optimization techniques contribute to ongoing advancements in sustainable energy solutions.
3) *Market competitiveness*: Biodiesel produced through optimized processes is positioned as a competitive and environmentally responsible alternative in the energy market. Enhanced competitiveness fosters the growth and adoption of sustainable biodiesel solutions.

17.6 Conclusion

In conclusion, this study has provided a detailed exploration of tobacco seed oil as a sustainable feedstock for biodiesel production, employing a rigorous parametric study facilitated by RSM and GWO. The research aimed to address the increasing demand for alternative and environmentally friendly sources of biofuels, and tobacco seed oil emerges as a promising and under explored feedstock in this context. The application of RSM and GWO allowed for a comprehensive parametric analysis, systematically varying key factors such as reaction temperature, catalyst concentration, and reaction time to optimize the biodiesel production process. The results highlight the intricate relationships between these factors and their impact on biodiesel yield. The use of tobacco seed oil not only provides a sustainable alternative but also addresses the potential utilization of a by-product from the tobacco industry. The efficiency and robustness of the optimization process were evident through the high precision of the models, as indicated by low values of various performance metrics such as RMSE, SEP, MAE, and AAD. The comparative analysis underscores the efficacy of GWO in enhancing the optimization process, resulting in superior statistical performance. This study contributes valuable insights into the sustainable utilization of tobacco seed oil for biodiesel production. The findings serve as a foundation for further research and development in the field of alternative energy sources, providing guidance for researchers, policymakers, and industry professionals. As the world seeks to transition toward greener and more sustainable energy solutions, the integration of unconventional feedstocks, such as tobacco seed oil, presents a viable and economically feasible avenue for the biofuel industry.

References

1 Ramachandran, B., Solomon, R., Sangwan, P. et al. (2023). An experimental analysis on performanceof tobacco seed oil as an alternative fuel for diesel engine. *Alexandria Eng. J.* 80: 408–416.

2 Carvalho, F.S., Fornasier, F., Leitão, J.O.M. et al. (2019). Life cycle assessment of biodiesel production from solaris seed tobacco. *J. Clean. Prod.* 230: 1085–1095.

3 Godwin, K.A., Albert, S., and Joseph, P. (2015). Effect of biodiesel production parameters on viscosity and yield of methyl esters: *Jatropha curcas, Elaeis guineensis* and *Cocos nucifera*. *Alex. Eng. J.* 54: 1285–1290.

4 Blasio, G.D., Ianniello, R., Beatrice, C. et al. (2023). Additive manufacturing new piston design and injection strategies for highly efficient and ultra-low emissions combustion in view of 2030 targets. *Fuel* 346: 128270.

5 R.Ianniello, G.Belgiorno, G.D.Luca, C.Beatrice, G.D.Blasio, Ethanol in dual-fuel and blend fueling modes for advanced combustion in compression ignition engines alcohol as an alternative fuel for internal combustion engines (2023). https://doi.org/10.1007/978-981-16-0931-2_2.
6 Sahu, T.K., Shukla, P.C., Belgiorno, G., and Maurya, R.K. (2022). Alcohols as alternative fuels in compression ignition engines for sustainable transportation: a review. *Energy Sources Part A* 44: 8736–8759.
7 Das, S. and Chowdhury, A. (2023). An exploration of biodiesel for application in aviation and automobile sector. *Energy Nexus* 10: 100204.
8 Gu, J., Zhang, X., Song, B. et al. (2022). Chemical composition of tobacco seed oils and their antioxidant, anti-inflammatory, and whitening activities. *Molecules* 27 (23): 8516.
9 Adepoju, T.F. (2020). Optimization processes of biodiesel production from pig and neem (*Azadirachta indica*Juss) seeds blend oil using alternative catalysts from waste biomass. *Ind. Crop. Prod.* 149: 112334. https://doi.org/10.1016/j.indcrop.2020.112334.
10 Wang, A., Li, H., Zhang, H. et al. (2019). Efficient catalytic production of biodiesel with acid-base bifunctional rod-like Ca-B oxides by the sol-gel approach. *Materials* 12: 83. https://doi.org/10.3390/ma12010083.
11 Tian, Y., Chen, K., Li, X. et al. (2020). Design of high oleic tobacco (*Nicotiana Tabacum* L.) seed oil by CRISPR-Cas9-mediated knockout of NtFAD2-2. *BMC Plant Biol.* 20: 233. https://doi.org/10.1186/s12870-020-02441-0.
12 Tian, Y., Liu, X., Fan, C. et al. (2021). Enhancement of tobacco (*Nicotiana Tabacum* L.) seed lipid content for biodiesel production by CRISPR-Cas9-mediated knockout of NtAn1. *Front. Plant Sci.* 11: 599474. https://doi.org/10.3389/fpls.2020.599474.
13 García-Martínez, N., Andreo-Martínez, P., Quesada-Medina, J. et al. (2017). Optimization of catalytic transesterification of tobacco (*Nicotiana tabacum*) seed oil using supercritical methanol to biodiesel production. *Energy Convers. Manage.* 131: 99–108. https://doi.org/10.1016/j.enconman.2016.10.078.
14 Berbeć, A.K. and Matyka, M. (2020). Biomass characteristics and energy yields of tobacco (*Nicotiana Tabacum* L.) cultivated in eastern Poland. *Agriculture* 10 (11): 551. https://doi.org/10.3390/agriculture10110551.
15 Wu, S., Guo, Y., Adil, M.F. et al. (2020). Comparative proteomic analysis by Itraq reveals that plastid pigment metabolism contributes to leaf color changes in tobacco (*Nicotiana tabacum*) during curing. *Int. J. Mol. Sci.* 21 (7): 2394. https://doi.org/10.3390/ijms21072394.
16 Hu, L., Wu, Z., Jiang, Y. et al. (2020). Recent advances in catalytic and autocatalytic production of biomass-derived 5-hydroxymethylfurfural. *Renew. Sustain. Energy Rev.* 134: 110317. https://doi.org/10.1016/j.rser.2020.11031.

17 Samuel, O.D., Okwu, M.O., Tartibu, L.K. et al. (2021). Modelling of *Nicotiana tabacum* L. oil biodiesel production: comparison of ANN and ANFIS. *Front. Energy Res.* 8: 612165. https://doi.org/10.3389/fenrg.2020.612165.

18 Grisan, S., Polizzotto, R., Raiola, P. et al. (2016). Alternative use of tobacco as a sustainable crop for seed oil, biofuel, and biomass. *Agron. Sustain. Dev.* 36: 55. https://doi.org/10.1007/s13593-016-0395-5.

19 Hu, T.W. and Lee, A.H. (2015). Tobacco control and tobacco farming in African countries. *J. Public Health Policy* 36: 41–51. https://doi.org/10.1057/jphp.2014.47.

20 Usta, N., Aydogan, B., Çon, A. et al. (2011). Properties and quality verification of biodiesel produced from tobacco seed oil. *Energy Convers. Manag.* 52: 2031–2039. https://doi.org/10.1016/j.enconman.2010.12.021.

21 Gao, X., Wu, X., Liu, G. et al. (2019). Characterization and mapping of a novel premature leaf senescence mutant in common tobacco (*Nicotiana tabacum* L.). *Plants* 8: 415.2.

22 Wang, L., Du, X., Feng, Y. et al. (2018). Ectopic expression of EuWRI1, encoding a transcription factor in *E. ulmoides*, changes the seeds oil content in transgenic tobacco. *Biotechnol. Prog.* 34: 337–346.

23 Zheng, C., Yoo, J., Lee, T. et al. (2005). Fatty acid synthesis is a target for antibacterial activity of unsaturated fatty acids. *FEBS Lett.* 579: 5157–5162.7.

24 Im, D.S. (2017). FFA4 (GPR120) as a fatty acid sensor involved in appetite control, insulin sensitivity and inflammation regulation. *Mol. Asp. Med.* 64: 92–108.

25 Ghalambaz, M., Yengejeh, R.J., and Davami, A.H. (2021). Building energy optimization using grey wolf optimizer (GWO). *Case Stud. Therm. Eng.* 27: 101250.

26 Mirjalili, S., Mirjalili, S.M., and Lewis, A. (2014). Grey wolf optimizer. *Adv. Eng. Softw.* 69: 46–61.

27 Faris, H., Aljarah, I., Al-Betar, M.A., and Mirjalili, S. (2018). Grey wolf optimizer: a review of recent variants and applications. *Neural Comput. & Applic.* 30 (2): 413–435.

28 Mirjalili, S., Aljarah, I., Mafarja, M. et al. (2020). Grey wolf optimizer: theory, literature review, and application in computational fluid dynamics problems. In: *Nature-Inspired Optimizers*, 87–105. Springer. M. Ghalambaz et al. Case Studies in Thermal Engineering 27 (2021) 101250 16.

29 Al-Tashi, Q., Rais, H.M., Abdulkadir, S.J. et al. (2020). A review of grey wolf optimizer-based feature selection methods for classification. In: *Evolutionary Machine Learning Techniques*, 273–286. in: Springer.

30 Chantar, H., Mafarja, M., Alsawalqah, H. et al. (2020). Feature selection using binary grey wolf optimizer with elite-based crossover for Arabic text classification. *Neural Comput. & Applic.* 32 (16): 12201–12220.

31 Qu, C., Gai, W., Zhang, J., and Zhong, M. (2020). A novel hybrid grey wolf optimizer algorithm for unmanned aerial vehicle (UAV) path planning. *Knowl.-Based Syst.* 105530.

32 Qu, C., Gai, W., Zhong, M., and Zhang, J. (2020). A novel reinforcement learning based grey wolf optimizer algorithm for unmanned aerial vehicles (UAVs) path planning. *Appl. Soft Comput.* 89: 106099.

33 Majumder, P. and Eldho, T. (2020). Artificial neural network and grey wolf optimizer based surrogate simulation-optimization model for groundwater remediation. *Water Resour. Manag.* 34 (2): 763–783.

34 Samuel, O.D., Okwu, M.O., Oyejide, O.J. et al. (2020). Optimizing biodiesel production from abundant waste oils through empirical method and grey wolf optimizer. *Fuel* 281: 118701.

35 Kharwar, P.K. and Verma, R.K. (2020). Exploration of nature inspired Grey wolf algorithm and Grey theory in machining of multiwall carbon nanotube/polymer nanocomposites. *Eng. Comput.* 1–22.

36 Miao, Z., Yuan, X., Zhou, F. et al. (2020). Grey wolf optimizer with an enhanced hierarchy and its application to the wireless sensor network coverage optimization problem. *Appl. Soft Comput.* 106602.

37 Zhao, M., Wang, X., Yu, J. et al. (2020). Optimization of construction duration and schedule robustness based on hybrid grey wolf optimizer with sine cosine algorithm. *Energies* 13 (1): 215.

38 Samadianfard, S., Darbandi, S., Salwana, E. et al. (2020). Predicting soil electrical conductivity using multi-layer perceptron integrated with Grey wolf optimizer. *J. Geochem. Explor.* 106639.

39 Yk, A. and Llsb, C. (2019). Silicon carbide and its composites for nuclear applications – historical overview. *J. Nucl. Mater.* 526: 151849.

40 Shuyu, D., Niu, D., and Li, Y. (2018). Daily peak load forecasting based on complete ensemble empirical mode decomposition with adaptive noise and support vector machine optimized by modified Grey wolf optimization algorithm. *Energies* 11 (1): 163–175. https://doi.org/10.3390/en11010163.

41 Xu, M., Girish, Y.R., Rakesh, K.P. et al. (2021). Recent advances and challenges in silicon carbide (SiC) ceramic Nanoarchitectures and their applications. *Mater. Today Commun.* 28: 102533.

42 Sodeifian, G., Ardestani, N.S.S.S.A., and Ghorbandoost, S. (2016). Application of supercritical carbondioxide to extract essential oil from Cleome coluteoides Boiss: experimental, response surface and grey wolf optimization methodology. *J. Supercrit. Fluids* 114: 55–63.

43 Okwu, M.O. and Tartibu, L.K. (2022). Metaheuristic optimization: nature inspired algorithms and swarm intelligence, theory and applications. In: *Studies in Computational Intelligence*, 927. Switzerland: SpringerInternational Publishing. https://www.springer.com/gp/book/978-3-030-61111-8.

18

Predictive Modeling of Biodiesel Production from Chrysobalanus Icaco: A Fuzzy Logic Approach

18.1 Introduction

The rising demand for sustainable and eco-friendly energy alternatives has spurred significant interest in biodiesel as a viable substitute for conventional diesel fuel [1, 2]. Among substitute fuels, biodiesel is considered the most promising and attractive alternative fuel because it is sulfur-free, nontoxic, biodegradable, eco-friendly, and more reliable [3]. Moreover, biodiesel can be applied to diesel engines without changing the engine systems because the combustion and emission characteristics of biodiesel are almost similar to conventional diesel [4]. In addition, compared to diesel, biodiesel has a higher cetane number [5], no aromatics, and 10%–11% oxygen content [6, 7]. These characteristics of biodiesel fuel can reduce CO, HC, and PM emissions in exhaust gases. However, there is no conclusive trend in oxygenated compounds such as aldehydes and ketones [8]. Recognized for its advantageous properties, including renewability, cleaner combustion, and a reduced carbon footprint, biodiesel presents a compelling solution to mitigate the environmental impact associated with fossil fuels. Chrysobalanus Icaco (CI), commonly known as cocoplum, emerges as a promising low-cost feedstock for biodiesel production due to its high oil yield [9]. As the global focus on renewable energy intensifies, it becomes imperative to explore innovative approaches that enhance the efficiency and predictability of biodiesel production processes. This study focused on the intricate system of biodiesel production from CI oil, with a distinctive emphasis on the application of fuzzy logic (FL) as a predictive tool.

Feedstocks for Sustainable Biodiesel Production: Characterization, Selection, and Optimization, First Edition. Chinwe P. Okonkwo and Modestus O. Okwu.

18.2 Review of Literature

Previous chapters focused more on biodiesel feedstocks, with an exposition of the diverse biodegradable materials utilized in biodiesel production. This entailed a thorough examination of existing research, articles, and publications pertaining to biodiesel feedstocks to gain insights into the range of biodegradable material options. The transesterification process has equally been discussed.The process is utilized in the production of biodiesel, employing either homogeneous (acidic or alkaline) or heterogeneous (acidic, alkaline, or enzymatic) catalysts. Notably, among these options, heterogeneous base-catalyzed transesterification emerges as the predominant method in biodiesel synthesis due to its rapidity and its ability to generate exceptionally high yields within short reaction periods [10]. The effectiveness of the transesterification reaction is contingent upon various variables, including the water content of oils or fats, the presence of free fatty acids (FFAs), the molar ratio of glycerides to alcohol, the nature of catalysts, reaction time, and reaction temperature [11]. A prerequisite for this process involves an acid pretreatment to convert FFAs into the corresponding methyl/ethyl esters before undergoing transesterification. The FFA content in triglycerides derived from diverse feedstocks exhibits significant variation, influenced by factors such as soil, environment, and chemical composition [12].

Cocoplum (*Chrysobalanus icaco*), a medium-sized shrub native to the coastal regions of South America, finds application in traditional medicine for treating conditions such as leucorrhoea, bleeding, and persistent diarrhea [13]. Recognized for its diuretic, hypoglycemic, and anti-antigenic properties, Cocoplum holds cultural significance. In Northern Brazil, the roots of Cocoplum are utilized in the management of diabetes. Phytochemical analyses of its fruits have identified myricetin and pomolic acid, with claims of possessing anti-hyperglycemic properties [14]. In addition, studies on Cocoplum leaves indicate their potential to restore insulin sensitivity, regulate blood glucose levels, and counteract weight gain in rats subjected to a high-fat diet [15]. The seed oil of Cocoplum is noteworthy for its composition, containing triacylglycerides and a significant concentration of unsaturated fats, particularly conjugated linoleic fatty acyls. Gas chromatography analysis of the hydrolyzed fats has revealed the presence of palmitic, stearic, oleic, and linoleic acids [16]. Despite these intriguing properties and their established applications in traditional medicine, there is currently no literature exploring the utilization of Cocoplum in biodiesel production.

18.2.1 Fuzzy Logic Models

The concept of FL, first introduced by Lotfi A. Zadeh [17], has developed into a linguistic methodology designed to tackle complex stochastic issues. FL, a computational paradigm inspired by human reasoning and decision-making processes, provides a flexible and robust framework for modeling intricate systems fraught with uncertainties. Its utility spans various knowledge domains, including the regulation of automatic electro-mechanical processes [18] and decision-making [19]. According to Nobre et al. [20], FL is perceived as a computational and mathematical framework well-suited for expressing approximate reasoning, accommodating everyday life concepts, experiences, and observations, all characterized by "fuzzy limits" [21]. Benson et al. [22] depict FL as an efficient methodology for constructing models, integrating intuition, and agent-related behavior.

The application of FL has been notably significant in addressing economic challenges, particularly within the energy sector. Spandagos et al. [23] argue that gaining an understanding of the factors shaping consumer energy behavior requires the creation of energy consumption models that consider economic factors. Consequently, Spandagos devised a model based on FL that incorporates notions of bounded rationality, time discounting of gains, and pro-environmental behavior. This methodology illustrates the effectiveness of employing FL to combine economic and physical quantitative data with qualitative concepts.

Furthermore, FL has been employed in appraising energy systems, as demonstrated in a study conducted by Gamalath et al. [24]. They introduced an evaluation framework for assessing the state of energy systems in multi-unit residential buildings (MURB). This assessment method utilizes FL to handle data uncertainty and imprecision, employing rules to amalgamate various performance categories and generate a rating that reflects the overall condition of a MURB. This research underscores the potential of FL in improving asset management and operational strategies for existing energy systems. Functioning as a heuristic technique, FL avoids the necessity for intricate mathematical models, relying instead on a basic grasp of linguistic terminologies. In contrast to Charles Boole's binary logic [25], where an element is either part of the whole or not, Zadeh's FL permits gradual variation between one and zero when an element is regarded as part of a fuzzy whole. The membership function (MF) assumes a critical role in problem-solving, indicating the extent to which an element is linked to or possesses a characteristic associated with a linguistic value.

By harnessing the capabilities of FL, this study aims to create a predictive model that not only anticipates biodiesel yield but also improves the comprehension of the intricate interactions among variables affecting the transesterification process in biodiesel production. The investigation places a significant emphasis on the use of natural heterogeneous catalysts, specifically waste periwinkle shells, selected for their environmentally friendly characteristics. These shells, typically discarded as waste, function as an innovative catalyst for biodiesel production, aligning with the principles of green chemistry. The research further delves into the physicochemical properties of CI oil, renowned for its high oil yield and cost-effectiveness, ensuring its adherence to established standards for biodiesel production. With the growing momentum in the pursuit of sustainable energy alternatives, the integration of advanced computational techniques, such as FL, becomes essential for enhancing the efficiency and predictability of biodiesel production processes. This research seeks to contribute to the evolving field of biofuel technology by proposing a robust prediction model tailored for CI, thereby advancing the potential for sustainable and environmentally conscious energy solutions.

18.3 Materials and Methods

The process of constructing a biodiesel production prediction model from CI using FL employs a methodical and multi-step strategy. The goal is to capitalize on the distinctive features of FL to develop a robust model capable of accurately forecasting biodiesel yield from the selected feedstock. Figure 18.1 illustrates the flowchart derived from the research methodology. This flowchart delineates a thorough research approach that amalgamates a literature review, data collection, FL modeling, and analysis to anticipate biodiesel yield based on input variables. The implementation of the fuzzy inference system within this methodology utilizes the FL Toolbox in MATLAB.

The steps include the identification of an appropriate feedstock for biodiesel production based on input variables. To facilitate this, a dataset for input variables and experimental results for a parametric study was acquired. This process involved the collection of experimental data related to the identified input variables and their influence on biodiesel yield. The data collection phase encompassed conducting new experiments. The final dataset includes values for each input variable alongside the corresponding experimental results, specifically biodiesel yield.

This dataset is fed into the FL Toolbox in MATLAB to develop a fuzzy inference system. Input and output variables in the fuzzy inference system are defined based on the dataset; MF are set up for each variable, and fuzzy rules are equally established to capture the relationships between the inputs and the desired output; fuzzy inference system is simulated using the dataset to analyze its performance so as to help evaluate the relationships between input variables and biodiesel yield; predicted biodiesel yield is obtained based on the learned relationships from the dataset; results obtained from the fuzzy inference system is analyzed, and conclusions are drawn about the effectiveness of the model in predicting biodiesel yield based on input variables.

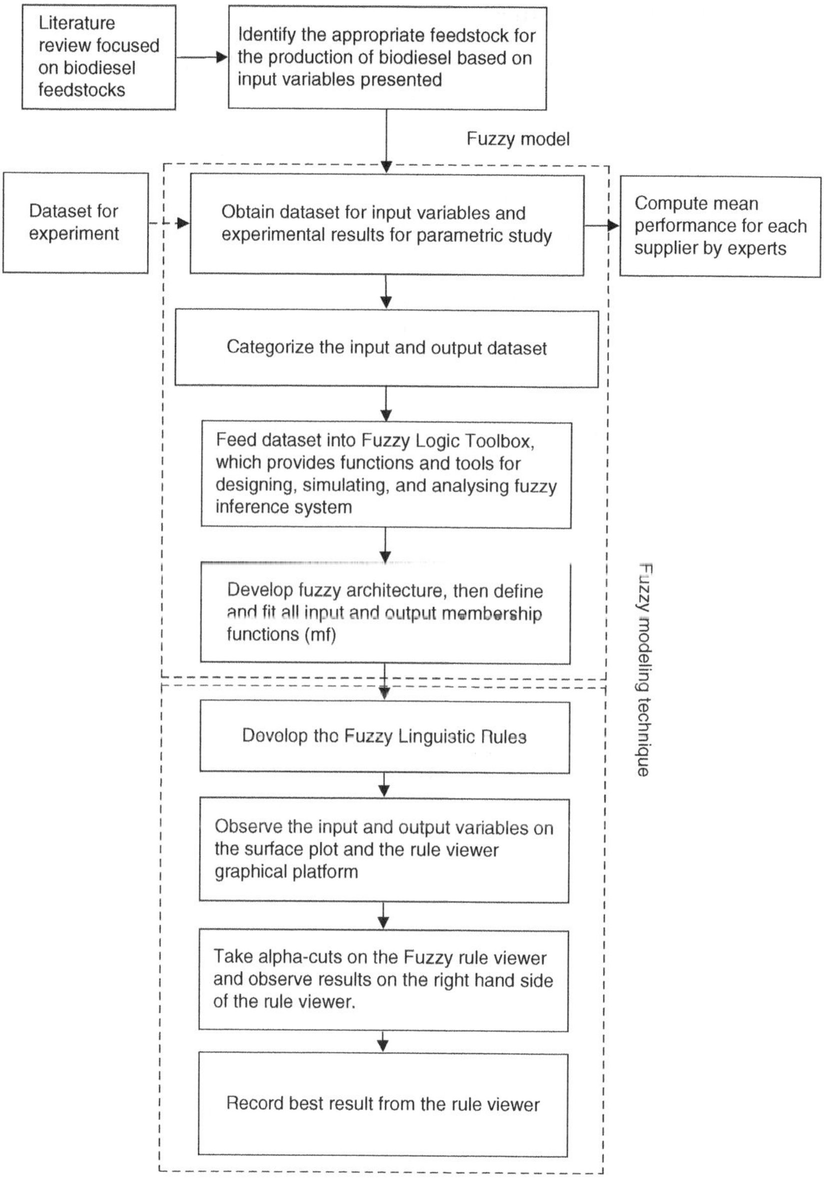

Figure 18.1 Flow chart based on research methodology.

18.3.1 Materials for the Study

Materials: The materials used for the study include the following: Stirrer, conical flask, fume cupboard, beaker, magnetic stirrer, point tester, pensky martens flash point tester, digital weighing balance, density bottle, separatory funnel, TECHNO R175A diesel engine, digital magnetic heater, powdered seeds, soxhlet extractor, viscosity bath, oven, muffle furnace, calcinator, Emission Scanning Electron Microscope (JSM-670IF), Fourier Transform infrared spectroscopy (FTIR) Spectrophotometer, GC–MS QP2010 Plus Shimadzu Japan, samples of feedstocks and diesel blends.

Reagents: The reagents used were of analytical grade, and they include concentrated sulfuric acid, distilled water, chloroform, potassium hydroxide, glacial acetic acid, hydrochloric acid, iodine monochloride, phenolphthalein, indicator, sodium hydroxide, *n*-hexane, potassium iodide, ethanol and methanol, sodium thiosulphate, hydrochloric acid, tetraoxosulphate (vi), sodium chloride.

18.3.2 Collecting Samples and Extracting Cocoplum Seed Oil

Chrysobalanus icaco, commonly known as cocoplum seeds, was procured in its raw form from a local market and subsequently identified by a taxonomist. To prepare the material for further processing, it underwent a preliminary step of sun drying. Following this, the dried material was finely crushed utilizing a standard industrial blender. The resultant mixture underwent a meticulous refinement process, passing through a sieve with a mesh size of 500 μm to attain the desired consistency. This careful sieving ensured that the final product met the specific standards required for subsequent stages in the biodiesel production process.

18.3.3 Oil Extraction

For the extraction of oil, the Soxhlet extraction method was utilized. Initially, the accurately weighed dried sample was enclosed in a semi-permeable cotton material. This encapsulated sample was then placed with precision into the thimble of a 500 mL Soxhlet extractor. Following this, 400 mL of *n*-hexane was introduced into a 500 mL round-bottom flask. The assembly of the Soxhlet apparatus involved connecting the thimble, which housed the sample in a semi-permeable membrane, to the condenser. The condenser was subsequently affixed to the round-bottom flask containing the *n*-hexane. This methodical setup ensured the efficient extraction of oil from the sample through the cyclical process of solvent circulation.

The initiation of the Soxhlet extraction system involved heating it on a heating mantle while circulating water was used to maintain the condenser's outer jacket. During the operation, *n*-hexane served as the solvent, extracting oil from the sample housed in the thimble. Upon completion of the extraction process, where all the oil was successfully extracted from the sample, the extraction procedure was concluded. The de-fatted sample was discarded, and the resultant mixture of oil and *n*-hexane in the round-bottom flask underwent separation through the use of a rotary evaporator device. Subsequent to the extraction and separation steps, the yield percentage was computed to assess the efficiency of the oil extraction method.

18.3.4 Catalyst Preparation and Characterization

The periwinkle shell sample was obtained from the market and underwent a meticulous cleaning process. Afterward, the cleaned sample was subjected to a drying process and subsequently finely crushed into a paste. As part of the additional preparation steps, a 5 g portion of the periwinkle shell was subjected to calcination at a temperature of 800 °C for a duration of two hours. Concurrently, another 10 g portion of the finely ground periwinkle shell underwent acidification using concentrated sulfuric acid (10 mL). This acidification process was followed by multiple rinses until all remnants of the acid were completely eliminated.

The processed samples, of acidified and calcined form, were carefully packed and subsequently delivered for supplementary assessment through scanning electron microscopy (SEM) and FTIR analyses.

18.3.5 Biodiesel Production (Transesterification)

The process of methyl ester production through trans-esterification was carried out using a 500 mL round-bottom glass reactor equipped with a spherical three-neck setup, as illustrated in Figure 18.2. In each experimental run, a predetermined amount of methanol and a modified catalyst were introduced, prestirred for 10 minutes to ensure proper dissolution, and then supplemented with 50 mL of oil. The reactor was equipped with a reflux condenser featuring cold water circulation in the outer jacket and positioned in the mid-neck. A mercury-in-glass thermometer, held in a plastic bung, was inserted into the right neck, while a capsular stirring knob, secured with a plastic bung, was inserted through the left neck.

The equipment was activated and positioned on a magnetic heating mantle. Upon initiation, the stirring mechanism maintained a consistent speed and temperature, indicating the commencement of the reaction. Each reaction was allowed to progress for the specified duration and temperature. Following the transesterification process, the resulting product underwent a settling period of 12 hours in a separating funnel to facilitate the separation of glycerol. The methyl ester (biodiesel) was retained after removing the crude glycerol through the funnel tap. To purify the biodiesel, it underwent a rinsing process with hot water,

Figure 18.2 Process required for the production of biodiesel. (a) represents the raw samples of Chrysobalanus icaco that have been washed and sundried. (b) illustrates the extraction of the oil using nhexane for Soxhlet extraction. (c) illustrates the separation process using a separatory funnel. (d) illustrates the transesterification process using a magnetic stirrer.

followed by transfer into a 250 mL beaker. The biodiesel was then heated to 105 °C to eliminate any remaining water molecules, subsequently cooled, and stored in calibrated specimen vials [12].

18.4 Results and Discussion

The overall characteristics indicate that the oil derived from *C. icaco* seeds holds significant promise as a biodiesel source, aligning well with the ASTM standards established for oils used in biodiesel production. The notably low acid value

Table 18.1 Physical and chemical characteristics of the oils obtained from *C. icaco*.

Properties	*C. icaco* seed oil
Yield(%)	21.90
Color	light brown
Refractive index at 29 °C	1.457
Moisture	3.5
Kinematic viscosity at 40 °C (mm^2/s)	24.16
The required value of energy (kJ/kg)	43 729
The required value of acid (mgKOH/g)	4.1
Peroxide value (meq/kg)	0.93
Saponification (mgKOH/kg)	185
Cloud point	12
Density(g/mL)	0.9278
Flashpoint (°C)	260
Iodine value (g/100 g of oil)	107
Molecular weight (g/mol)	771.66

implies that the oil can undergo direct processing into biodiesel through a base-catalyzed transesterification method [26]. Elevated levels of FFAs in the oil have the potential to deactivate the base catalyst, leading to soap formation and subsequently reducing the biodiesel yield. The acidity observed in biodiesel may stem from residual mineral acids in the production process, lingering FFAs resulting from hydrolysis, or post-hydrolysis esterification processes. In addition, it may arise from the oxidation of by-products, giving rise to other organic acids [27] (Table 18.1).

The iodine value findings, registering at 107, signify that *C. icaco* seed oil is categorized as nondrying. A higher iodine content in the oil indicates elevated levels of unsaturation, rendering it more prone to oxidation and polymerization. Biodiesel possessing a heightened iodine content may undergo polymerization, leading to the formation of deposits on injector nozzles, piston rings, and piston ring grooves. Furthermore, the stability of biodiesel at higher temperatures diminishes as the iodine content increases. It is noteworthy that the Euro norm for iodine value in biodiesel, as outlined in EN 14214, is set at 120, while the German standard (DIN 5160) designates a limit of 115 [28].

It is essential to recognize that the iodine value is also influenced by the type and ester content of the feedstocks used in biodiesel production. The choice of feedstock plays a pivotal role in determining the iodine value, impacting the susceptibility of biodiesel to oxidation and its overall stability under various conditions. Therefore, understanding and managing the iodine value is crucial for optimizing biodiesel production processes and ensuring compliance with established standards.

The kinematic viscosity of the oil surpasses that of conventional diesel, primarily attributed to the specific fatty acid composition and molecular weight of the oil [29]. Kinematic viscosity plays a direct role in fuel performance, impacting atomization and combustion quality. Despite the oil's kinematic viscosity being higher than that of conventional diesel, it remains within the acceptable range defined by ASTM standards [30].

Table 18.2 provides a breakdown of the prevalent fatty acids found in *C. icaco*, highlighting stearic acid as the dominant saturated fatty acid, constituting 25%, followed by palmitic acid at 6%. The sole unsaturated fatty acid identified is oleic acid, accounting for 20% of the composition. In addition, linolenic acid emerges as the singular polyunsaturated fatty acid, making up 13% of the overall composition. The presence of palmitic acid, recognized for its natural antioxidant properties, suggests that *C. icaco* may inherently exhibit high oxidation stability. This inherent quality could potentially obviate the need for supplementary additives aimed at enhancing oxidation stability in biodiesel production. The composition's richness in stearic and oleic acids also contributes to the overall fatty acid profile, impacting the fuel's characteristics and potential applications (Table 18.3).

Table 18.2 Composition of fatty acids in methyl ester from *C. icaco* seed oil.

S/no	Common name	Systematic name	Shorthand	R.A (%)
1	Palmitic acid	Hexadecanoic acid	C16:0	6
2	Arachidic acid	5,8,11,14-Eicosatetraenoic acid	C20:1	Trace
3	Stearic acid	Octadecanoic acid	C18:0	25
4	Almonic acid	4-oxo-Octadeca-cis-9-trans-11-trans-13-cis-15-tetraenoic acid	C18:4	18
5	Linoleic acid	cis-9,12,15-Otacatrienoic acid	C18:3	13
6	Oleic acid	cis-9-Octadeaoic acid	C18:1	20
7	α-Licanic	4-Oxooctadeca-9,11,13-trienoic acid	C18:3	10
8	α-Parinaric	Octadeca-9,11,13,15-tetraenoic acid	C18:2	11

Table 18.3 Catalysts characterization of PSA.

Parameter	Raw	Acid activated	Calcined
Moisture (%)	0.05	0.06	0.04
Bulk density (g/cm^3)	0.729	0.571	0.539
pH	6.74	5.73	6.07
Surface area (m^2/g)	755	821	856
Carbon (%)	4.08	1.24	2.17
Organic matter (%)	12.09	3.68	6.44
Loss on ignition (%)	9.29	3.96	0.85
Particle density (g/cm^3)	1.52	1.39	0.83
Total porosity (%)	52.04	58.92	35.06
Ash (%)	3.09	2.44	1.76

18.4.1 Surface Morphology of the Catalyst

The examination of the catalyst's surface morphology was conducted using a SEM, as illustrated in Figure 18.3. The SEM images unveil that the pore widths within the carbon structures of the PSA catalysts increase in correlation with the duration of carbonization, peaking before eventually collapsing. Among the carbonized catalysts, those subjected to calcination displayed the most extensive and well-organized pores. The substantial diameter of these pores contributes to an augmented surface area of the catalyst, creating favorable conditions for the transformation of bio-oil. This observation implies that, when deriving a catalyst from PSA, achieving significant pore diameters may necessitate a carbonization period exceeding 60 minutes. The observed breakdown of the carbon structure after one hour of carbonization may be attributed to the conversion of biomass into graphite [31]. Understanding these structural changes in the catalyst is vital for optimizing the conditions for bio-oil transformation and enhancing the efficiency of the catalytic process.

The examination of the catalyst's activity and stability included a comprehensive study of its reusability. The calcined form was obtained by exposing the raw PSA to a temperature of 800 °C for a duration of three hours. On the contrary, the acidified form was prepared by treating 5 g of the raw PSA with 5 mL of sulfuric acid, followed by thorough washing until no residual acid was detectable, as confirmed by litmus paper testing. In each successive run, both the calcined and acidified samples underwent their respective processes to maintain their structural and chemical characteristics. This reusability study is crucial in assessing the

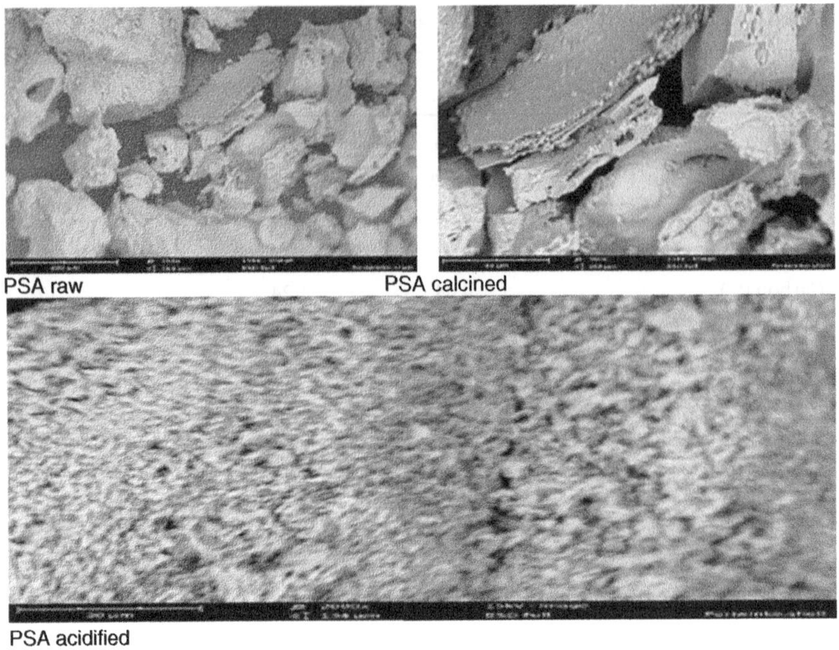

Figure 18.3 Investigating the reusability of PSA catalyst through scanning electron microscope (SEM) imaging in its original, calcined, and acidified states.

long-term performance and durability of the catalyst, providing valuable insights into its practical application in biodiesel production.

The reaction was carried out under the specified conditions:

Oil ratio of methanol: 10:1
Catalyst quantity: 0.75 g
Reaction time: 1 hour 30 minutes

This experimental configuration facilitated an investigation into the reusability of the catalyst, providing an opportunity to evaluate its performance and stability across multiple reaction cycles. The set parameters, including the oil ratio, catalyst quantity, and reaction time, were intentionally chosen to mimic realistic biodiesel production conditions and to comprehensively assess the catalyst's efficacy over an extended duration of use (Table 18.4).

Upon completing the transesterification process under the specified optimal conditions, characterized by a 10:1 methanoltooil molar ratio, 0.75 g of catalyst, and a reaction time of 1 hour 30 minutes, the focus of the investigation shifted toward assessing the reusability of the catalyst for subsequent transesterification

Table 18.4 PSA catalyst regeneration.

Runs	Remains of the catalyst after each run	Conversion yield CSO (vol. %) (acid activated)	Conversion yield CSO (vol. %) (calcined)
1	1.8	89.26	92.25
2	1.3	84.50	87.40
3	0.9	82.80	85.90
4	0.5	80.70	82.10
5	0.2	78.50	80.60

runs while maintaining the same operating conditions. In the initial regeneration cycle, both the acidactivated and calcined catalysts demonstrated promising conversion rates, achieving 89.2% and 92.25%, respectively, in the conversion of Chrysobalanus Icao Seed Oil (CSO) to biodiesel.

However, as the regeneration cycles progressed, a gradual decline in conversion rates was observed. After the fifth regeneration run, the conversion rates decreased to 78.50% for the acid-activated catalyst and 80.60% for the calcined catalyst. This diminishing trend in conversion rates over successive regeneration cycles suggests the need for a more in-depth understanding of catalyst behavior and potential strategies for improving or maintaining performance across multiple reaction cycles. It also underscores the importance of considering catalyst reusability in the design and optimization of biodiesel production processes.

The assessment of catalyst reusability, as illustrated in Figure 18.4, involved a meticulous examination of various process parameters. A particularly noteworthy observation was that the percentage conversion achieved with the calcined (carbon-based) catalysts surpassed that of the acid-activated catalysts. However, a discernible decline in catalytic activity manifested in the fourth cycle, signaling the deactivation of specific active sites, potentially influenced by impurities within the catalyst. It is further inferred that the leaching of calcium oxide content in the catalyst serves as the primary factor contributing to the abrupt decrease in conversion rates, aligning with findings from previous studies [32, 33]. This comprehensive assessment underscores the dynamic nature of catalyst performance during multiple regeneration cycles, providing valuable insights into the factors influencing reusability. The identification of potential impurities and their impact on catalyst deactivation highlights key considerations for sustainable and efficient biodiesel production processes. Understanding these dynamics contributes to the ongoing efforts to optimize catalyst design and enhance the overall reliability of biodiesel production systems.

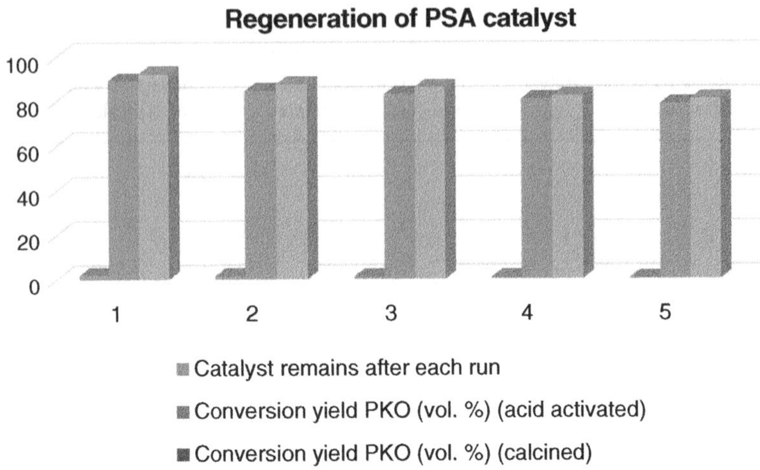

Figure 18.4 PSA catalyst regeneration.

18.4.2 Fuzzy Mamdani Model (FMM)

The Fuzzy Mamdani Model (FMM) is a popular method in the field of FL control and decision-making. It was introduced by Lotfi Zadeh in the 1970s [34]. The FMM is particularly useful when dealing with systems that involve uncertainty and imprecision. Important terms to note:

Fuzzy sets: The first step in the FFM is defining fuzzy sets. Fuzzy sets are used to represent linguistic variables that have imprecise boundaries. A linguistic variable is a variable whose values are words or sentences rather than numerical values. For example, in the context of biodiesel prediction, variables like "temperature,""pressure," and "reactor time" might be considered linguistic variables.

MF: For each fuzzy set, anMF is defined. MF describe the degree to which an element belongs to a fuzzy set. These functions map the input values to membership values between 0 and 1. There are various types of MF, such as triangular, trapezoidal, Gaussian, and more. The choice of MF depends on the characteristics of the variables being modeled.

Fuzzy rules: Fuzzy rules are the core of the FFM. These rules express relationships between the fuzzy sets of input variables and the fuzzy sets of output variables.

Each rule consists of an antecedent (if-part) and a consequent (then-part). The antecedent uses FL to determine the degree to which each rule is satisfied, and the consequent defines the fuzzy set for the output variable.

Inference: The inference engine combines the fuzzy rules to determine the overall fuzzy set for each output variable. This is typically done using FL operators

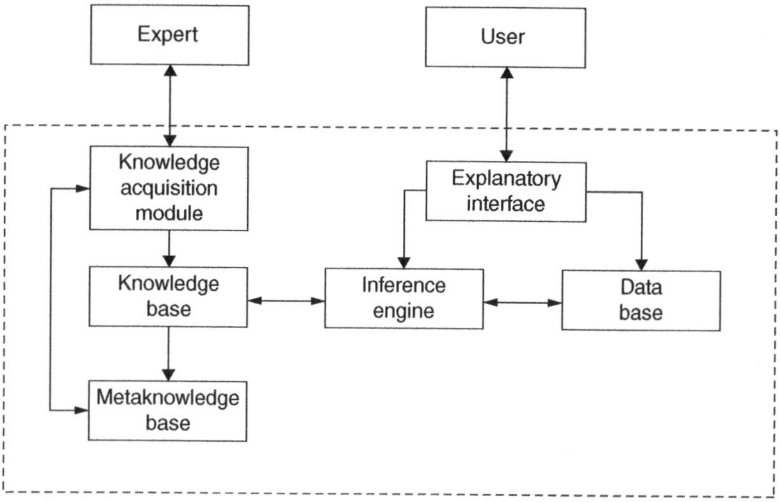

Figure 18.5 Fuzzy logic flowchart.

such as AND, OR, and NOT. The result is a fuzzy output that represents the degree of membership of the output variables.

Defuzzification: The final step is to convert the fuzzy output into a crisp, nonfuzzy value. This process is called defuzzification and is necessary for obtaining a concrete and usable result. Common defuzzification methods include centroid, mean of maximum, and others. Figure 18.5 illustrates the flow chart of FL, depicting the process of acquiring expert opinions and knowledge.

18.4.3 FFM in the Context of Biodiesel Production

In the framework of biodiesel prediction, the input parameters, including temperature, mole ratio, catalyst concentration, and yields from *C. icaco* as raw seed oil, *C. icacos* seed oil acid activated, and *C. icaco* oil calcined, are clearly defined. Simultaneously, the output variable is biodiesel yield. Subsequently, utilizing either expert knowledge or experimental data, relationships between these variables are modeled through fuzzy sets, MF, and fuzzy rules. The implementation of a FFM in MATLAB 2023 is achieved, leveraging the FLtoolbox, which offers functionalities and tools for the design, simulation, and analysis of fuzzy inference systems. The process involves creating a fuzzy inference system, defining input and output variables, setting up MF, and specifying fuzzy rules using the toolbox functions. The architecture of the fuzzy system is illustrated in Figure 18.6, depicting four input variables and one output. This structured approach establishes a foundation for effective modeling and analysis within the Fuzzy Mamdani framework.

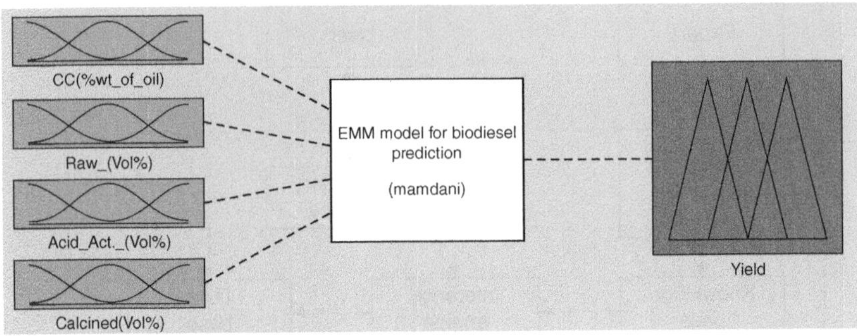

Figure 18.6 FMM architecture for four inputs and one output.

18.4.4 Membership Function

Computational studies involving Fuzzy analysis are incomplete without involving adjustments to input parameters. The main parameters considered in this study include catalyst concentration, *C. icaco* as raw seed oil, *C. icaco* seed oil acid activated, and *C. icaco* oil calcined. Concurrently, the resulting experimental biodiesel yield was imported as the expected output variable. In Figure 18.7, the MF for the input variables, encompassing catalyst concentration, *C. icaco* as raw seed oil, *C. icaco* seed oil acid activated, and *C. icaco* oil Calcined, are graphically represented in orange font. This section on the MF plot signifies categories such as very very low (VVL), very low (VL), low (L), high (H), very high (VH), and very very high (VVH). Correspondingly, the MF for the output variable, which is the biodiesel yield, is depicted in Figure 18.8, categorized as VL, low (L), high (H), and VH.

These MF support the FL system to interpret input data using linguistic terms, facilitating adaptable and imprecise reasoning. Through the categorization of input data into these linguistic variables, the system can make well-informed decisions and generate more precise predictions of biodiesel yield, considering the diverse compositions of biodegradable feedstocks. In summary, Figure 18.7 elucidates the MF associated with the FL pertaining to the input variables, while Figure 18.8 presents the plot for the MF concerning the output mapping.

The biodiesel yield is intricately tied to the input variables, with varying compositions of these components exerting a significant influence on the final biodiesel output. To encapsulate the linguistic variables associated with the input parameters, four to five MFs are defined as listed: VVL, VL, L, H, and VH.

1) *VVL*: This MF represents an exceedingly minimal proportion of catalyst concentration, *C. icaco* as raw seed oil, *C. icaco* seed oil acid activated, and *C. icaco* oil calcined, as outlined in Table 18.5. The values associated with this MF indicate an exceptionally low presence of these variables in the mixture.

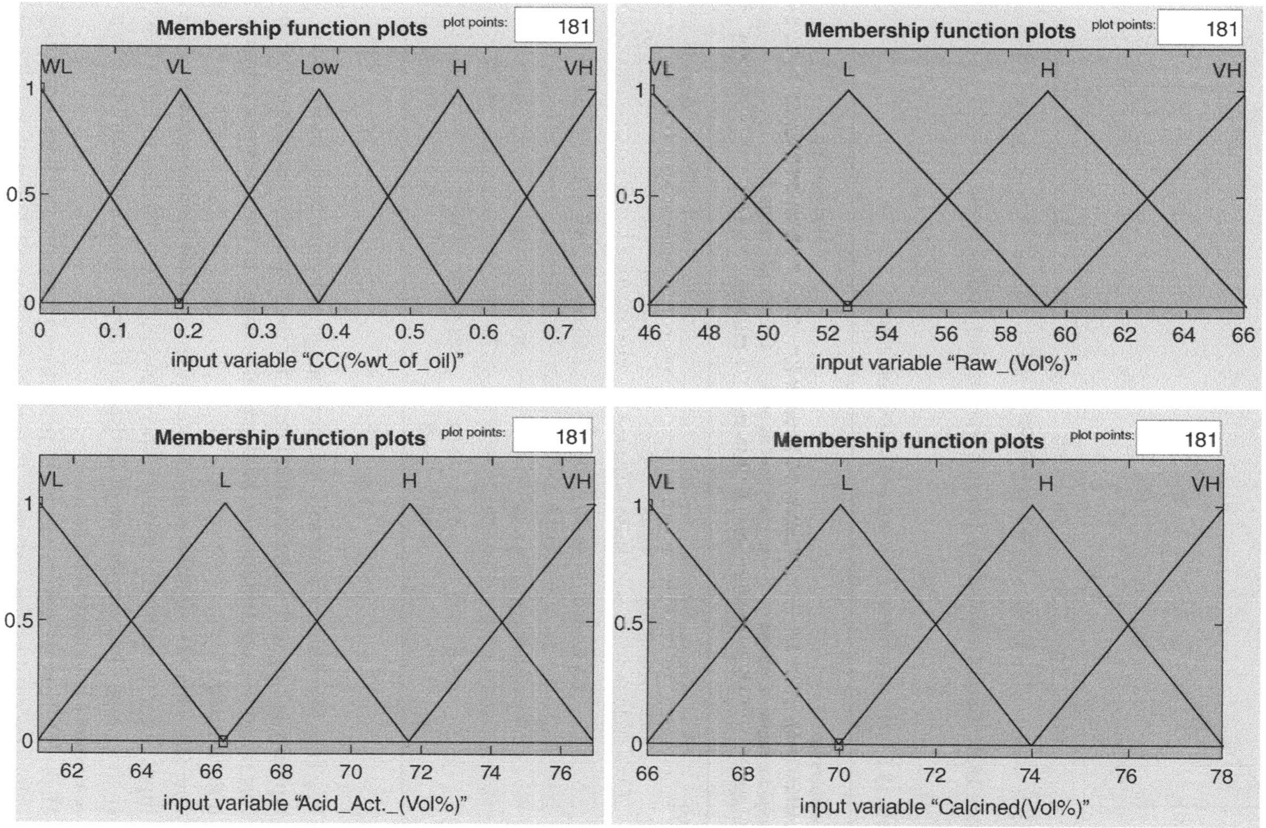

Figure 18.7 Input membership function based on four input variables.

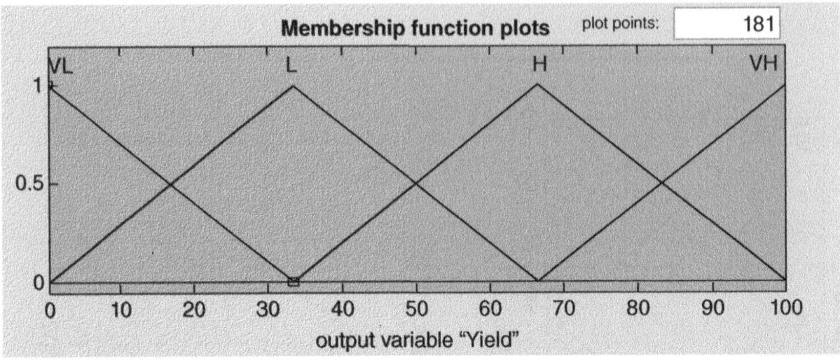

Figure 18.8 Output membership function representing biodiesel yield.

Table 18.5 MF parameters for input and output variables.

Membership function (MF) range for input variables	Catalyst concentration	*C. icaco* as raw seed oil	*C. icaco seed* oil as acid activated	*C. icaco* oil calcined	Membership function (MF) range for output	Yield(%)
VVL	0.0-0.1-0.2	—	—	—	VLP	0-10-20
VL	0.0-0.2-0.4	46-48-50-52	60-62-64-66	66-68-70	LP	20-25-30
L	0.4-0.5-0.6	46-50-54-58	66-68-70-72	66-68-72-78	AP	30-40-50
H	0.4-0.6-0.8	52-56-62-66	66-70-74-78	70-72-76-78	HP	50-55-60-70
VH	0.6-0.7.0.8	60-62-64-66	72-74-76-78	74-76-78	VHP	70-80-90-100

2) *VL*: This MF denotes a notably low percentage of hybrid waste in the blend, typically falling within the specified range for the four input variables. Within this range, the contribution of biodegradable waste to the feedstock remains quite limited.
3) *Low (L)*: This MF signifies a moderate presence of oil seed waste and other input variables within the mixture, encompassing values detailed in Table 18.6. In this range, the contribution of feedstock is appreciable and contributes to a reasonably balanced mixture.
4) *High (H)*: This MF covers a range of percentages for input variables, providing values for catalyst concentration, *C. icaco* as raw seed oil, *C. icaco* seed oil acid

Table 18.6 Fuzzy Mamdani model solution for the predicted values of *C. icaco* using the three forms of PSA in different forms (raw, acid activated, and calcined).

Std run	A: A	Catalyst. conc. (wt. % of oil)	Raw (vol. %)	Activated (vol. %)	Calcined (vol. %)	FMM predicted value yield 1 (vol. %)	FMM predicted value yield 2 (vol. %)	FMM predicted value yield 3 (vol. %)
1	9	0.75	57.6666	68.9999	73.0000	62.5000	76.5000	75.2500
2	6	3.75	56.7143	69.5715	71.5714	61.6667	76.0000	77.5000
3	2	0.75	57.6666	68.9999	73.0000	62.5000	76.5000	75.2500
4	14	3.75	56.7143	69.5715	71.5714	61.6667	76.0000	77.5000
5	10	0.75	57.6666	68.9999	73.0000	62.5000	76.5000	75.2500
6	8	3.75	56.7143	69.5715	71.5714	61.6667	76.0000	77.5000
7	4	0.75	57.6666	68.9999	73.0000	62.5000	76.5000	75.2500
8	19	3.75	56.7143	69.5715	71.5714	61.6667	76.0000	77.5000
9	1	0.75	57.6666	68.9999	73.0000	75.2500	76.5000	75.2500
10	13	3.75	56.7143	69.5715	71.5714	61.6667	76.0000	77.5000
11	23	0.75	57.6666	68.9999	73.0000	62.5000	76.5000	75.2500
12	11	3.75	56.7143	69.5715	71.5714	61.6667	76.0000	77.5000
13	12	0.75	57.6666	68.9999	73.0000	62.5000	76.5000	75.2500
14	24	3.75	56.7143	69.5715	71.5714	61.6667	76.0000	77.5000
15	18	0.75	57.6666	68.9999	73.0000	62.5000	76.5000	75.2500
16	3	3.75	57.6667	69.0000	73.0000	61.6667	76.0000	77.5000
17	7	−0.75	57.6667	69.0000	73.0000	60.0000	75.0000	77.5000
18	20	5.25	57.6667	69.0000	73.0000	60.0000	75.0000	77.5000

(Continued)

Table 18.6 (Continued)

Std run	A: A	Catalyst. conc. (wt. % of oil)	Raw (vol. %)	Activated (vol. %)	Calcined (vol. %)	FMM predicted value yield 1 (vol. %)	FMM predicted value yield 2 (vol. %)	FMM predicted value yield 3 (vol. %)
19	15	2.25	57.6667	69.0000	73.0000	62.0000	73.8000	77.5000
20	5	2.25	57.6667	69.0000	73.0000	62.0000	73.8000	77.5000
21	25	2.25	57.6667	69.0000	73.0000	62.0000	73.8000	77.5000
22	17	2.25	57.6667	69.0000	73.0000	77.5000	73.8000	77.5000
23	21	2.25	57.6667	69.0000	73.0000	62.0000	73.8000	77.5000
24	22	2.25	57.6667	69.0000	73.0000	62.0000	73.8000	77.5000
25	16	2.25	57.6667	69.0000	73.0000	62.0000	73.8000	77.5000

activated, and *C. icaco* oil calcined. This range indicates a notable presence of these variables in the mixture.

5) *VH*: This MF corresponds to a VH percentage of the input variables, including catalyst concentration, *C. icaco* as raw seed oil, *C. icaco* seed oil acid activated, and *C. icaco* oil calcined. Detailed values and representations can be found in Table 18.6, Figures 18.7 and 18.8, providing insights into the significant presence of these variables in the mixture.

18.5 Result and Discussion

18.5.1 Rule Base Creation

The architecture of the FFM is illustrated in Figure 18.6, while the graphical view of the MF construction of the FL system is presented in Figures 18.7–18.9, followed by the development of a comprehensive rule base. This rule base effectively defines the intricate relationship between the input variables and the output variable in linguistic terms. These rules were systematically generated using the input dataset for variable mapping within the FL system, ensuring that the model was adept at handling the complexity of the input data. Each rule serves to effectively connect linguistic variables to the output variable, establishing a set of instructions for the model's decision-making process.

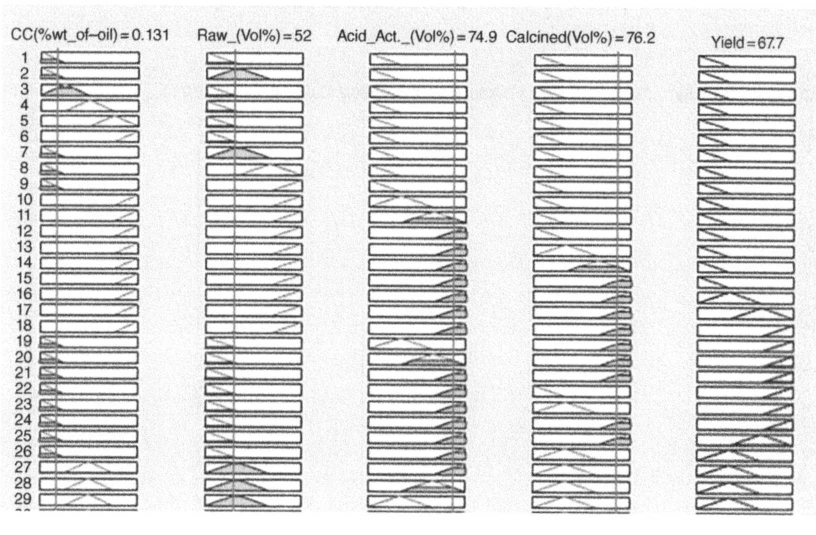

Figure 18.9 Rule viewer lfor four input variables to output response.

Subsequently, the intricate web of fuzzy rules is applied using the rule viewer depicted in Figures 18.10–18.12 to generate a fuzzy output. This fuzzy output forms the basis for biodiesel yield predictions presented in Table 18.6. This process underscores the effectiveness of the FL model in providing valuable insights for optimizing biodiesel production processes in waste-to-energy systems.

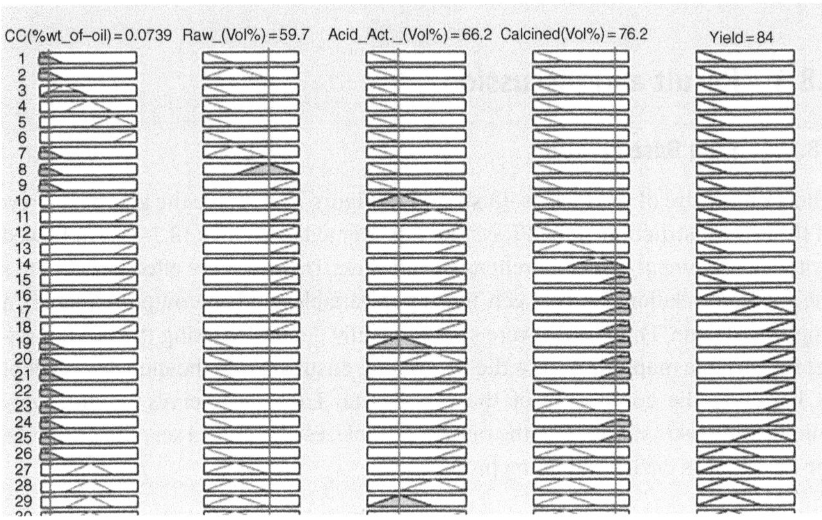

Figure 18.10 Rule viewer II for four input variables to output response.

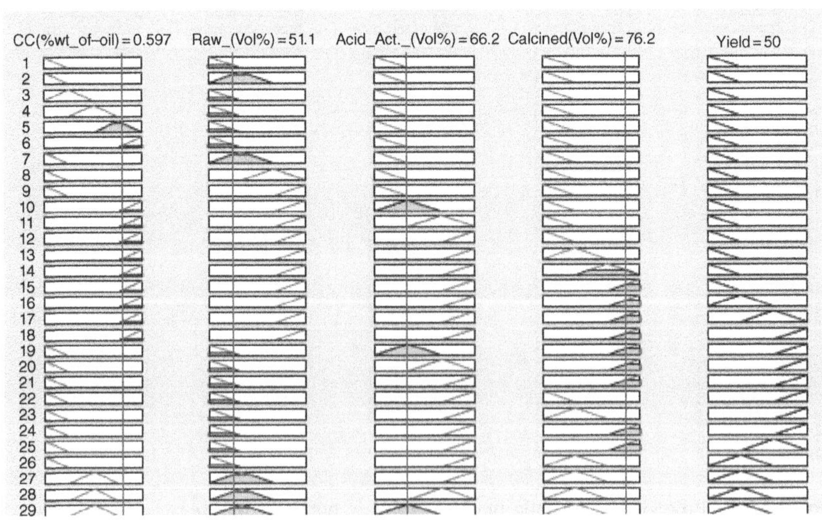

Figure 18.11 Rule viewer III for four input variables to output response.

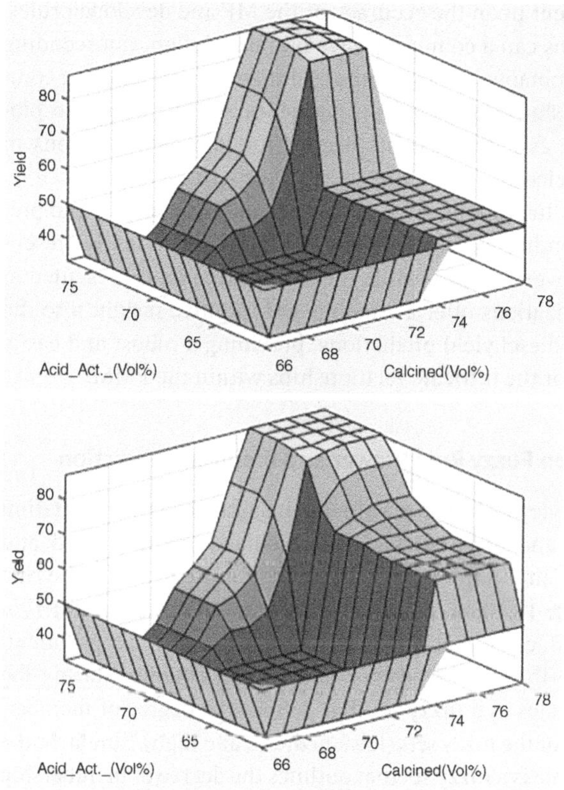

Figure 18.12 3D surface plot.

18.5.2 Analytical Exploration

This section is focused on a comprehensive examination of the FMM applied to forecast biodiesel production from unconventional feedstocks. A critical element of this study involves the development of a standard fuzzy rule viewer – a crucial tool for harnessing the FMM's capabilities in providing precise and dependable predictions. The standard fuzzy rule viewer employs the alpha cut concept, a technique facilitating the exploration of boundaries and subtleties within input and output MF. This exploration is crucial for extracting the most accurate and meaningful predictions from the model.

Once the FMM architecture is established, the rule viewer facilitates the application of alpha cuts. In this context, the alpha cut sets a threshold or confidence level within the MF, allowing the investigation of various degrees of certainty and imprecision. Adjusting this threshold effectively fine-tunes the model's predictions to align with actual conditions. A standard rule viewer is essential in this analytical process, contingent upon the accuracy of the MF and developed rules. Only under these conditions can a comprehensive output solution, transcending mere numerical values, be obtained. These output solutions signify a deeper comprehension of the relationships between input variables and the predicted biodiesel yield, providing the essential insights needed for informed decisions in optimizing the process efficiently.

Furthermore, graphical illustrations serve as complements to analysis and prediction under stochastic conditions. Figures 18.9–18.11 present a visual representation of the fuzzy rule viewer's output, while the 3D surface plot is presented in Figure 18.12. These visualizations offer a tangible and intuitive insight into the fuzzy rules utilized for biodiesel yield predictions, providing a robust and easily understandable depiction of the intricate relationships within the FMM.

18.5.3 Further Details on Fuzzy Rule Viewer and Biodiesel Prediction

Analysis of the fuzzy rule viewer provides valuable insights into the correlation between input parameters and output variables, particularly biodiesel yield, and sheds light on the mechanisms employed by the model for predictions. The rule viewer illustrated in Figure 18.10 offers a clear interpretation of the prediction outcomes, indicating a satisfactory biodiesel yield achieved during the experiment.

The input MF, based on the novel feedstock and dataset obtained during the experimental process, operates as a fuzzy set that defines the degree of membership of a data point in each of the fuzzy sets (low, medium, and high). Similarly, the MF for the first input variable is a fuzzy set that outlines the degree of membership of a data point in each corresponding fuzzy set. These MF play a crucial role in defining the input variables for the FFM. In addition, the output MF for biodiesel yield is also a fuzzy set, specifying the degree of membership of a data point in each of its fuzzy sets.

Upon comparison, discrepancies between the predicted and actual biodiesel yield become evident, as detailed in Table 18.6. Fundamentally, this study emphasizes the pivotal significance of precise biodiesel yield prediction through the FMM in optimizing waste-to-energy conversion processes. This precision has the potential to make significant contributions to the advancement of renewable energy production from biodegradable waste materials.

18.6 Conclusion

In conclusion, this research underscores the promising potential of utilizing waste periwinkle shells as a green catalyst in CI seed oil-based biodiesel production. The comprehensive evaluation of CI seed oil's physicochemical properties confirms its suitability for biodiesel production, aligning with

ASTM standards. The characterization of periwinkle shells in raw, calcined, and acidified forms using advanced techniques establishes their viability as an environmentally friendly, reusable, and cost-effective catalyst for large-scale biodiesel synthesis. The transesterification process, incorporating methanol as the alcohol, explored various parameters such as methanol–oil molar ratio, catalyst concentration, agitation speed, reaction temperature, and reaction time. The investigation into the catalyst's reusability revealed the superiority of the calcined periwinkle shell form, demonstrating higher percentage conversion compared to the raw and acidified counterparts. Moreover, the incorporation of FL as an innovative approach for predicting biodiesel yield showcased notable success. Despite marginal differences between experimental and predicted yields, the FL model exhibited accuracy, as evidenced by the result obtained from the rule viewer. This emphasizes the flexibility and enhanced predictive performance of the FL model, particularly in handling stochastic datasets associated with biodiesel transesterification processes. The findings of this study not only contribute to the understanding of an environmentally sustainable approach to biodiesel production but also hint at the potential for cost reduction through the utilization of waste periwinkle shells. The observed success with the calcined periwinkle shell catalyst suggests a pathway for the future commercialization of biodiesel production from CI seed oil. This research opens avenues for further exploration and optimization of the proposed methodology, paving the way for a greener and economically viable biodiesel production process.

References

1 Zhang, Z., Jiagiang, E., Deng, Y. et al. (2018). Effects of fatty acid methyl esters proportion on combustion and emission characteristics of a biodiesel fueled marine diesel engine. *Energy Convers. Manage.* 159: 244–253.

2 Zhang, Z., Tian, J., Li, J. et al. (2022). The development of diesel oxidation catalysts and the effect of sulfur dioxide on catalysts of metal-based diesel oxidation catalysts: a review. *Fuel Process. Technol.* 233: 107317.

3 Yan, Z.M., Gainey, B., and Lawler, B. (2022). A parametric modeling study of thermal barrier coatings in low-temperature combustion engines. *Appl. Therm. Eng.* 200: 117687.

4 Tamilselvan, P., Nallusamy, N., and Rajkumar, S. (2017). A comprehensive review on performance, combustion and emission characteristics of biodiesel fuelled diesel engines. *Renewable Sustainable Energy Rev.* 79: 1134–1159.

5 Mohd Noor, C.W., Noor, M.M., and Mamat, R. (2018). Biodiesel as alternative fuel for marine diesel engine applications: a review. *Renewable Sustainable Energy Rev.* 94: 127–142.

6 Zhang, B., Jiagiang, E., Gong, J. et al. (2017). Influence of structural and operating factors on performance degradation of the diesel particulate filter based on composite regeneration. *Appl. Therm. Eng.* 121: 838–852.

7 Zhong, C., Tan, J., Zuo, H. et al. (2021). Synergy effects analysis on CDPF regeneration performance enhancement and NOx concentration reduction of NH3–SCR over Cu–ZSM–5. *Energy* 230: 120814.

8 Zuo, H., Zhang, B., Huang, Z. et al. (2022). Effect analysis on SOC values of the power lithium manganate battery during discharging process and its intelligent estimation. *Energy* 238: 121854.

9 Okonkwo, C.P., Ajiwe, V.I.E., Obiadi, M.C. et al. (2023). Modeling and optimization of biodiesel production from Cocoplum seed oil (*Chrysobalamus icaco*) using artificial neural network. *J. Cleaner Prod.* 385: 135631.

10 Karaa, K., Ouanjib, F., Mohammed El Mahia, E.L. et al. (2019). *Biodiesel Synthesis from Vegetable Oil Using Eggshell Waste as a Heterogeneous Catalyst*, 1–7. Taylor and Francis ISSN: 1759-7269 (Print) 1759–7277. https://doi.org/10.1080/17597269.2019.1580972.

11 Ikbal, B.L., Kalyani, R., Rajat, G. et al. (2018). Waste ‘snail shell derived heterogeneous catalyst for biodiesel production by the transesterification of soybean oil. *R. Soc. Chem.* 8: 20131–20142.

12 Okonkwo, C.P., Ajiwe, V.I.E., Obiadi, M.C., and Okwu, M. (2021). A systematic review of biodiesel production with sustainable feedstock using assorted catalyst. *Am. J. Appl. Chem.* 9 (5): 154–163. https://doi.org/10.11648/j.ajac.20210905.1.

13 Konstantin, P., Alexander, Z., Antonios, S. et al. (2020). The advances and limitations in biodiesel production: feedstocks, oil extraction methods, production, and environmental life cycle assessment. *Green Chem. Lett. Rev.* 13 (4): 275–294. https://doi.org/10.1080/17518253.2020.1829099.

14 Feitosa, E.A., Xavier, H.S., and Randau, K.P. (2012). Chrysobalanaceae: traditional uses, phytochemistry and pharmacology. *Braz. J. Pharmacognosy* 22: 1181–1186.

15 White, P.A.S., Araujo, J.M.D., Cercato, L.M. et al. (2016). *Chrysobalanus icaco* L. leaves normalizes insulin sensitivity and blood glucose and inhibits weight gain in high-fat diet-induced obese mice. *J. Med. Food* 19: 155–160.

16 Medeiros, T., de Aguiara, R., Luob, A.A. et al. (2017). Regulatory science chemical characterization of Cocoplum (*Chrysobalanus icaco*, L.) seed oil and seeds. *J. Regul. Sci.* 5 (2): 15–28.

17 Zadeh, L. (1965). Fuzzy sets. *Inf. Control* 8: 338–353.

18 Gutierrez, S.M.C. and Trujillo, O.E. (2005). Design and implementation of a speed and position controller for a conveyor belt applying fuzzy logic and digital techniques. *Rev. Colomb. Tecnol. Av.* 2: 27–33.

19 Shang, K. and Hossen, Z. (2013). Applying fuzzy logic to risk assessment and decision-making. *Casualty Actuar. Soc. Can. Inst. Actuar. Soc. Actuar.* 2: 1–59.

20 Nobre, F.S., Tobias, A.M., and Walker, D.S. (2009). The impact of cognitive machines on complex decisions and organizational change. *AI Soc.* 24: 365–381.

21 Tron, E. and Margaliot, M. (2003). Modeling observed natural behaviour using fuzzy logic. In: *Proceedings of the 12th IEEE International Conference on Fuzzy Systems (FUZZ '03)*, vol. II, 74–78, St Louis, MO, USA (25–28 May 2003).

22 Bernstein, D.A., Clarke-Stewart, A., Roy, E.J., and Wickens, C.D. (1997). Boston: Houghton Mifflin Company. *Psychology* 4.

23 Spandagos, C. and Ng, T.L. (2018). Fuzzy model of residential energy decision-making considering behavioural economic concepts. *Appl. Energy* 213: 611–625.

24 Gamalath, I., Hewage, K., Ruparathna, R. et al. (2018). Energy rating system for climate conscious operation of multi-unit residential buildings. *Clean Technol. Environ. Policy* 20: 785–802.

25 Gossett, E. (2008). *Discrete Mathematics with Proof*, 2e, 17–23. Ander Hills, MN, USA: Bethel University.

26 Sivanesan, S. (2009). Kinetic study of the esterification of free fatty acids in non-edible *Pongamia pinnata* oil using acid catalyst. *Indian J. Sci. Technol.* 2 (12): 20–24.

27 Berthiaume, D. and Tremblay, A. (2006). Study of the Rancimat test method in measuring the oxidation stability of biodiesel ester and blends NRCan project No. CO414CET-327O, LEOTEK Inc Quebec, Canada. http://www.technopolethetford.ca/Industrialolechemistry/info_observatoiredeleochimieetudes-etrecherches_187-ang.cfm (accessed 16 February 2016).

28 Vargončík, M. (2007). Waste vegetable oil as a diesel replacement fuel.University ofWest Bohemia. Faculty of Electrical Engineering. Department of Electrical Power Engineering and Environmental Engineering. http://home.zcu.cz/~tesarova/IP/Proceedings/Proc_2007/index.htm http://hdl.handle.net/11025/22626.

29 Ezekwe, C.C. and Ajiwe, V.I.E. (2011). The effects of biodiesel's composition on its viscosity. *Anachem* 5 (1): 1016–1028.

30 Knothe, G. and Steidley, K.R. (2018). The effect of metals and metal oxides on biodiesel oxidative stability from promotion to inhibition. *Fuel Process. Technol.* 177: 75–80.

31 Wang, L., Dong, X., Jiang, H. et al. (2014). Preparation of a novel carbon- based solid acid from cassava stillage residue and its use for the esterification of free fatty acids in waste cooking oil. *Bioresour. Technol.* 158: 392–395.

32 Milan, D.K., Ana, V.V., Nataša, M.J. et al. (2016). Optimization and kinetic modeling of esterification of the oil obtained from waste plum stones as a pretreatment step in biodiesel production. *Waste Manage. (Oxford)* 48: 619–629.

33 Sai Bharadwaj A., Subramaniapillai N., Meera Sheriffa B. and Anantharaman N. (2019) Catalyst reusability and kinetic modeling of biodiesel produced from rubber seed oil, Energy Sources https://doi.org/10.1080/15567036.2020.1785056.

34 Machesa, M.G.K., Tartibu, L.K. & Okwu, M.O, (2023). Performance analysis of Stirling engine using computational intelligence techniques (ANN & Fuzzy Mamdani Model) and hybrid algorithms (ANN-PSO & ANFIS). Neural Comput. Appl. https://doi.org/10.1007/s00521-022-07385-0.

19

Parametric Investigation of Biodiesel Production from Biodegradable Feedstock Using Artificial Neural Network (ANN) Models

19.1 Introduction

The rise in energy demand across key sectors – transportation, heating, cooling, and electricity – due to sustained economic growth has prompted the exploration of alternative energy sources with lower greenhouse gas emissions than fossil fuels. Biodiesel has emerged as a promising alternative to traditional fuels in the transportation sector, recognized for its environmental benefits and potential [1]. In the pursuit of environmentally sustainable practices in biodiesel production, repurposing biomass waste has gained prominence [2].

Biodiesel stands out as a carbon-neutral, nontoxic, biodegradable fuel that is environmentally friendly. It is derived from feedstocks, whether edible or inedible, including various triglyceride sources. The quantity of biodiesel that can be produced is influenced by factors such as the cost of raw materials and the availability of specific feedstocks. Despite its widespread commercialization, a significant challenge remains, particularly concerning the cost of biodiesel feedstocks [3].

The exponential growth in fossil fuel-based energy production has been driven by the escalating global demand for transportation, heating, cooling, and electricity. The extensive reliance on fossil fuels has given rise to environmental challenges, including resource depletion, heightened carbon dioxide emissions, and an acceleration of global warming [4]. The utilization of petroleum products and fossil fuels has seen a substantial increase. In a span of just 10 years, the global consumption of diesel derived from petroleum has risen from 3.5 million tonnes in 2010 to 3.9 million tonnes in 2019. Notably, diesel consumption in the Asia-Pacific region, encompassing China, Japan, and Malaysia, witnessed an

Feedstocks for Sustainable Biodiesel Production: Characterization, Selection, and Optimization, First Edition. Chinwe P. Okonkwo and Modestus O. Okwu.

uptick from 1.1 million tonnes in 2010 to 1.4 million tonnes in 2019 [5]. This underscores the imperative of identifying a substitute for diesel to mitigate the global consumption impact.

Waste cooking oil (WCO) proves to be a valuable raw material for biodiesel synthesis. Unfortunately, a significant portion of WCO is often improperly disposed of, ending up in soil, dustbins, or drainage systems, leading to environmental challenges. Utilizing WCO as a feedstock for biodiesel production offers a sustainable alternative to its disposal. The cost-effectiveness of the biodiesel production process is greatly enhanced by the affordability of WCO. In contrast to edible oils, WCO is abundantly available, does not pose harm to the environment, and avoids competition with food production during times of fuel crisis. While WCO presents a promising feedstock, its high water and free fatty acid (FFA) content poses challenges in the transesterification reaction. To address this issue, a necessary step involves pretreatment or esterification using an acid before initiating the transesterification reaction [6].

Various methods employed in the production of biodiesel from WCO include supercritical techniques, transesterification, and pyrolysis. Among these, the most versatile process is transesterification, wherein feedstock oil is utilized to generate biodiesel along with glycerol as a by-product. A catalyst plays a crucial role in expediting the transition state reaction by reducing the activation energy. Catalysts can be broadly classified into three types: bases, acids, and enzymes. The use of acidic catalysts has been associated with corrosion issues while employing alkaline catalysts has been linked to saponification problems. Catalysts, based on their composition, fall into categories such as homogeneous, heterogeneous, or enzymatic.

Due to their prolonged lifespan, sustainability, ease of separation, resistance to corrosion, and capacity for reuse and recycling, heterogeneous catalysts are favored over homogeneous catalysts. However, several other factors also impact biodiesel production, including stirring speed, oil-to-alcohol ratio, catalyst dose, reaction time, and temperature [7]. An artificial neural network (ANN) is a computational system designed to analyze and process data in a manner akin to the human brain. This technology, a subset of artificial intelligence (AI), provides solutions to a wide range of problems that have proven challenging or unsolvable using statistical or human standards. Representing a contemporary computational approach, it leverages complex datasets and nonlinear modeling. In numerous fields, establishing connections between theoretical principles and experimental data is crucial. The statistical tool emulates the structure and functions of biological neural networks through a modeling formulation style. At the core of every ANN is an artificial neuron, a fundamental mathematical model (function) [8–12].

This model incorporates three fundamental sets of operations: multiplication, summation, and activation. ANNs have gained widespread popularity for their ability to accurately analyze datasets ranging from small to large, offering a cost-effective and time-efficient alternative. Various fields, including chemical and environmental engineering, neuroscience, computational analysis, engineering design, and applications, can benefit from their utilization [13]. In the artificial neuron, inputs at the entrance are weighted, implying that each input value is multiplied by its specific weight. The central region of the artificial neuron encompasses the summation function, which aggregates all the weighted inputs and biases. The activation function, also known as the transfer function, is situated at the artificial neuron's output. It receives the sum of the previously weighted inputs plus bias and processes it to generate a response.

An excellent resource for conducting predictive analysis of this nature is the MATLAB ANN toolbox [14–16]. This study aims to assess the sustainability and environmental implications of biodiesel production from used cooking oil waste across various control variables. The process is optimized through ANNs, showcasing a notable transformation of waste into a valuable resource. The primary objective of employing ANN in this research is to assess the effectiveness of utilizing WCOs, specifically in terms of their biodiesel yield in the transesterification process. In addition, the goal is to develop a system that establishes a relationship between process variables and biodiesel yield using the ANN modeling technique.

19.1.1 Artificial Neural Network

ANNs possess the capability to tackle problems that are often considered challenging or impossible by human or statistical standards [17]. Comprising three layers – input, hidden, and output – the modeling of ANNs holds significant potential to contribute to the advancement of renewable energy systems, particularly in the context of accelerating research on biodiesel [18]. The experimental design serves as a guide for achieving the optimal response [14]. The popularity of ANN has surged due to its precision in analyzing datasets of varying sizes, offering a cost-effective and time-efficient solution. Its application extends to diverse fields, encompassing chemical and environmental engineering, neuroscience, various forms of computational analysis, and engineering design and application [19].

These computational systems are designed to replicate the biological neural networks present in the brains of animals. At the core of these systems are artificial neurons, which emulate the neurons found in biological brains. Artificial neurons consist of interconnected units or nodes, enabling the transmission of signals to other neurons through each connection [15]. These synthetic neurons function by receiving input, processing it, and subsequently transmitting signals to other neurons within their network. The output of each neuron is determined by a nonlinear function applied to the sum of its inputs, and the "signal" at a

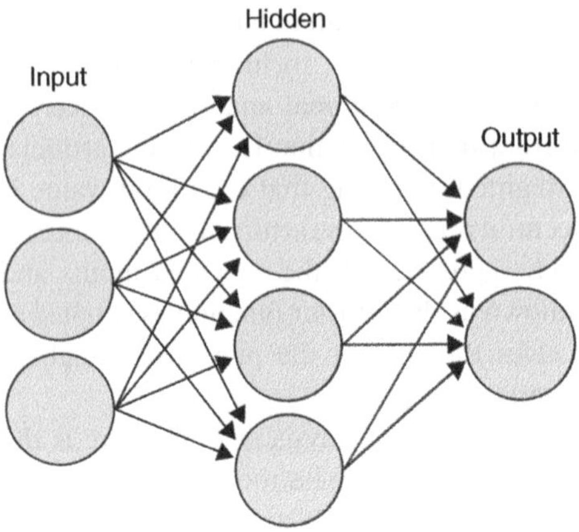

Figure 19.1 Artificial neural networks structural model.

connection is represented by a real number. The connections are referred to as edges. Typically, the weights of neurons and edges undergo adjustments throughout the learning process, affecting the signal strength at a connection. Neurons may have a threshold, and a signal is transmitted only if the cumulative signal surpasses this threshold. Neurons often organize into layers, and different layers may modify their inputs in distinct manners. Signals progress from the initial layer – the input layer – to the final layer – the output layer, potentially traversing through the layers multiple times [15].

These network architecture as presented in Figure 19.1 is composed of interconnected nodes, simplified versions of brain neurons. In this illustration, each circular node represents an artificial neuron, and arrows depict connections between the input and output of an artificial neuron. Functioning as a computational tool, this statistical instrument operates through modeling formulations that simulate the structure and functions of biological neural networks [12]. The foundation of any ANN is an artificial neuron, which serves as a fundamental mathematical model or function. This model incorporates three essential operations – multiplication, summation, and activation. At the entrance of the artificial neuron, inputs are weighted, signifying that each input value is multiplied by its specific weight. The sum function, situated within the artificial neuron, aggregates all the weighted inputs plus bias. Finally, the activation function, also referred to as the transfer function, processes the sum of previously weighted inputs and bias at the artificial neuron's exit, generating the desired response [14, 20].

19.1.2 Selection of Raw Materials for Biodiesel Production

Selecting the optimal oil for biodiesel production involves a nuanced consideration of both chemical and economic factors. The chemical decision-making process hinges on the type of fats and oils chosen, with attention directed toward assessing the quantity of FFAs associated with triglycerides. In addition, the chemical reaction's complexity can be influenced by the presence of contaminants in diverse feedstocks, necessitating varying levels of preparation. To facilitate the industrialization of biodiesel production, a critical challenge lies in identifying a raw oil that is not only abundant but also meets criteria such as cost-effectiveness, environmental friendliness, and high utility. The quest for a large-scale, cost-effective, environmentally sustainable, and highly versatile raw oil becomes paramount in overcoming this hurdle. Countries worldwide navigate this decision-making process based on their unique national circumstances, selecting raw materials for biodiesel production that align with their specific economic, environmental, and agricultural considerations. This dynamic interplay of chemistry and economics underscores the multifaceted nature of the biodiesel production setting.

19.1.3 Raw Materials for Biodiesel Production

Vegetable oils (herbal and woody oils), microbial oils, animal oils, and waste of no use oils are the raw materials which are of primary use in the production of biodiesel.

19.1.3.1 Vegetable Oils and Fats

Biodiesel production can benefit from a diverse range of vegetable oils derived from various sources, encompassing both herbal oil crops and woody oil crops. Herbal oil crops, such as, peanut, soybean, sesame, rapeseed, cottonseed and corn, offer valuable feedstocks for biodiesel synthesis. In addition, woody oil crops like the *Jatropha curcas*, tallow tree, palm oil, coptis chinensis etc. contribute to the array of excellent ingredients available for biodiesel production. The utilization of such diverse oil sources underscores the versatility and adaptability of biodiesel manufacturing, allowing for the exploration of different agricultural and botanical resources to meet the growing demand for sustainable and eco-friendly fuel alternatives.

19.1.3.2 Animal Fat

Animal fats, derived from sources such as animal slaughter, waste generated during fur treatment, and residuals from edible meat processing like butter, lard, lamb oil, and fish oil, present a variety of potential sources for biodiesel production. While animal fats may not serve as the primary raw material due to the necessity of multiple imports, they can be effectively utilized as an additional resource in biodiesel manufacturing. Various techniques, including enzyme synthesis, transesterification, and microemulsion, can be employed to produce biodiesel from animal fats. A practical example involves the use of tuna oil as the initial material, where a step-by-step alcoholysis procedure with immobilized lipase resulted in a biodiesel yield of 90%.

19.1.3.3 Microbial Grease

The predominant fatty acids found in microbial oils are typically C16 and C18. The exploration of microbial grease production dates back to the 1940s when scientists initially delved into the mechanisms employed by microbes to generate oils. Recent research has particularly focused on advancements in microalgae oils and other microbial oils. It has been known since the late 19th century that microorganisms such as microalgae, fungi, bacteria, and yeast produce a diverse array of unique oils. Microalgae, in particular, stand out for their primitive photosynthetic efficiency. They exhibit the potential for large-scale cultivation, offering high oil yields, rich oil content, and the advantage of not requiring arable land. Using seawater as a natural medium presents an opportunity to conserve agricultural resources, which can be significantly more than those required for terrestrial plants. This advantage is particularly notable in the context of employing engineered microalgae for diesel oil production. Microalgae typically contain oil in the range of 20%–50%. Some specific microalgae species even boast an oil content exceeding 80% of their dry weight, a characteristic that often leads to their neglect by many industrial facilities.

19.1.3.4 Waste Useless Oils and Fats

Waste oils and fats, notably those derived from cooking processes such as fried food oil and residual cooking oil, present promising options as alternative fuels. Despite the shortcomings associated with petrochemical diesel – such as elevated viscosity, restricted volatility, inadequate air mixing, and vulnerability to thermal polymerization – transesterification emerges as a solution that meets the performance requirements for optimal diesel oil substitutes. Each year, the food, animal, and plant processing industries collectively generate millions of tons of waste oils and fats.

Acidified oils obtained from the processing of vegetable oils are characterized by a predominance of FFAs. While waste oils and fats stand out as cost-effective raw materials for biodiesel production, they come with certain drawbacks, including limited availability, complex compositions, and demanding processing requirements.

19.2 Review of Related Research Focused on Artificial Neural Network for Biodiesel Production

Detailed information and findings from pertinent studies that specifically explore the application of ANNs in the context of biodiesel production are systematically presented in a tabular format within Table 19.1. The table serves as a comprehensive repository of key insights, methodologies, and outcomes derived from various

Table 19.1 Review of related research using artificial neural network for biodiesel production.

Citations	Authors	Research title	Applications	Year
[2]	Fangfang et al.	A research conducted in this field is focused on sustainable biodiesel production from waste cooking oil with the application of ANN modelling, though a deep exposition of ANN was not presented as showcased in this study.	B	2021
[20]	Matei et al.	This researchers focused on the yield optimization using artificial neural networks in biodiesel production from soybean oil	B	2020
[8]	Okonkwo et al.	Production of biodiesel from the novel nonedible seed of *Chrysobalanus icaco* using natural heterogeneous catalyst: modeling and prediction using artificial neural network	B	2023
[21]	Okonkwo et al.	Production and performance evaluation of biodiesel from *Elaeis guineensis* using natural snail shell-based heterogeneous catalyst: kinetics, modeling and optimisation by artificial neural network	B	2023
[6]	Aghbashlo et al.	The study focused on machine learning technology in biodiesel research: a review	B	2021
[22]	Karacan et al.	Artificial neural networks modelling for biodiesel production from waste cooking oil	B	2020
[23]	Ayoola et al.	Data on artificial neural network and response surface methodology analysis of biodiesel production	B	2020
[24]	Hoanget al.	A review on application of artificial neural network (ANN) for performance and emission characteristics of diesel engine fueled with biodiesel-based fuels	B	2021
[25]	Treve et al.	Performance evaluation of artificial neural networks in sustainable modelling biodiesel synthesis	B	2022
[26]	Soltani et al.	The optimization of biodiesel production from transesterification of sesame oil via applying ultrasound-assisted techniques: Comparison of RSM and ANN-PSO hybrid model	B	2020

B, biodiesel.

research endeavors within this domain. Each entry in the table encapsulates specific aspects such as experimental setups, data parameters, ANN architectures employed, and notable findings, offering a structured and accessible overview of the broader setting of research activities that probe into the utilization of ANNs for enhancing biodiesel production processes.

19.3 Materials and Method

Biodiesel production was initiated using residual cooking oil, with an initial washing step designed to eliminate approximately 50% of water-soluble fatty acids. The reactor employed mechanical stirring, and the transesterification reaction involved methanol and an alkaline catalyst [27]. The experiment applied a controlled variable approach, with the following five variables manipulated during the experiment [27]:

Methanol to used cooking oil molar ratio (6:1–12:1)
Reaction time (30–90 minutes)
Agitation speed (300–700 rpm)
Temperature range: 30–70 °C
Catalyst concentration (0.75–3.75 g)

Following the completion of the transesterification reaction, a separating funnel was employed to segregate biodiesel from glycerol. Glycerol settled at the bottom, while crude biodiesel was situated at the top. Upon discarding the glycerol layer, the crude biodiesel underwent heating in an open pan above 100 °C to eliminate residual water. Subsequently, it underwent a three-time water-washing process to eliminate alcohol, catalyst residues, and unreacted oil. The resulting purified biodiesel was obtained after the heating process. The refined biodiesel was then collected for further analysis and characterization.

The viscosity of both raw and esterified oil materials was determined using a redwood viscometer (ASTM D445). Density and calorific value were measured using a hydrometer (ASTM D1298) and a bomb calorimeter (ASTM D420), respectively. The flash and ignition points were assessed through the open cup method (ASTM D93). Fuel properties analysis indicated a high percentage of monounsaturated fatty acids in the produced biodiesel. In comparison to leftover cooking oil, biodiesel exhibited a 4% decrease in specific gravity and an 85% decrease in kinetic viscosity. In addition, biodiesel demonstrated a 4% increase in heating value compared to leftover cooking oil [27].

The unique properties of WCO are a result of the chemical reactions that take place during frying. Nevertheless, WCO exhibits similar properties to fresh oil in terms of producing biodiesel that meets quality standards, including density, acid

Table 19.2 Physicochemical properties of the extracted oil.

Properties	WCO
Color	Light brown
Yield (%)	92.000
Moisture (%)	2.000
Kinematic viscosity at 40 °C (mm^2/s)	126.000
Refractive index at 29 °C	1.457
Energy value (kJ/kg)	43 938
Acid value (mgKOH/g)	35.400
Saponification (mgKOH/kg)	243.000
Peroxide value (meq/kg)	8.45
Iodine value (g/100 g of oil)	121.000
Molecular weight (g/mol)	824.640
Flashpoint (°C)	296.000
Cloud point (°C)	13.500
Pour point (°C)	7.000

value, kinematic viscosity, and iodine value [28]. The elevated FFA content in WCO compared to vegetable oil is a crucial factor when choosing a catalyst for the biodiesel production process [29]. The presence of soap, a byproduct of high FFA concentrations, poses a limitation to large-scale biodiesel production [30]. Research indicates that WCO has significantly high values for kinematic viscosity, saponification, flash point, moisture content, and FFAs [31]. See Table 19.2 for the physiochemical properties of the extracted oil.

These chemical characteristics play a role in determining the quantity and quality of the biodiesel produced. In addition, high oil viscosity can impede fuel injection from the tank and hinder atomization. WCO offers substantial cost savings, being two to three times more economical than fresh vegetable oil [32]. Given that raw materials can constitute 75%–90% of the total biodiesel production cost, their affordability significantly impacts the commercial price of the fuel [33]. Utilizing WCO as a biodiesel feedstock avoids concerns related to the environment and potential conflicts between fuel and food resources.

19.3.1 ANN Modeling

ANN, as a soft computing tool, has been developed to emulate the functioning of the human brain. The ANN model comprises interconnected processing units called neurons [28]. Input data is processed between two connected neurons through weights in the ANN. The ANN determines the relationship between input and output values based on the weights among various interconnected neurons through the matrix. The model undergoes training with a substantial amount of pertinent data, and iterations are performed until the optimal model for the dataset is achieved. The objective is to minimize the model's generalization error. As ANNs operate as linear systems, they depend on weight prediction parameters to replicate the excitation of biological neurons. Activation functions are also employed by ANNs to mimic the excitation of biological neurons, with the neuron becoming active when the excitation value surpasses its threshold. This research employs ANN architecture for the prediction of biodiesel production.

19.3.2 Experimental Design

Table 19.3 is presented, which provides experimental data showcasing the fatty acid methyl ester (FAME) yield across varying catalyst amounts, agitation speeds, molar ratios, and temperatures. The MATLAB-based ANN toolbox is utilized for receiving datasets to facilitate kinetic modeling. The experimental design encompassed 25 samples for each variable under consideration. The primary focus of the study was the dependent variable, representing the percentage of FAME yield, while the independent variables included temperature, catalyst amount, agitation speed, and the methanol-to-oil molar ratio. The comprehensive dataset is crucial for the subsequent modeling and analysis using the ANN framework.

Table 19.3 Experimental results to illustrate the FAME yield from waste cooking oil.

		Input variable	Exp. value	Exp. value	Exp. value
Std run	**A:A**	**Catalyst. conc.**	**Yield 1**	**Yield 2**	**Yield 3**
1	9	0.75	55	72	73
2	13	3.75	55	72	73
3	15	0.75	65	72	73
4	23	3.75	65	72	73
5	5	0.75	55	78	73
6	1	3.75	55	78	73
7	12	0.75	65	78	73
8	21	3.75	65	78	73
9	2	0.75	55	72	82

Table 19.3 (Continued)

		Input variable	Exp. value	Exp. value	Exp. value
Std run	A:A	Catalyst. conc.	Yield 1	Yield 2	Yield 3
10	4	3.75	55	72	82
11	7	0.75	65	72	82
12	17	3.75	65	72	82
13	19	0.75	55	78	82
14	18	3.75	55	78	82
15	10	0.70	65	78	82
16	22	3.70	65	78	82
17	24	−0.75	60	75	77.5
18	8	5.25	60	75	77.5
19	6	2.25	50	75	77.5
20	20	2.25	70	75	77.5
21	14	2.25	60	69	77.5
22	25	2.25	60	81	77.5
23	11	2.20	60	75	68.5
24	3	2.20	60	75	86.5
25	16	2.20	60	75	77.5

19.4 ANN Model Validation

The dataset obtained through experimental trials conducted in a laboratory reactor served as the foundation for developing the ANN model. MATLAB 2020, equipped with the neural network toolbox, was employed for this purpose. The ANN model adopted a three-layered feedforward (FF) backpropagation (BP) network architecture to capture the intricacies of the biodiesel production process. The ANN Network's architecture as presented in Figure 19.2 played a crucial role in the validation, testing, and training phases. The three-layered structure facilitated the propagation of information through the network, allowing for the development of a robust model. This sophisticated network architecture ensured that the ANN model could effectively analyze and predict the FAME yield based on the experimental data collected from the lab reactor tests. The utilization of MATLAB, a widely recognized computational tool, added precision and efficiency to the implementation of the ANN model, contributing to its reliability in capturing the underlying relationships in the biodiesel production system.

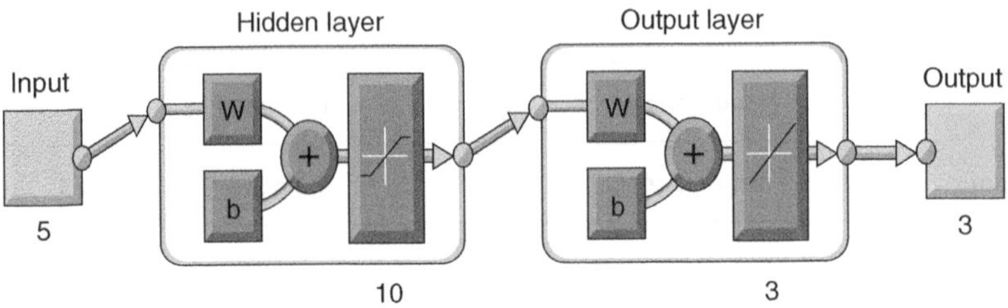

Figure 19.2 ANN architecture with hidden and output layer.

Table 19.4 Shows the MSE and R value for training validation and testing.

	Results		
	Samples	**MSE**	***R***
Training	17	1.36617	8.8914×10^{-1}
Validation	4	3.247646	9.1165×10^{-1}
Testing	4	3.060481	9.4661×10^{-1}

The mean squared error (MSE) value is remarkably low, as indicated by the *R* values of 0.875, 0.941, and 0.929 obtained during the training, testing, and validation phases, respectively. The result obtained is shown in Table 19.4. The overall *R* value, combining these results, stands at 0.882. This low MSE suggests a high degree of accuracy in the correlation between the experimental biodiesel yield and the predicted solution model derived from the ANN. The regression plot is presented in Figure 19.3.

The final training outcome highlights a robust correlation between the observed biodiesel yield in experiments and the predictions generated by the ANN model. The exceptionally small error further underscores the reliability of the model. The correlation coefficient, denoted by *R*, serves as an indicator of the closeness of the relationships between variables. It is crucial to note that the correlation coefficient's range spans from zero (0) to one (1). A value of one (1) signifies a perfect relationship between the variables, while zero indicates no relationship. In this context, a coefficient of determination falling between 0.7 and 1.0 is considered more than satisfactory, reflecting a highly accurate solution. Conversely, a value below 0.7 is deemed outside the acceptable solution range [34].

The comprehensive error histogram for the entire process as presented in Figure 19.4 highlights the distinct phases of training, validation, testing, and minimal error. In the histogram, the training set is represented by blue lines, the testing dataset is depicted in red, and the validation dataset is shown in green font. A thin, straight orange line on the graph denotes the point where the error tends to zero. In this scenario, the BP

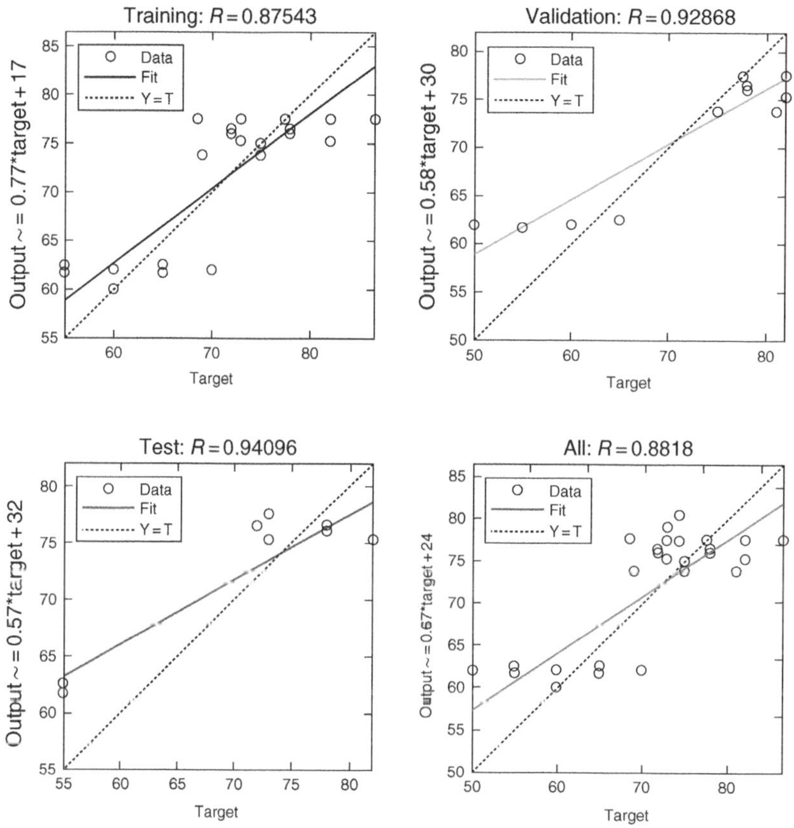

Figure 19.3 Regression plot for training, testing, and validation.

algorithm is the employed technique. Analyzing the error histogram with 20 bins reveals the minimum error to be –0.003125. This indicates that the BP algorithm successfully minimized the error during the ANN modeling process.

19.4.1 Validation Performance

As discussed in the preceding section, the outcomes of the experiment with PSA raw, calcined, and acidified catalysts were utilized to generate regression plots for training, validation, testing, and an overall plot. This plot is presented in Figure 19.5. The satisfactory point is reached when there is a discernible trend toward establishing a perfect connection between the identified variables. See the predicted result in Table 19.5.

There was a noticeable concurrence between the empirical findings and the outcomes obtained from the predicted values. This alignment underscores the efficacy of employing an ANN predictive tool for simulating the production of

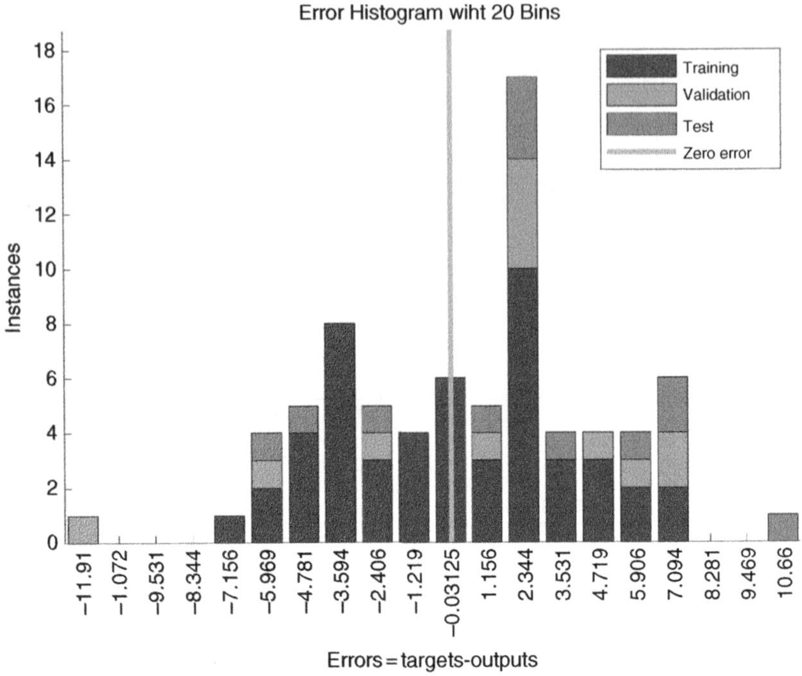

Figure 19.4 Error histogram of the model.

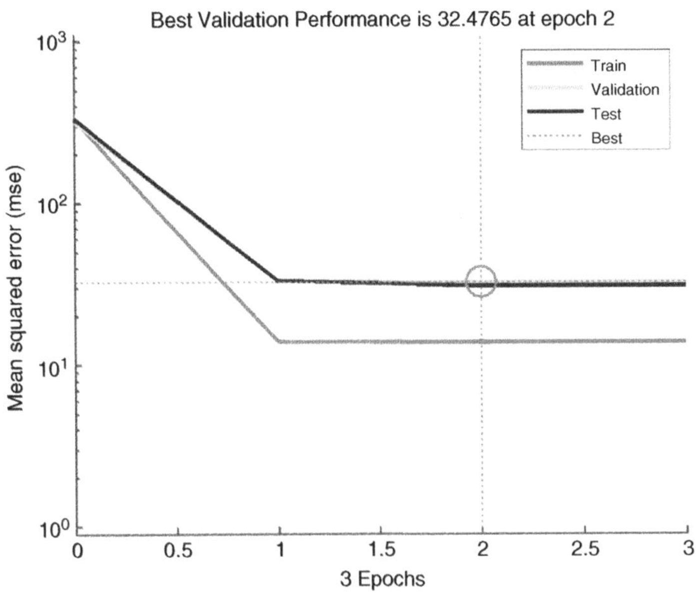

Figure 19.5 Validation performance at epoch 2.

Table 19.5 ANN solution for predicted values for wastecooking oil.

		Input variable	Output value	Output value	Output value	ANN predicted value	ANN predicted value	ANN predicted value
Std run	A:A	Catalyst. conc.	Yield 1	Yield 2	Yield 3	Yield	Yield	Yield
1	9	0.75	60.0003	75.9994	76.0004	60.0003	75.9994	76.0004
2	13	3.75	61.0001	75.6001	74.8002	61.0001	75.6001	74.8002
3	15	0.75	60.0003	75.9994	76.0004	60.0003	75.9994	76.0004
4	23	3.75	61.0001	75.6001	74.8002	61.0001	75.6001	74.8002
5	5	0.75	60.0003	75.9994	76.0004	60.0003	75.9994	76.0004
6	1	3.75	61.0001	75.6001	74.8002	61.0001	75.6001	74.8002
7	12	0.75	60.0003	75.9994	76.0004	60.0003	75.9994	76.0004
8	21	3.75	61.0001	75.6001	74.8002	61.0001	75.6001	74.8002
9	2	0.75	60.0003	75.9994	76.0004	60.0003	75.9994	76.0004
10	4	3.75	61.0001	75.6001	74.8002	61.0001	75.6001	74.8002
11	7	0.75	60.0003	75.9994	76.0004	60.0003	75.9994	76.0004
12	17	3.75	61.0001	75.6001	74.8002	61.0001	75.6001	74.8002
13	19	0.75	60.0003	75.9994	76.0004	60.0003	75.9994	76.0004
14	18	3.75	61.0001	75.6001	74.8002	61.0001	75.6001	74.8002
15	10	0.75	60.0003	75.9994	76.0004	60.0003	75.9994	76.0004
16	22	3.75	61.0001	75.6001	74.8002	61.0001	75.6001	74.8002
17	24	−0.75	60.0002	75.0002	77.5001	60.0002	75.0002	77.5001

(Continued)

Table 19.5 (Continued)

		Input variable	Output value	Output value	Output value	ANN predicted value	ANN predicted value	ANN predicted value
Std run	**A:A**	**Catalyst. conc.**	**Yield 1**	**Yield 2**	**Yield 3**	**Yield**	**Yield**	**Yield**
18	8	5.25	60.0001	75.0014	77.5011	60.0001	75.0014	77.5011
19	6	2.25	62.5000	73.5000	75.2500	62.5000	73.5000	75.2500
20	20	2.25	62.5000	73.5000	75.2500	62.5000	73.5000	75.2500
21	14	2.25	62.5000	73.5000	75.2500	62.5000	73.5000	75.2500
22	25	2.25	62.5000	73.5000	75.2500	62.5000	73.5000	75.2500
23	11	2.25	62.5000	73.5000	75.2500	62.5000	73.5000	75.2500
24	3	2.25	62.5000	73.5000	75.2500	62.5000	73.5000	75.2500
25	16	2.25	62.5000	73.5000	75.2500	62.5000	73.5000	75.2500

biodiesel from WCO. The congruence between the experimental data and the predictions substantiates the reliability and accuracy of the ANN model in forecasting biodiesel production from WCO.

19.5 Conclusion and Future Direction

The global setting focused on biofuel production, demand, and capacity is experiencing a consistent and upward trajectory, with expectations for continued growth in the coming years. This study focused on the parametric investigation of biodiesel production from biodegradable feedstock with exposition to environmental implications associated with the production of biodiesel from used cooking oil. The study aims to predict biodiesel yield and assess its properties, contributing valuable insights to the evolving field of biofuel research. The utilization of WCO in biodiesel production emerges as an economically viable option, effectively reducing production costs. However, the heating process introduces adverse changes to the oil's properties, leading to increased levels of FFA, higher water content, and the introduction of various impurities. These challenges in biodiesel production from WCO contribute to elevated costs in the purification and separation processes. The translation of laboratory-scale processes, which often involve immiscible phases, to commercial scale encounters limitations in mass and heat transfer, presenting obstacles to scalability. Future research activities should focus on key areas to enhance the sustainability of biodiesel production from WCO. Addressing the environmental impact necessitates the exploration of eco-friendly catalysts and integration with renewable energy sources, such as solar energy. These avenues of research aim to propel the development of more sustainable and scalable processes for biodiesel production from WCO.

References

1 Rocha-Meneses, L., Hari, A., Inayat, A. et al. (2023). Recent advances on biodiesel production from waste cooking oil (WCO): a review of reactors, catalysts, and optimization techniques impacting the production. *Fuel* 348: 128514.

2 Fangfang, F., Alagumalai, A., and Mahian, O. (2021). Sustainable biodiesel production from waste cooking oil: ANN modeling and environmental factor assessment. *Sustainable Energy Technol. Assess.* 46: 101265.

3 Brahma, S., Nath, B., Basumatary, B. et al. (2022). Biodiesel production from mixed oils: a sustainable approach towards industrial biofuel production. *Chem. Eng. J. Adv.* 10: 100284.

4 Ambar, G., Mishra, S., Chowdhury, S. et al. (2021). A review on factor affecting biodiesel production from waste cooking oil: an Indian perspective. *Mater Today: Proc.* 46: 5594–5600.

5 BP (2020). Statistical review of world energy. https://www.bp.com/content/dam/bp/businesssites/en/global/corporate/pdfs/energyeconomics/statistical-review/bp-stats-review-2020-full-report.pdf (accessed June 2020).

6 Aghbashlo, M., Peng, W., Tabatabaei, M. et al. (2021). Machine learning technology in biodiesel research: a review. *Prog. Energy Combust. Sci.* 85: 100904. https://doi.org/10.1016/j.pecs.2021.100904.

7 Banga, S. and Pathak, V.V. (2023). Biodiesel production from waste cooking oil: a comprehensive review on the application of heterogeneous catalysts. *Energy Nexus* 100209.

8 Okonkwo, C.P., Ajiwe, V.I.E., Obiadi, M.C. et al. (2023). Production of biodiesel from the novel non-edible seed of *Chrysobalanus icaco* using natural heterogeneous catalyst: modeling and prediction using artificial neural network. *J. Clean. Prod.* 385: 135631.

9 Bravo-Moncayo, L., Lucio-Naranjo, J., Chavez, M. et al. (2019). A machine learning approach for traffic-noise annoyance assessment. *Appl. Acoust.* 156: 262–270.

10 Hamad, K., Khalil, M.A., and Shanableh, A. (2017). Urban road traffic noise monitoring, mapping, modelling, and mitigation: A thematic review. *Transp. Res. Part D: Transp. Environ.* 53: 161–177.

11 Steinbach, L. and Altinsoy, M.E. (2019). Prediction of annoyance evaluations of electric vehicle noise by using artificial neural networks. *Appl. Acoust.* 145: 149–158.

12 Okwu, M.O. and Olufemi, A. (2018). A comparative study of artificial neural network (ANN) and adaptive neuro-fuzzy inference system (ANFIS) model in distribution system with non-deterministic inputs. *Int. J. Eng. Bus. Manage.* 10: 1–17. https://doi.org/10.1177/1847979018768421.

13 Khudsange, C.R. and Wasewar, K.L. (2017). Experimental and Modeling Study of Esterification Reaction for Synthesis of Butyl Butyrate: Desirability Function Approach for Optimization and Prediction Comparative Study of RSM and ANN. *Int. J. Chem. React. Eng.* 15 (3): 1542–6580. https://doi.org/10.1515/ijcre-2016-0176

14 Ayoola, A., Hymore, F., Omonhinmin, C. et al. (2020). *Chem. Data Collect.* 28: 100478.

15 Okwu, M.O., Chukwu, V.U., and Oguoma, O. (2019). *International Work-Conference on Artificial Neural Networks*, 539–554. Springer.

16 Ewim, D.R.E., Okwu, M.O., Onyiriuka, E.J. et al. (2022). *Eng. Appl. Sci. Res.* 49: 444–458.

17 Yang, G.R. and Wang, X.J. (2020). Artificial neural networks for neuroscientists: a primer. *Neuron* 107 (6): 1048–1070.

18 Mohammed, M.L., Patel, D., Mbeleck, R. et al. (2013). Optimisation of alkene epoxidation catalysed by polymer supported Mo (VI) complexes and application of artificial neural network for the prediction of catalytic performances. *Appl. Catal. A Gen.* 466: 142–152.

19 Alkım, E., Gürbüz, E., and Kılıç, E. (2012). A fast and adaptive automated disease diagnosis method with an innovative neural network model. *Neural Netw.* 33: 88–96.

20 Matei, D.A.N.U.T.A., Doicin, B.O.G.D.A.N., Cursaru, D.I.A.N.A., and Ezeanu, D.S. (2020). Yield optimization using artificial neural networks in biodiesel production from soybean oil. *Rev. Chim.* 71 (5): 132–140.

21 Okonkwo, C.P., Ajiwe, V.I., Ikeuba, A.I. et al. (2023). Production and performance evaluation of biodiesel from Elaeis guineensis using natural snail shell-based heterogeneous catalyst: kinetics, modeling and optimisation by artificial neural network. *RSC Adv.* 13 (28): 19495–19507.

22 Karacan, S., Gedikaslan, B., and Çağatay, M. (2020). Artificial neural networks modelling for biodiesel production from waste cooking oil. *Int. J. Energy Appl. Technol.* 7 (3): 74–80.

23 Ayoola, A.A., Hymore, F.K., Omonhinmin, C.A. et al. (2020). Data on artificial neural network and response surface methodology analysis of biodiesel production. *Data Brief* 31: 105726.

24 Hoang, A.T., Nižetić, S., Ong, H.C. et al. (2021). A review on application of artificial neural network (ANN) for performance and emission characteristics of diesel engine fueled with biodiesel-based fuels. *Sustainable Energy Technol. Assess.* 47: 101416.

25 Treve, M., Patra, I., Prabu, P. et al. (2022). Performance evaluation of artificial neural networks in sustainable modelling biodiesel synthesis. *Sustainable Energy Technol. Assess.* 52: 102098.

26 Soltani, H., Karimi, A., and Falahatpisheh, S. (2020). The optimization of biodiesel production from transesterification of sesame oil via applying ultrasound-assisted techniques: comparison of RSM and ANN–PSO hybrid model. *Chem. Prod. Process. Model.* 17 (1): 55–67.

27 Pugazhendhi, A., Alagumalai, A., Mathimani, T., and Atabani, A.E. (2020). Optimization, kinetic and thermodynamic studies on sustainable biodiesel production from waste cooking oil: an Indian perspective. *Fuel* 273. https://doi.org/10.1016/j.fuel.2020.117725.

28 Cordero-Ravelo, V. and Schallenberg-Rodriguez, J. (2018). Biodiesel production as a solution to waste cooking oil (WCO) disposal. Will any type of WCO do for a transesterification process? A quality assessment. *J. Environ. Manag.* 228: 117–129. https://doi.org/10.1016/J.JENVMAN.2018.08.106.

29 Gaur, A., Mishra, S., Chowdhury, S. et al. (2020). A review on factor affecting biodiesel production from waste cooking oil: an Indian perspective. *Mater. Today: Proc.* 46: 5594–5600. https://doi.org/10.1016/j.matpr.2020.09.432.

30 Zahan, K.A. and Kano, M. (2018). Biodiesel production from palm oil its by-products and mill effluent: a review. *Energies* 11 (8): 2132. https://doi.org/10.3390/EN11082132.

31 Ganesan, K., Sukalingam, K., and Xu, B. (2017). Impact of consumption of repeatedly heated cooking oils on the incidence of various cancers – a critical review. 59 (3): 488–505. https://doi.org/10.1080/10408398.2017.1379470.

32 Said, N.H., Ani, F.N., and Said, M.F.M. (2015). Review of the production of biodiesel from waste cooking oil using solid catalysts. *J. Mech. Eng. Sci.* 8 (June): 1302–1311. https://doi.org/10.15282/jmes.8.2015.5.0127.

33 Nomanbhay, S., Hussein, R., and Ong, M.Y. (2018). Sustainability of biodiesel production in Malaysia by production of bio-oil from crude glycerol using microwave pyrolysis: are view. *Green Chem. Lett. Rev.* 11 (2): 135–157. https://doi.org/10.1080/17518253.2018.1444795.

34 Okwu, M.O., Chukwu, V.U., and Oguoma, O. (2019). *Application of Artificial Neural Network Model for Cost Optimization in a Single, Multi-Destination System with Non-Deterministic Inputs*. Advances in Computational Intelligence, Lecture Note in Computer Science book series (LNCS), vol. 11, 507. Springer.

20

Modeling Sustainable Biofuel Production from Tobacco Seeds by Implementing Response Surface Methodology (RSM) and Adaptive Neuro-Fuzzy Inference System (ANFIS)

20.1 Introduction

In the ever-evolving process of sustainable energy, the quest for renewable and environmentally friendly alternatives to conventional fuels remains a paramount focus. Biodiesel, derived from diverse feedstocks, stands out as a promising avenue for addressing the global energy challenge [1]. Among the myriad sources explored for biodiesel production, underutilized tobacco seeds emerge as a novel and intriguing field. The utilization of underexplored resources not only addresses the imperative of finding sustainable sources of energy but also contributes to the efficient management of agricultural by-products. In this context, this research endeavors to pioneer a new frontier in biofuel production by investigating the potential of underutilized tobacco seeds for biodiesel synthesis.

At the forefront of the methodological approach is the adaptive neuro-fuzzy inference system (ANFIS), a sophisticated computational tool celebrated for its adeptness in handling nonlinearities and intricate stochastic datasets. ANFIS has demonstrated its efficacy in diverse fields, making it a preferred choice for modeling complex systems [2]. The present study represents a paradigm shift to leverage the power of ANFIS and RSM in modeling the biodiesel production process using underutilized tobacco seeds in tropical climates.

The dataset under scrutiny originates from a preceding study focused on the design of experiments for the production of tobacco seed oil methyl ester (TSOME) [3]. Building upon this foundation, this investigation is focused on the nuances of transesterification parameters, with a particular emphasis on the influence of reaction duration (T), methanol/oil molar ratio (M:O), and CAT dosage on biodiesel yield. Through rigorous analysis and modeling, the study aims to unravel the intricate relationships governing the biodiesel production process, shedding light on the optimal conditions for maximizing yield and efficiency.

Feedstocks for Sustainable Biodiesel Production: Characterization, Selection, and Optimization, First Edition. Chinwe P. Okonkwo and Modestus O. Okwu.

This pioneering study not only contributes to the expanding body of knowledge in biodiesel research but also introduces a novel application of ANFIS to the under-explored domain of biofuel synthesis from tobacco seeds. The findings hold the promise of not only advancing the understanding of sustainable energy solutions but also unlocking the untapped potential of underutilized resources in the pursuit of a greener and more sustainable energy future.

20.2 Computational Models for Biodiesel Production

20.2.1 ANFIS Model for Biodiesel Production

It is simple to use biodiesel as a substitute fuel in diesel engines. The main barrier facing the industrial production of biodiesel at the moment is meeting ASTM criteria and optimizing biodiesel yield [4]. The ANFIS is a hybrid system that solves issues by combining an artificial neural network (ANN) with fuzzy logic. There are numerical advantages to the combination [2]. It is a widely established and reliable model for estimating biodiesel yield. ANFIS is a data-driven, adaptive computational method for mapping nonlinear and complex data [5]. This model is used to estimate biodiesel yield based on temperature, pressure, reaction time, and alcohol/oil ratio. The ANFIS method, which is based on the implementation of a fuzzy inference system within an adaptive neural network framework, is one of the most innovative hybrid systems aimed at reducing the individually disadvantageous utilization of neural networks and FIS [6].

ANFIS modeling has also been used to forecast biodiesel characteristics. Al-Shanableh [7] examined the ANFIS approach for simulating the performance parameters and exhaust emissions of a diesel engine using different diesel fuels, biodiesel, and their blends. The cetane number of biodiesel fuels was estimated using the ANFIS modeling technique based on FAME double bonds and molar weight [8]. In a more recent study, the same author used the ANFIS approach to estimate the iodine value, kinematic viscosity, cloud point, and pour point of several biodiesel fuels based on the carbon number, amount of double bonds, temperature, and molar weight of FA [9]. Aminian and ZoreNezhad [10] proposed the ANFIS model, which demonstrated the superiority of their constructed model in predicting the viscosity of 18 types of biodiesels with a high correlation coefficient of 0.9964.

An ANFIS model is made up of five separate layers: The first layer (Layer1) is the fuzzification, the second layer (Layer 2) is the product, the third layer (Layer 3) is the rule, the fourth layer (Layer 4) is the defuzzification, and the fifth layer (Layer 5) is the overall output summation. The primary goal of the first layer is to convert the input variables into adaptive fuzzy sets. To achieve the targeted goal in the study, ΣMUFA, ΣSFA, and ΣPUFA were transformed into fuzzy sets with linguistic nomenclature like "very low,""low," high, and "very high" and membership functions (MF) like a generalized bell, sigmoid function, or triangular MF. The second layer combines sets of fuzzy from each input and evaluates the weight of each MF. The layer employs a Π-norm operator with fuzzy concurrence (and) to generate a nonadaptive output. Layer 3 estimates the ratio of a rule's firing power to the sum of all specified "if-then" fuzzy rules. The fourth layer applies the Sugeno fuzzy rule to multiply the output of the previous levels. In this layer, the centroid area is calculated under MF to defuzzify it. The last layer, made up of a single node, computes the sum of all outputs using the Σ-norm operator. Finally, the weighted average approach transforms the fuzzy output into a crisp one [7]. ANFIS is capable of recognizing patterns, whether there is a linear or nonlinear relationship in the data available.

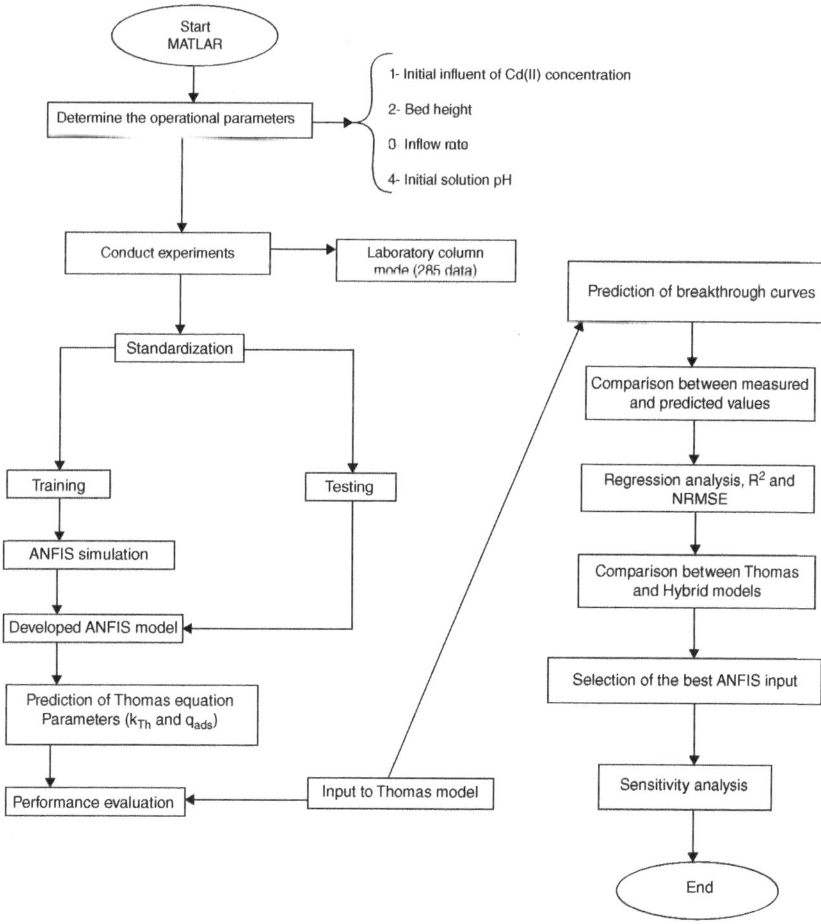

Figure 20.1 Flow chart of ANFIS modeling.

ANFIS models produce more accurate estimates than other models documented in the literature and studied by the current researchers. Several research has been conducted to model and optimize biodiesel production from various sources. Mostafaei et al. [11] evaluated and compared the results of a response surface methodology (RSM) and an ANFIS for modeling the transesterification yield achieved in an ultrasonic reactor, whereas Naderloo et al. [12] developed an ANFIS model to estimate the energy production ratio from biodiesel production. The ideal ANFIS model for prediction is based on the reaction temperature, CAT CAT weight percentage, mixing intensity, and reaction time, which are typically created in MATLAB [13] (Figure 20.1).

20.2.2 Further Review of Related Research

Table 20.1 showcases research works that have been utilised using ANFIS across different fields.

Table 20.1 Related research conducted using ANFIS in diverse analysis.

Citations	Authors	Title	Applications	Year
[7]	Al-Shanableh et al.	Estimation of cold flow properties of biodiesel using ANFIS-based models	B	2023
[14]	Najafi et al.	Application of ANNs, ANFIS and RSM to estimating and optimizing the parameters that affect the yield and cost of biodiesel production	B	2018
[2]	Okwuet al.	A comparative study of artificial neural network (ANN) and adaptive neuro-fuzzy inference system (ANFIS) models in distribution system with nondeterministic inputs	DS	2018
[15]	Kumar et al.	Performanceevaluation of ANFIS, ANN and RSM in biodiesel synthesis from Karanja oil with Domestic Microwave set up	B	2023
[16]	Kikon et al.	ANFIS-based soft computing models for forecasting effective drought index over an arid region of India	GP	2023
[17]	Oladipo and Sun	Enhanced adaptive neuro-fuzzy inference system using genetic algorithm: a case study in predicting electricity consumption	GP	2023
[18]	Yazdani et al.	Using an adaptive network-based fuzzy inference system for prediction of successful aging: a comparison with common machine learning algorithms	GP	2023

Table 20.1 (Continued)

Citations	Authors	Title	Applications	Year
[19]	Suparta and Samah	Rainfall prediction by using ANFIS times series technique in South Tangerang, Indonesia	GP	2020
[20]	Esmaili et al.	Assessment of adaptive neuro-fuzzy inference system (ANFIS) to predict production and water productivity of lettuce in response to different light intensities and CO_2 concentrations	GP	2021
[21]	Prasad et al.	Development of ANFIS models for air quality forecasting and input optimization for reducing the computational cost and time	GP	2016

ANFIS, adaptive neuro-fuzzy inference system; B, biodiesel; DS, distribution system; GP, general purpose.

20.2.3 Response Surface Methodology

RSMrepresents a statistical and mathematical approach employed in experiment design, data analysis, and process optimization. Originally, it was crafted for the field of experimental design and optimizing responses; RSM serves as a pivotal tool in unraveling the intricate connections between various input variables and their corresponding output responses. The core objective of RSM is to construct models elucidating the intricate relationships amid input variables and responses, enabling the anticipation and enhancement of these responses within specified ranges of input values. This methodology proves particularly advantageous in scenarios involving numerous factors in experiments, where the goal is to pinpoint the optimum combination of these factors to attain desired outcomes.

20.2.4 Key Components of RSM

20.2.4.1 Key Elements of RSM

The key elements of RMS include experimental design, mathematical modeling, analysis of variance, optimization, and graphical representation. It is important to discuss the key elements further.

Experimental design: In the framework of RSM, careful planning of experiments is essential. Various factors are systematically manipulated to observe their impact on the response variable of interest. These experiments are meticulously designed to efficiently gather relevant data, forming the foundation for constructing response surface models.

Mathematical modeling: Response surfaces, integral to RSM, are mathematical models that depict the relationship between input variables and the response. Typically, these models take the form of polynomial equations, capturing the curvature and interactions among factors.

Analysis of variance (ANOVA): The application of ANOVA is pivotal in assessing the statistical significance of the model and individual factors. It aids in identifying factors that significantly influence the response and those that can be deemed negligible.

Optimization: Once the response surface model is validated, optimization techniques come into play. These techniques are employed to identify the optimal set of input variables that produce the desired response. This step allows researchers and practitioners to pinpoint the most efficient and effective conditions for a given process.

Graphical representation: Visualization tools, including contour plots and three-dimensional response surface plots, are commonly utilized in RSM. These tools offer a graphical depiction of the relationships between variables, facilitating the interpretation and communication of results.

The objective of this study is to apply soft computing techniques to the tobacco seed dataset with a further exposition of the potential of the tobacco plant as a feedstock. The experimental analysis will demonstrate the effectiveness of the ANFIS and RSM for biodiesel production using a dataset obtained from tobacco seeds.

20.3 Materials and Method

20.3.1 Modeling Technique Using ANFIS

ANFIS is a modeling technique that combines the principles of neural networks and fuzzy logic to create a powerful and adaptive system. ANFIS was employed to model the methylic biodiesel production from TSO based on the available dataset by predicting biodiesel yield based on various input parameters like temperature, reaction time, and CAT concentration. Relevant data were obtained elsewhere [3] for training and testing the ANFIS model. This data includes input parameters such as retention time (RT), M:O, and CAT and corresponding biodiesel yield as output. To adopt the ANFIS model involves MF design, rule base formation, training the ANFIS model, model validation, optimization, model testing, and performance evaluation

20.3.1.1 ANFIS Model Solution for Dataset

The hybrid algorithm demonstrates rapid computational speed. It integrates qualitative techniques involving fuzzy and linguistic rules from the fuzzy inference system (FIS) with quantitative techniques encompassing the mathematical and graphical expressions of the Neural Network (NN). This hybridization merges two innovative global learning techniques into a unified framework system. The structure is illustrated in Figure 20.2.

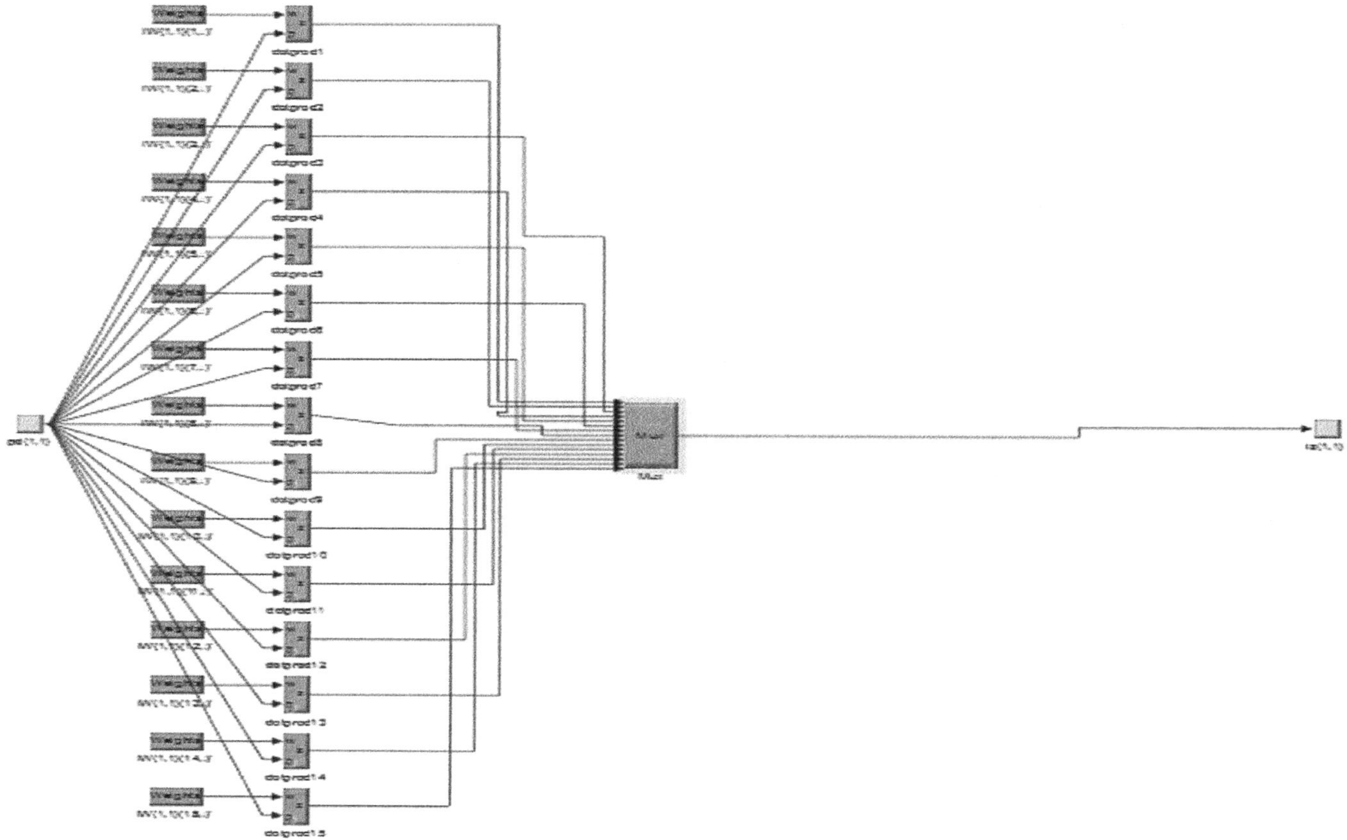

Figure 20.2 ANFIS architecture.

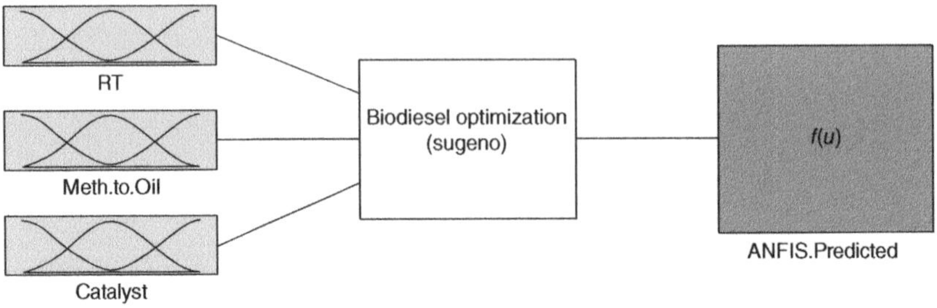

Figure 20.3 General ANFIS architecture for solar radiation dataset with input variables.

Figure 20.3 illustrates the structure of the ANFIS model, showcasing the MF designed for three key input variables: reaction time, methanol/oil molar ratio, and catalyst amount. This graphical representation provides insight into how these variables are characterized within the model, offering a visual understanding of their influence on the biofuel production process. The ANFIS architecture serves as a valuable reference for comprehending the model's complexity and the interplay between input parameters, laying the groundwork for further analysis and optimization of the biofuel production system. The yellow color toward the left-hand side of the structure depicts the MF design for the input parameter (RT, MR, and CAT), while the right-hand side shows the MF design for the corresponding biodiesel yield as output.

The diagrammatic representation of the ANFIS model with three (3) input variables is presented in Figure 20.3. All the functions have been presented in the range of the functional Table (see Table 20.2). The presented Tables serve as a support to facilitate the development of the MF plot. The decision variables are sectioned for the minimum and maximum range of dataset as presented in Table 20.3. The decision variables include RT, MR, CAT.

20.3.1.2 Membership Functions and Plots

These input variables underwent fuzzification using the ANFIS toolbox accessed from the latest version of MATLAB. The output MF was generated and subsequently defuzzified. The ANFIS MF for all variables was segmented to serve as input values for the simulation model. Figures 20.4–20.7 show a high-level graphical representation of the initial MF for the ANFIS model. Aligning MF for all variables is essential for constructing a rule-based engine, through the random application of alpha cuts while adhering to the constraints established by the MF of the various criteria (Table 20.2). The result of the ANFIS model is a set of rules, providing an ANFIS rule viewer indicating the best coefficient of correlation and MSE value. MF were defined for all input variables.

Retention Time (RT): Low (L), Medium (M), High (H)
Methanol to Oil Mole Ratio (MR): Low (L), Medium (M), High (H)
Catalyst (CAT): Low (L), Medium (M), High (H)
Biodiesel Yield (BY): Low (L), Medium (M), High (H)

Table 20.2 Rule base system for input variables.

Run	Reaction time (min)	Methanol to oil molar ratio	Catalyst (w/w. %)	Experimental yield
1	L	L	L	VL
2	H	L	L	L
3	L	H	L	M
4	H	H	L	M
5	L	L	H	M
6	H	L	H	M
7	L	H	H	M
8	H	H	H	M
9	L	M	M	H
10	H	M	M	VH
11	M	L	M	H
12	M	H	M	H
13	M	M	L	L
14	M	M	H	M
15	M	M	M	VH
16	M	M	M	VH
17	M	M	M	VH
18	M	M	M	VH
19	M	M	M	VH
20	M	M	M	VH

Table 20.3 Parameters for membership for input and output for TSOME.

Membership function/range	Reaction time (min)	Methanol/oil molar ratio	Catalyst amount	Yield (%)
Low	40–45–50	4–4.5–5	0.5–0.65–0.8	45–55–60
Medium	55–60–65	5.5–6–6.5	0.9–1–1.1	55–65–75
High	70–75–80	7–7.5–8	1.2–1.35–1.5	75–85–95

As presented in Table 20.3, the range of values of retention time expressed as low, medium, and high range for the experimental dataset is presented in orange color in Figure 20.4. The plot is well defined showing low range between 40 and 50, medium range between 55 and 65 and high range between 70 and 80.

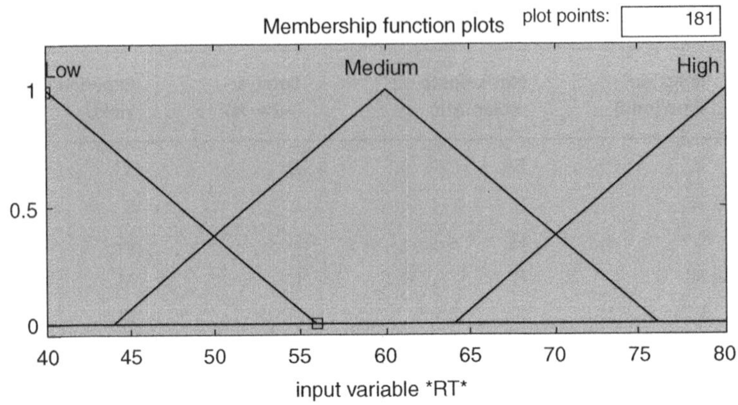

Figure 20.4 Membership function for retention time.

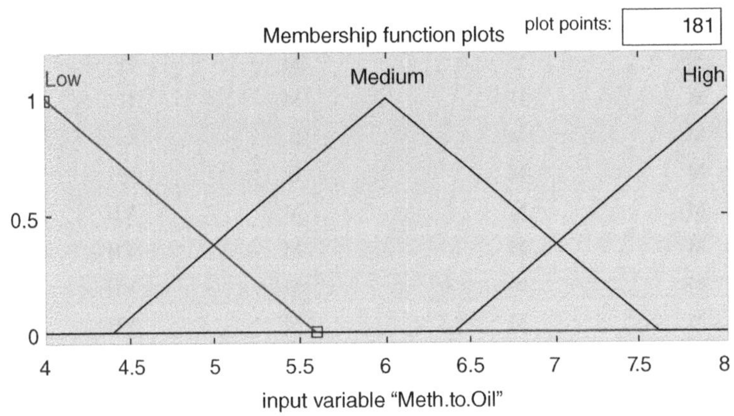

Figure 20.5 Membership function for methanol to oil molar ratio.

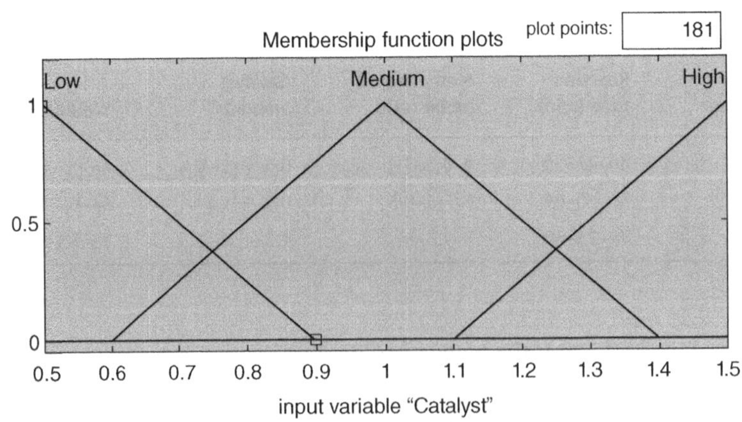

Figure 20.6 Membership function for catalyst.

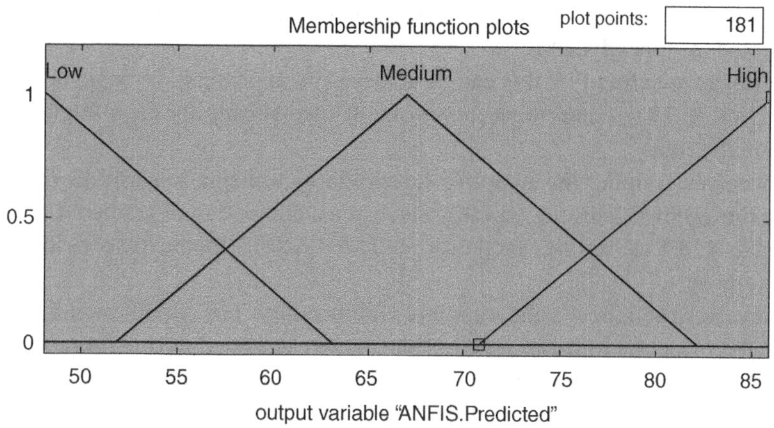

Figure 20.7 Membership function for biodiesel yield.

As presented in Table 20.3, the range of value for M:O expressed as low, medium, and high range for the experimental dataset is presented in orange color in Figure 20.5. The plot is well defined showing low range between 4 and 5, medium range between 5 and 6.5 and the highest range of 6.

Also from Table 20.3, the range of value for input variable in the form of catalyst expressed is as low, medium, and high range for the experimental dataset is presented in orange color in Figure 20.6. The plot is well-defined showing low range between 0.5 and 0.8, medium range between 0.9 and 1.1 with highest range of 1.5.

The last membership function plot is the output variable in terms of Biodiesel Yield. The graphical representation is shown in green font the range of value for output data expressed as low, medium, and high range for the experimental design. The plot is well defined showing low range between 45 and 60, medium range between 60 and 75 with the highest range of 95.

20.3.1.3 ANFIS Linguistic Rules

Creating linguistic rules for an ANFIS involves defining fuzzy sets, linguistic terms, and the relationships between input variables and output predictions. For the prediction of biodiesel experimental results from tobacco seed oil, considering three input variables (retention time, methanol to oil mole ratio, and catalyst), the following steps are required:

Step 1: This involves defining the Input and Output Variables, which include Inputs: Retention Time (RT), MR, Catalyst (CAT), and Output: Biodiesel Yield (BY).

Step 2: Define Fuzzy Sets and Linguistic Terms for MF as presented in Table 20.2. The Figures representing the MFs are also presented (Figures 20.4–20.7). The range could be expressed as low, medium, and high.

Step 3: This step is to establish linguistic rules based on the defined parameters as presented in Appendix 20.1.

Step 4: This is to define FIS, this can be achieved by applying fuzzy logic operations (AND, OR) to determine the degree of membership for each linguistic term in the rules.

Step 5 involves Training the Adaptive Neuro FIS by utilizing a hybrid learning algorithm. Hybrid learning typically involves a combination of gradient-based methods and least-squares methods. MATLAB ANFIS Toolbox was used for this purpose.

Step 6 involves the model Validation and Optimization. The ANFIS model was validated using the remaining part of the dataset sectioned during training. It was ensured that the model generalizes well to new, unseen data. The optimization aspect involves fine-tuning the MF and adjusting the rule base, and optimizing the training parameters.

20.3.1.4 ANFIS Rule Viewer

The next Step 7 involves the use of the trained and validated ANFIS model to predict biodiesel yield from the input combinations. The rule viewer helps to showcase the output results by taking alpha cut for rule base that relates the input variables to the output. This is typically done through a set of fuzzy if-then rules. The structure and number of rules depend on the complexity of the system. See the rule viewer presented in Figure 20.8.

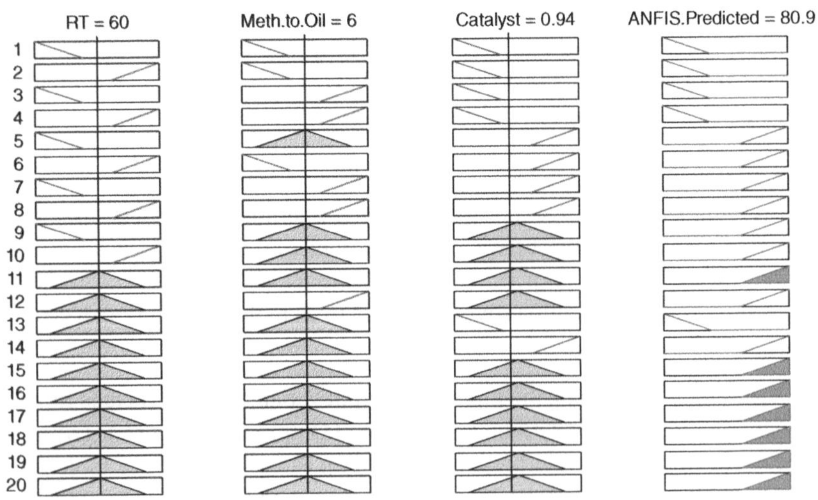

Figure 20.8 Rule viewer for ANFIS predicted result.

20.3.2 Response Surface Methodology (RSM)

This section focuses on identifying the optimal conditions for maximizing biodiesel yield from tobacco seed oil. The optimization of independent variables was carried out using RSM. The RSM model, incorporating a response variable (Y) and several independent variables (x_1, x_2, ...x_n), is defined by Eq. (20.1):

$$Y = f\left(X_1, X_2, \ldots X_n\right) + \varepsilon \qquad (20.1),$$

where f is the independent function, and ε denotes the error associated with the experiments. In this study, the central composite design (CCD) method was employed to assess the impact of input parameters, namely, reaction time, methanol-to-oil mole ratio, and catalyst, on biodiesel yield. The combination of CCD with RSM was utilized to analyze biodiesel yield. Design Expert was used to evaluate the suitability and significance of the design of experiments (DoE) model in establishing the correlation between input variables and the response. Furthermore, analysis of variance (ANOVA) was employed to illustrate the influence of independent variables on the response. The predicted response ($Y_{\text{predicted}}$) is presented in the system of Eq. (20.2).

$$Y_{\text{Predicted}} = \beta_0 + \sum_{i=1}^{k} \beta_{ij} x_i + \sum_{i=1}^{k} \beta_{ij} x_i^2 + \sum_{j>1}^{k} \sum_{j=1}^{k} \beta_{ij} x_i x_j + e \qquad (20.2),$$

where, $Y_{\text{predicted}}$ is defined as the predicted response in terms of Biodiesel yield; β_0; β_{ii}; β_{ij} are the regression coefficients; k is the number of factors optimized and considered in the experiment, and e is presented as the random error.

20.4 Result and Discussions

20.4.1 Data for Experimental Investigation

In this study, 20 experimental runs were obtained from a study conducted elsewhere [3], and used as the Experimental runs matrix. Each experimental run comprising the reaction time (min), methanol to oil mole ratio, and catalyst (%) was used as input parameters, while the TSOME Yield (%) as the output parameters, respectively, as presented in Table 20.4.

20.4.2 Modeling and Prediction Using the RSM

In this study, an attempt is made to develop a second-order mathematical relationship between selected input variables, namely; reaction time (min), MR and catalyst (%) coupled with one response variables, namely; TSOME Yield (%) using the RSM.

Table 20.4 Experimental dataset.

Run	Reaction time (min)	Methanol to oil mole ratio	Catalyst (%)	TSOME Yield (%)
1	40	4	0.5	48.06
2	80	4	0.5	55.63
3	40	8	0.5	56.52
4	80	8	0.5	56.04
5	40	4	1.5	62.97
6	80	4	1.5	76.01
7	40	8	1.5	65.13
8	80	8	1.5	73.45
9	40	6	1	79.38
10	80	6	1	89.75
11	60	4	1	81.15
12	60	8	1	78.18
13	60	6	0.5	59.71
14	60	6	1.5	74.71
15	60	6	1	90.07
16	60	6	1	86.04
17	60	6	1	86.04
18	60	6	1	90.07
19	60	6	1	85.13
20	60	6	1	85.13

20.4.3 Modeling and Prediction of TSOME Yield (%)

To assess the suitability and effectiveness of the quadratic model for a comprehensive analysis of experimental data, a thorough validation process was conducted. This process include calculating the sequential model sum of squares, a crucial statistical metric employed to evaluate the model's adequacy in capturing the variation in TSOME Yield (%), as detailed in Table 20.4. This analysis plays a crucial role in ensuring the robustness and reliability of the quadratic model, confirming its capability to accurately model and predict the TSOME Yield (%) response. The sequential model sum of squares not only validates the model's fit but also offers valuable insights into the contribution of each term, thereby bolstering precision and confidence in the model's predictive capacities.

A noteworthy observation surfaced during the examination of the data presented in Table 20.5, indicating that the cubic polynomial was found to be aliased.

Table 20.5 Sequential model sum of squares.

Source	Sum of squares	*df*	Mean square	*F*-value	*p*-Value	
Mean vs.total	1.094E+05	1	1.094E+05			
Linear vs.mean	736.05	3	245.35	1.45	0.2661	
2FI vs.linear	56.58	3	18.86	0.0923	0.9629	
Quadratic vs. 2FI	**2602.65**	**3**	**867.55**	**165.87**	**<0.0001**	**Suggested**
Cubic vs. quadratic	16.03	4	4.01	0.6627	0.6403	Aliased
Residual	36.27	6	6.05			
Total	1.128E+05	20	5642.24			

This inherent aliasing suggests that the complexity and potentially confounding nature of the cubic polynomial render it ineffective for contributing to the final model. In response to this realization, a judicious decision was made, redirecting the analysis toward the quadratic and 2FI (Two-Factor Interaction) models as more practical and suitable alternatives. This careful selection not only simplifies the intricacies associated with the modeling process but also reinforces the use of the quadratic polynomial as the preferred analytical tool in this comprehensive investigation.

To assess the quadratic model's ability to elucidate the inherent variation in the experimental data, a critical diagnostic step involved estimating the lack of fit for the response variable. The lack of fit test serves as a robust metric, evaluating how well the quadratic model accounts for the residual discrepancies between the predicted and observed values. In Table 20.3, the meticulously presented results of the computed lack of fit for TSOME Yield (%) offer valuable insights into the adequacy and reliability of the quadratic model. This analysis sheds light on its effectiveness in capturing and explaining the intricacies present in the experimental data. The lack of fit analysis significantly contributes to the overall validation of the chosen quadratic model, ensuring a comprehensive understanding of its predictive capabilities and suitability for the analytical task.

The examination of the outcomes provided in Table 20.6 highlighted a pivotal criterion for model selection – an ideal model should demonstrate an insignificant lack of fit to be considered suitable for predictive purposes. Adhering to this criterion, any model exhibiting a substantial lack of fit is deemed unsuitable for reliable predictions. In our scrutiny, it was reaffirmed that the quadratic polynomial displayed a nonsignificant lack of fit, establishing its suitability for model analysis and subsequent predictions. In contrast, the cubic polynomial, characterized by a noteworthy lack of fit, was considered significant and consequently aliased, thereby disallowing its use in the predictive modeling process.

Table 20.6 Lack of fit tests.

Source	Sum of squares	*df*	Mean square	*F*-value	*p*-Value	
Linear	2683.88	11	243.99	44.12	0.0003	
2FI	2627.30	8	328.41	59.39	0.0002	
Quadratic	**24.65**	**5**	**4.93**	**0.8917**	**0.5485**	**Suggested**
Cubic	8.63	1	8.63	1.56	0.2669	Aliased
Pure error	27.65	5	5.53			

Table 20.7 Model summary statistics.

Source	Std. dev.	R^2	Adjusted R^2	Predicted R^2	PRESS	
Linear	13.02	0.2135	0.0660	−0.3622	4696.44	
2FI	14.29	0.2299	−0.1255	−4.1064	17604.57	
Quadratic	**2.29**	**0.9848**	**0.9712**	**0.9369**	**217.62**	**Suggested**
Cubic	2.46	0.9895	0.9667	−2.0848	10635.04	Aliased

To further validate the appropriateness of the chosen model, comprehensive model statistics were computed for the response variable of centrifugal force. These statistics, originating from the model sources, are intricately presented in Table 20.7. This table serves as a repository of crucial information, providing insights into the performance and reliability of the selected model in elucidating the variation in the centrifugal force response. The model statistics, encompassing regression coefficients and associated *p*-values, contribute to a comprehensive understanding of the model's efficacy in representing the intricate relationships inherent in the experimental data. The significance of each model term is meticulously examined, facilitating a nuanced evaluation of the model's predictive capability and its potential application in practical scenarios.

Based on the information extracted from Table 20.7, the quadratic polynomial model was recommended for consideration, while the cubic polynomial model was identified as aliased. As a result, the quadratic polynomial model emerged as the preferred option for this analytical investigation. To assess the effectiveness of the quadratic model in minimizing TSOME Yield (%), a one-way ANOVA table was generated and is outlined in Table 20.5. This table serves as a comprehensive tool for evaluating the statistical significance of the quadratic model's terms and offers valuable insights into the model's effectiveness in explaining variations in the TSOME Yield (%). The ANOVA results contribute to the assessment of the

Table 20.8 ANOVA for quadratic model.

Source	Sum of squares	*df*	Mean square	*F*-value	*p*-Value	
Model	3395.27	9	377.25	72.13	< 0.0001	Significant
A: Reaction time	150.70	1	150.70	28.81	0.0003	
B: Methanol to oil mole ratio	3.03	1	3.03	0.5784	0.4645	
C: Catalyst	582.32	1	582.32	111.34	< 0.0001	
AB	20.38	1	20.38	3.90	0.0766	
AC	25.45	1	25.45	4.87	0.0519	
BC	10.74	1	10.74	2.05	0.1823	
A^2	2.22	1	2.22	0.4242	0.5295	
B^2	92.45	1	92.45	17.68	0.0018	
C^2	916.24	1	916.24	175.18	<0.0001	
Residual	52.30	10	5.23			
Lack of fit	24.65	5	4.93	0.8917	0.5485	Not significant
Pure error	27.65	5	5.53			
Cor total	3447.58	19				

robustness and reliability of the chosen quadratic model, providing support for its suitability for the specific analytical objectives at hand.

Reviewing the information in Table 20.8, highlights a crucial step in validating the model's significance and thoroughly assessing the substantial contributions of individual variables, along with their combined and quadratic effects on each response. This involved employing ANOVA, a statistical method serving a dual purpose: determining the overall significance of the model and investigating the specific influences exerted by each variable and their interactions.

The Model *F*-value, a key metric derived from the ANOVA, registers an impressive 72.13, indicating a noteworthy level of significance in the model. The magnitude of the *F*-value implies a mere 0.01% probability that such a value could arise solely due to random noise in the data. This statistical insight underscores the model's robustness, suggesting that the observed effects are highly unlikely to be a result of chance or random variability. Therefore, the Model *F*-value provides compelling evidence supporting the model's significance, enhancing confidence in its capacity to effectively capture and explain variations in the considered responses.

P-values below 0.0500 indicate the significance of model terms. In this case, model terms A, C, B^2, and C^2 exhibit significance, with *p*-values below this threshold. Conversely, values surpassing 0.1000 suggest that corresponding model terms lack significance. If numerous model terms are found insignificant, excluding those crucial for maintaining hierarchy, reducing the model may enhance its effectiveness. The Lack of Fit *F*-value, standing at 0.89, is noteworthy in its nonsignificance relative to pure error. This implies a 54.85% probability that a Lack of Fit *F*-value of this magnitude could arise from random noise. A lack of significance in the Lack of Fit is advantageous for model fitting, suggesting that observed discrepancies are likely not due to unaccounted factors or noise.

To assess the quadratic model's effectiveness in minimizing centrifugal force, a detailed examination of fit statistics is presented in Table 20.9. This analysis represents a crucial step in validating the adequacy of the quadratic model and its ability to effectively capture and minimize variations in centrifugal force.

The predicted R^2, achieving a value of 0.9369, demonstrates a commendable level of agreement with the adjusted R^2, which stands at 0.9712. Importantly, the difference between these two values is less than 0.2, indicating a robust agreement. This alignment emphasizes the model's reliability in predicting responses, highlighting that its predictive capabilities remain strong even when accommodating variations in the number of terms.

An essential metric, Adeq Precision, provides insights into the signal-to-noise ratio. A ratio exceeding 4 is considered desirable, and in this case, your ratio of 25.334 signifies a highly adequate signal. This indicates that the model possesses a substantial signal component relative to the noise, reinforcing its reliability for navigating within the design space. The high Adeq Precision ratio adds credibility to the model's predictive capabilities, supporting its practical utility.

In the pursuit of the optimal solution, careful attention is devoted to coefficient statistics and their corresponding standard errors. These statistics, crucial for determining the significance and reliability of each coefficient, play a pivotal role in understanding the impact of each variable on the response.

Table 20.9 Fit statistics.

Std. dev.	2.29	R^2	0.9848
Mean	73.96	Adjusted R^2	0.9712
C.V. %	3.09	Predicted R^2	0.9369
		Adeq precision	25.3337

Table 20.10 Coefficients in terms of coded factors.

Factor	Coefficient estimate	*df*	Standard error	95% CI low	95% CI high	VIF
Intercept	86.43	1	0.7862	84.68	88.19	
A: Reaction time	3.88	1	0.7232	2.27	5.49	1.0000
B: Methanol to oil mole ratio	0.5500	1	0.7232	−1.06	2.16	1.0000
C: Catalyst	7.63	1	0.7232	6.02	9.24	1.0000
AB	−1.60	1	0.8086	−3.40	0.2053	1.0000
AC	1.78	1	0.8086	−0.0178	3.59	1.0000
BC	−1.16	1	0.8086	−2.96	0.6428	1.0000
A^2	−0.8982	1	1.38	−3.97	2.17	1.82
B^2	−5.80	1	1.38	−8.87	−2.73	1.82
C^2	−18.25	1	1.38	−21.33	−15.18	1.82

Table 20.10 thoroughly presents the coefficient statistics for the TSOME Yield (%) response variable. This thorough analysis aids in refining the model and fine-tuning the coefficients to achieve an optimized and reliable solution within the experimental design.

Gleaning insights from Table 20.10, it is essential to grasp the significance of the coefficient estimate, representing the anticipated change in the response for each unit shift in factor value, assuming all other factors remain constant. In the context of an orthogonal design, the intercept takes on the role of the overall average response across all experimental runs. The coefficients, in turn, serve as adjustments around this average, depending on the specific factor settings. In cases where factors are orthogonal, variance inflation factors (VIFs) maintain a value of 1, indicating an absence of multi-collinearity. Conversely, VIFs surpassing 1 signal the presence of multicollinearity, with the severity of correlation among factors increasing with higher VIF values. Generally, VIFs below 10 are considered acceptable, suggesting a manageable level of multi-collinearity. This metric proves valuable in assessing the extent of interdependence among factors, thereby influencing the reliability and stability of the model.

The optimal equation, which delineates the individual effects and combined interactions of the selected input variables against the response, is presented based on the coded variables in Eq. (20.3).

Equation in terms of coded factors (20.3)

TSOME Yield (%) =

$$
\begin{aligned}
&+86.43\\
&+3.88\ \mathrm{A}\\
&+0.5500\ \mathrm{B}\\
&+7.63\ \mathrm{C}\\
&-1.60\ \mathrm{AB}\\
&+1.78\ \mathrm{AC}\\
&-1.16\ \mathrm{BC}\\
&-0.8982\ \mathrm{A}^2\\
&-5.80\ \mathrm{B}^2\\
&-18.25\ \mathrm{C}^2
\end{aligned}
$$

The equation expressed in coded factors is applicable for making predictions regarding the response at specified levels of each factor. In this context, the high levels of factors are conventionally coded as +1, while the low levels are coded as −1. The coded equation serves a practical purpose in discerning the relative impact of factors by facilitating a comparison of the coefficients assigned to each factor.

Equation in terms of actual factors (20.4)

TSOME Yield (%) =

$$
\begin{aligned}
&-86.02098\\
&+0.524617\ \text{Reaction time}\\
&+21.22267\ \text{Methanol to oil mole ratio}\\
&+157.53745\ \text{Catalyst}\\
&-0.039906\ \text{Reaction time} \times \text{methanol to oil mole ratio}\\
&+0.178375\ \text{Reaction time} \times \text{catalyst}\\
&-1.15875\ \text{Methanol to oil mole ratio} \times \text{catalyst}\\
&-0.002245\ \text{Reaction time}^2\\
&-1.44955\ \text{Methanol to oil mole ratio}^2\\
&-73.01273\ \text{Catalyst}^2
\end{aligned}
$$

The equation expressed in terms of actual factors (Eq. 20.4) is applicable for predicting the response at specific levels of each factor. It is important to note that the levels must be specified in the original units for each factor. However, this equation is not suitable for determining the relative impact of each factor because the coefficients are scaled to accommodate the units of each factor, and the intercept is not positioned at the center of the design space.

The diagnostic case statistics, which present the observed values of centrifugal force against their predicted values, are detailed in Table 20.11. These diagnostic case statistics offer insights into the strength of the model and the adequacy of the optimal second-order polynomial equation.

To accept any model, its satisfactoriness must first be checked by an appropriate statistical analysis output. To diagnose the statistical properties of the response surface model, the normal probability plot of residual for the TSOME Yield (%) is represented in Figure 20.9.

The normal probability plot of studentized residuals stands out as the crucial assumption for assessing the adequacy of a statistical model. The findings indicate that the calculated residuals follow an approximately normal distribution, affirming the satisfactory nature of the developed model. In order to identify the presence of significant patterns or expanding variance, a plot of residuals against the predicted values for the TSOME Yield (%) was generated, as depicted in Figure 20.10.

Figure 20.10 visually presents a graph where residuals are plotted against increasing response values, aiming to assess the assumption of consistent variance within the model. The objective is to examine the dispersion of residuals across the spectrum of response values and determine if the variability remains relatively uniform.

In a complementary endeavor to unveil values or groups that might escape easy model detection, Figure 20.11 is showcased. This figure includes a plot illustrating the relationship between predicted and actual values, specifically for TSOME Yield (%). By comparing predicted and actual values on this plot, insights into the model's alignment with observed data are gained, aiding in the identification of any disparities or outliers that could impact overall model performance.

Based on careful examination of the graph, it is evident that the data points closely follow the fitted line, indicating a robust correlation between predicted values and actual observations. The model demonstrates a notable ability to accurately predict the majority of the data points, highlighting its effectiveness in capturing the inherent patterns in the experimental data.

To further scrutinize the dataset for potential outliers or influential observations, a Cook's distance plot was generated for the responses. Cook's distance serves as a statistical metric evaluating the impact of individual data points on the regression model. In this context, Figure 20.12 presents the Cook's distance specifically for centrifugal force. This plot offers valuable insights into the influence of each data point on the model, aiding in the identification of any observations that might significantly affect the model's performance or assumptions.

Table 20.11 Diagnostics case statistics report of observed vs predicted centrifugal force.

Run order	Actual value	Predicted value	Residual	Leverage	Internally studentized residuals	Externally studentized residuals	Cook's distance	Influence on fitted value DFFITS	Standard order
1	48.06	48.45	−0.3895	0.793	−0.374	−0.358	0.054	−0.701	1
2	55.63	55.84	−0.2085	0.793	−0.200	−0.191	0.015	−0.373	2
3	56.52	55.06	1.46	0.793	1.404	1.487	0.756	2.912	3
4	56.04	56.06	−0.0235	0.793	−0.023	−0.021	0.000	−0.042	4
5	62.97	62.46	0.5085	0.793	0.489	0.469	0.092	0.919	5
6	76.01	76.99	−0.9755	0.793	−0.938	−0.932	0.337	−1.825	6
7	65.13	64.44	0.6935	0.793	0.667	0.647	0.171	1.267	7
8	73.45	72.58	0.8745	0.793	0.841	0.827	0.271	1.621	8
9	79.38	81.65	−2.27	0.491	−1.393	−1.472	0.187	−1.446	9
10	89.75	89.42	0.3329	0.491	0.204	0.194	0.004	0.190	10
11	81.15	80.09	1.06	0.491	0.653	0.633	0.041	0.621	11
12	78.18	81.19	−3.01	0.491	−1.842	−2.149	0.327	−2.110	12
13	59.71	60.55	−0.8391	0.491	−0.514	−0.494	0.025	−0.486	13
14	74.71	75.81	−1.10	0.491	−0.675	−0.655	0.044	−0.643	14
15	90.07	86.43	3.64	0.118	1.693	1.902	0.038	0.696	15
16	86.04	86.43	−0.3933	0.118	−0.183	−0.174	0.000	−0.064	16
17	86.04	86.43	−0.3933	0.118	−0.183	−0.174	0.000	−0.064	17
18	90.07	86.43	3.64	0.118	1.693	1.902	0.038	0.696	18
19	85.13	86.43	−1.30	0.118	−0.607	−0.587	0.005	−0.215	19
20	85.13	86.43	−1.30	0.118	−0.607	−0.587	0.005	−0.215	20

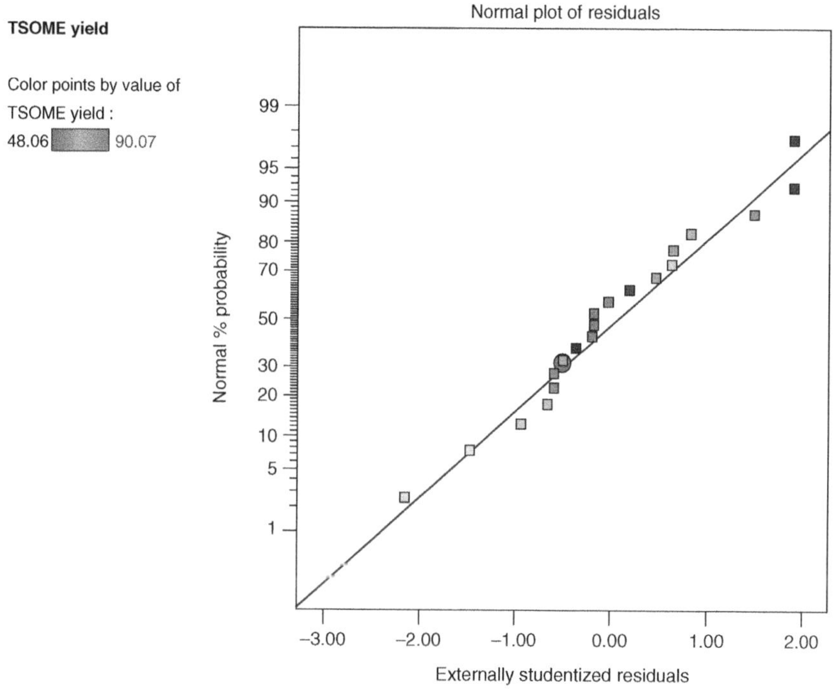

Figure 20.9 Normal probability plot of studentized residuals for TSOME Yield (%).

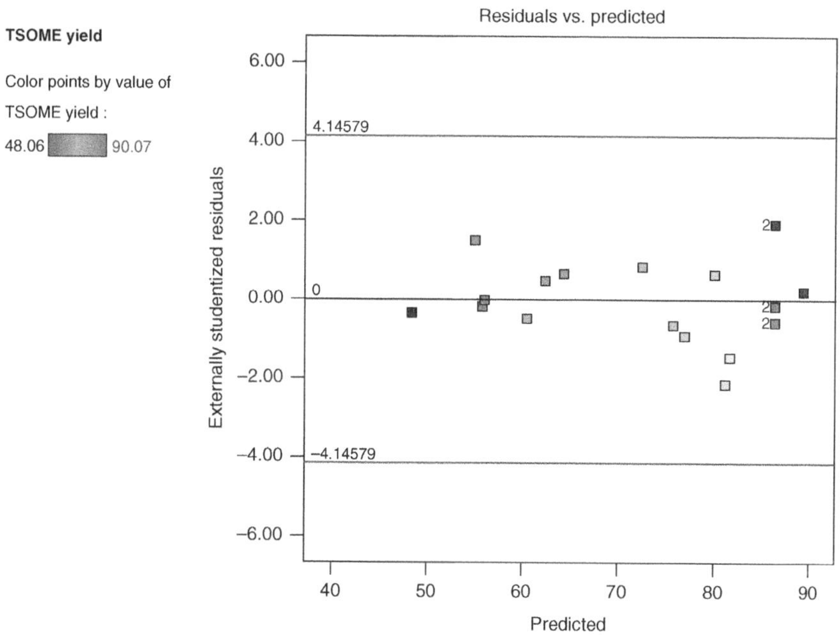

Figure 20.10 Plot of residual vspredicted for TSOME Yield (%).

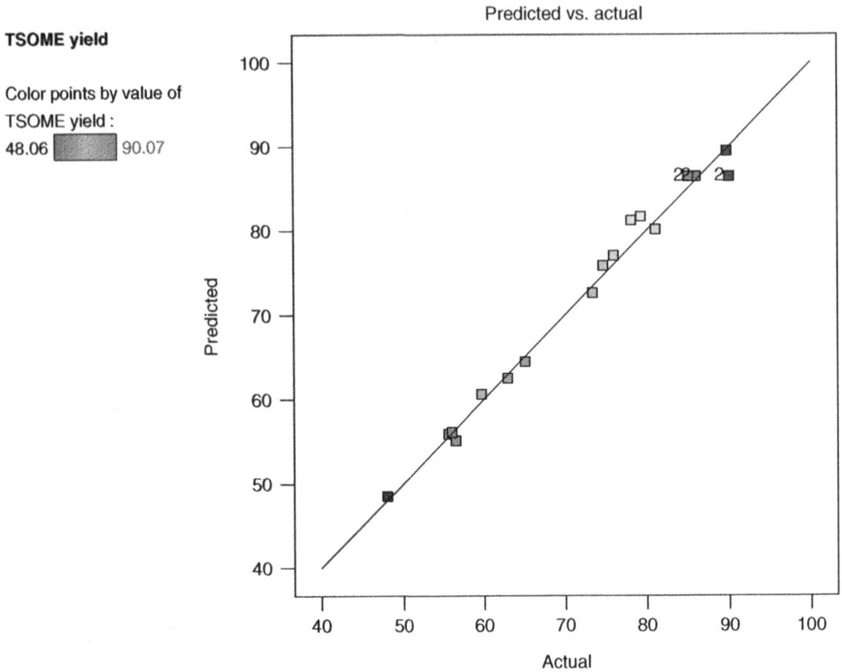

Figure 20.11 Plot of predicted vs actual for TSOME Yield (%).

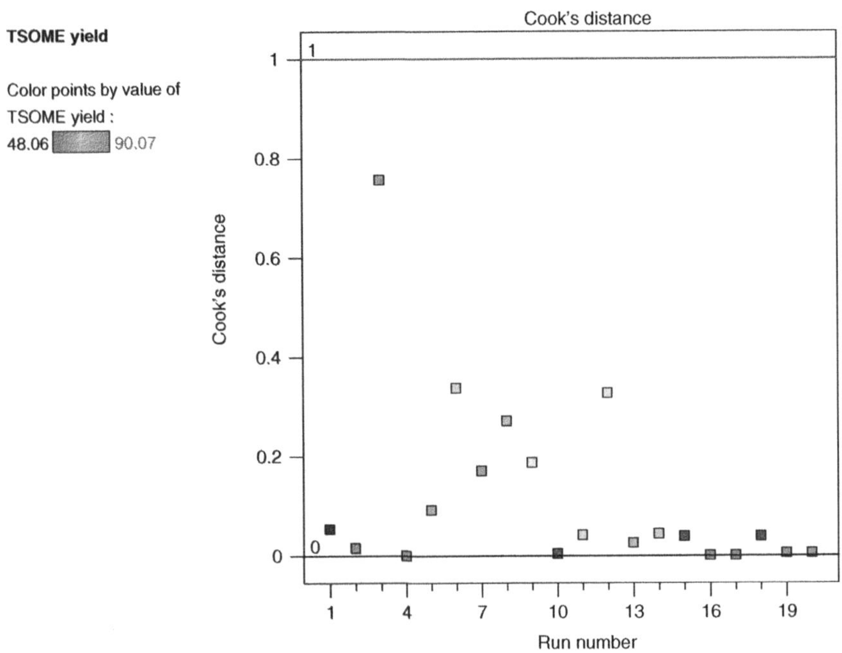

Figure 20.12 Generated cook's distance for TSOME Yield (%).

The Cook's distance plot is characterized by an upper limit of 1.00 and a lower limit of 0.00. In this context, experimental values that fall below the lower bound or exceed the upper bound are considered potential outliers, necessitating thorough investigation and scrutiny. Identifying and understanding such outliers is crucial to ensure the robustness and reliability of the analysis.

To thoroughly assess the combined influence of input variables on TSOME Yield (%), a series of 3D surface plots is presented in Figure 20.13. These plots visually represent the relationships and interactions among multiple input variables and their impact on TSOME Yield. Exploring these surface plots provides valuable insights into how variations in the input variables collectively contribute to observed changes in TSOME Yield, aiding in the interpretation of the overall system behavior.

To examine the impacts of the combined input variables, Catalyst and Reaction Time, on TSOME Yield, the 3D surface plots shown in Figure 20.14 were created in the following manner.

By carefully representing the multidimensional dataset on a three-dimensional surface, these plots offer an intuitive visualization of how changes in the levels of

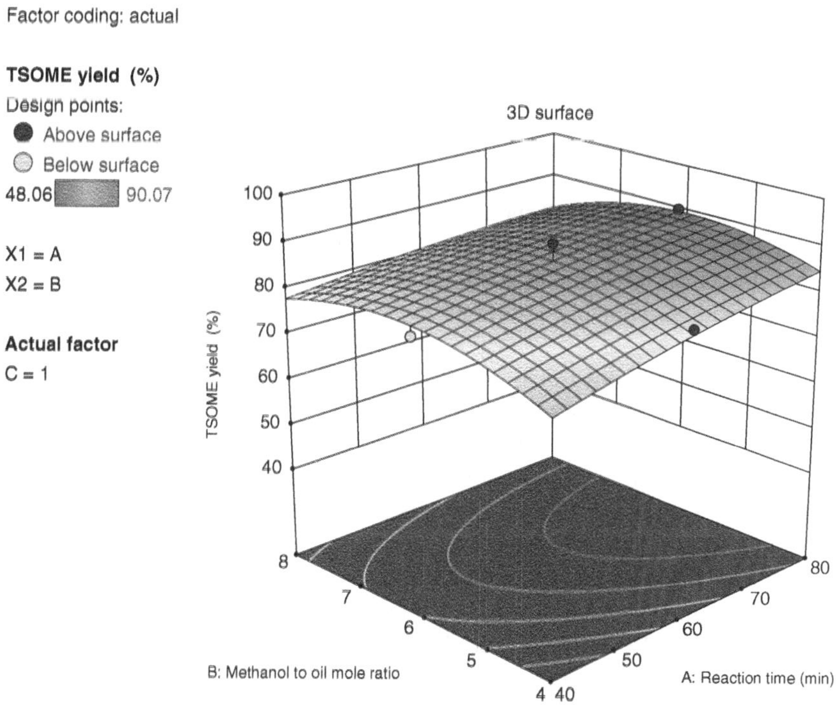

Figure 20.13 3D surface plot of TSOME Yield, methanol to oil mole ratio, and reaction time.

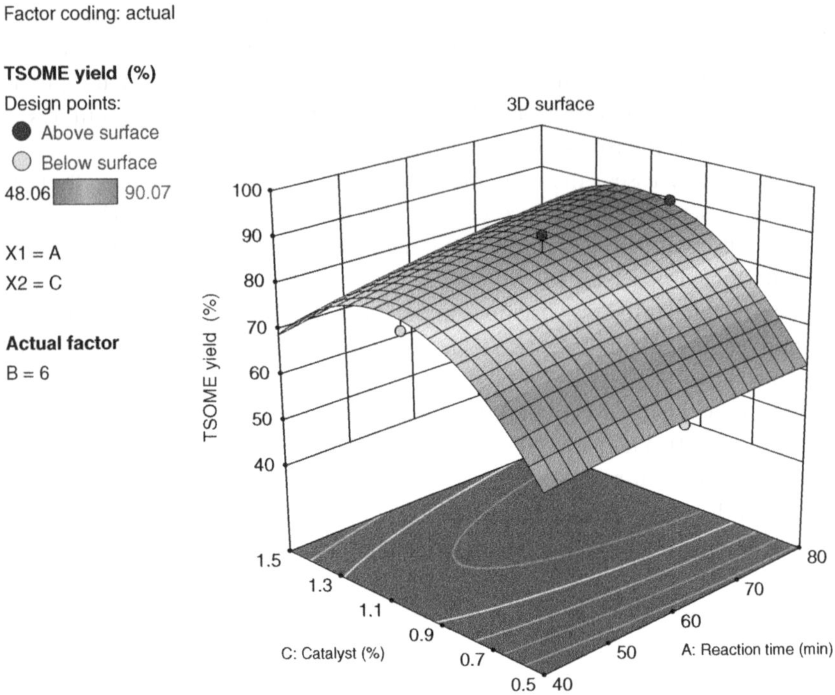

Figure 20.14 3D surface plot of TSOME Yield, catalyst, and reaction time.

catalyst and methanol to oil mole ratio collectively influence the variations in TSOME Yield, as depicted in Figure 20.15. This approach enables a nuanced comprehension of the interconnections between the input variables and the resulting TSOME Yield, assisting in the interpretation of the intricate dynamics governing the system.

The contour plots illustrated in Figure 20.16 vividly illustrate the changes in the TSOME Yield response variable in response to variations in both catalyst and methanol to oil mole ratio. These plots visually represent the intricate relationships and interactions between the two input variables, providing valuable insights into their combined effects on TSOME Yield. By displaying contours of constant TSOME Yield levels on the catalyst–MR plane, these plots facilitate an intuitive understanding of how changes in these input variables collectively impact the response variable. The contour lines connect points with equal TSOME Yield values, allowing for the identification of regions where the response is optimized or minimized based on different combinations of catalyst and methanol to oil mole ratio. This graphical representation assists in discerning the optimal conditions for achieving desired TSOME Yield levels, contributing to a more comprehensive interpretation of the experimental outcomes.

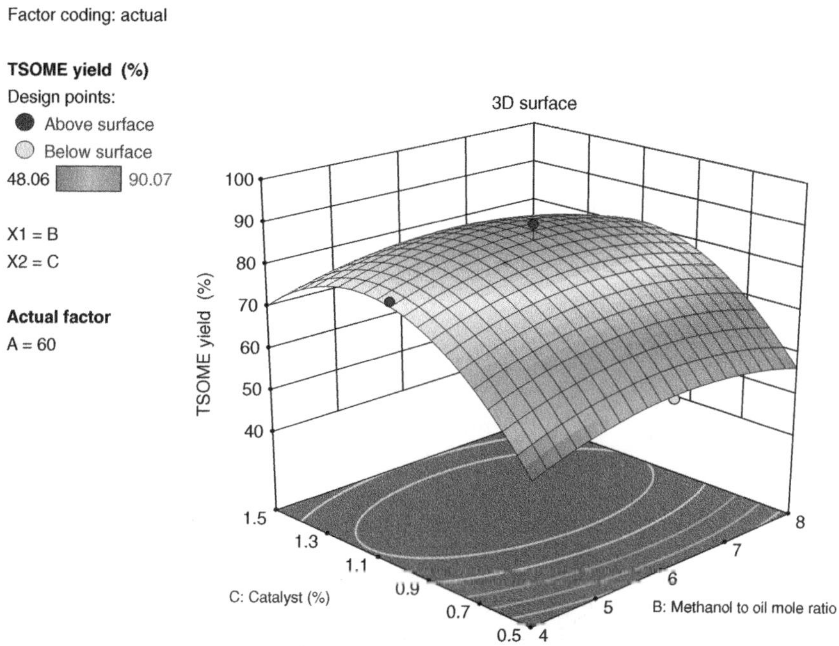

Figure 20.15 3D surface plot of TSOME Yield, catalyst, and methanol to oil mole ratio.

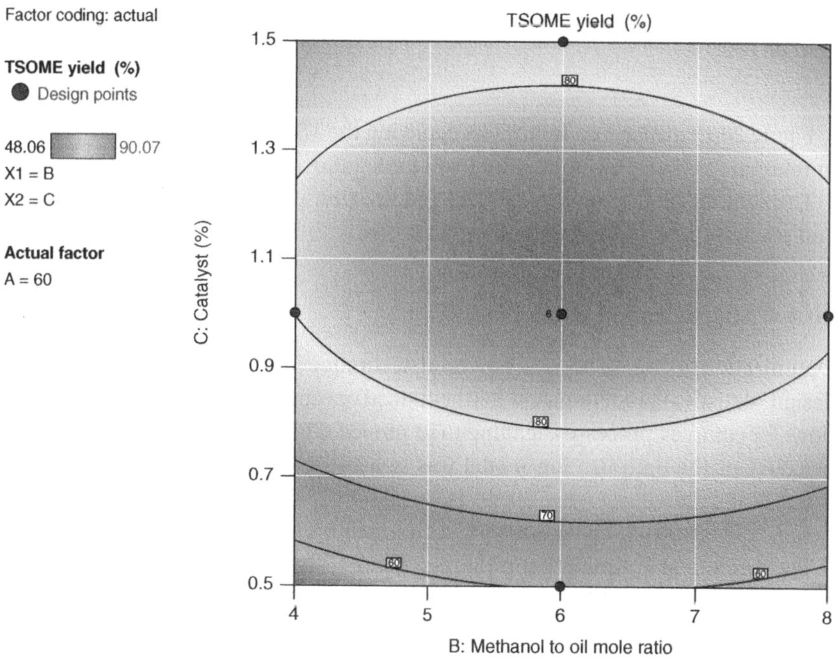

Figure 20.16 The contour plots of TSOME Yield, catalyst, and methanol to oil mole ratio.

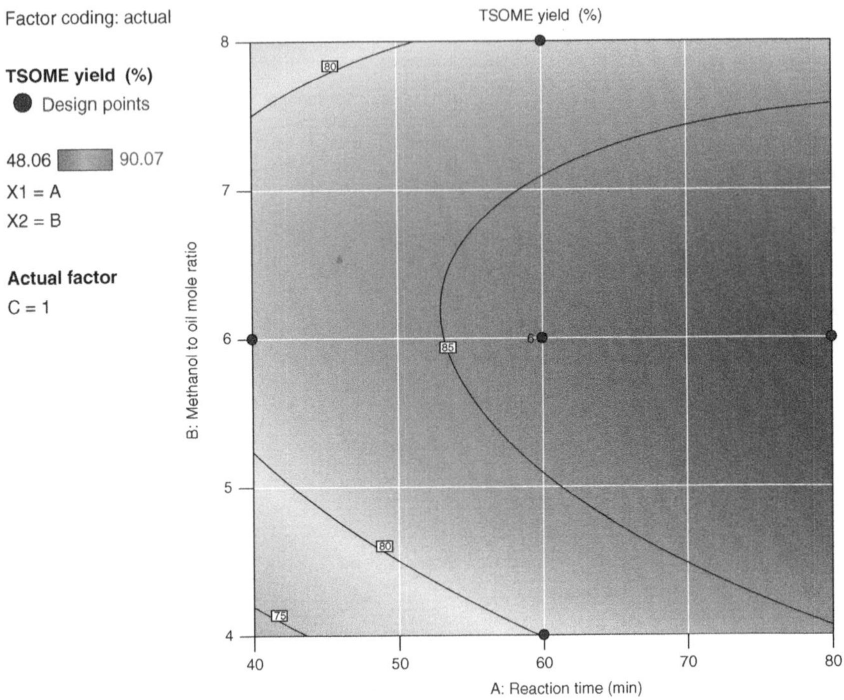

Figure 20.17 The contour plots of TSOME Yield against methanol to oil mole ratio and reaction time.

Figure 20.17 displays contour plots depicting the TSOME Yield response variable in relation to changes in methanol to oil mole ratio and reaction time.

The contour plots showing TSOME Yield response variable against the catalyst and reaction time is presented in Figure 20.18.

20.4.4 Optimization Using the RSM

Optimization using RSM is a powerful technique that combines experimental design, statistical modeling, and mathematical optimization to find optimal conditions for complex processes, leading to improved efficiency and effectiveness. The objective of the optimization model was to achieve the highest possible TSOME Yield, as illustrated in Figure 20.19. The peak TSOME Yield is observed within the range of 48.06–90.07. The optimization constraint is shown in Table 20.12.

The numerical optimization produces about 15 optimal solutions which a represented in Table 20.13. The experimental result and predicted solution using RSM and ANFIS is presented in Table 20.14.

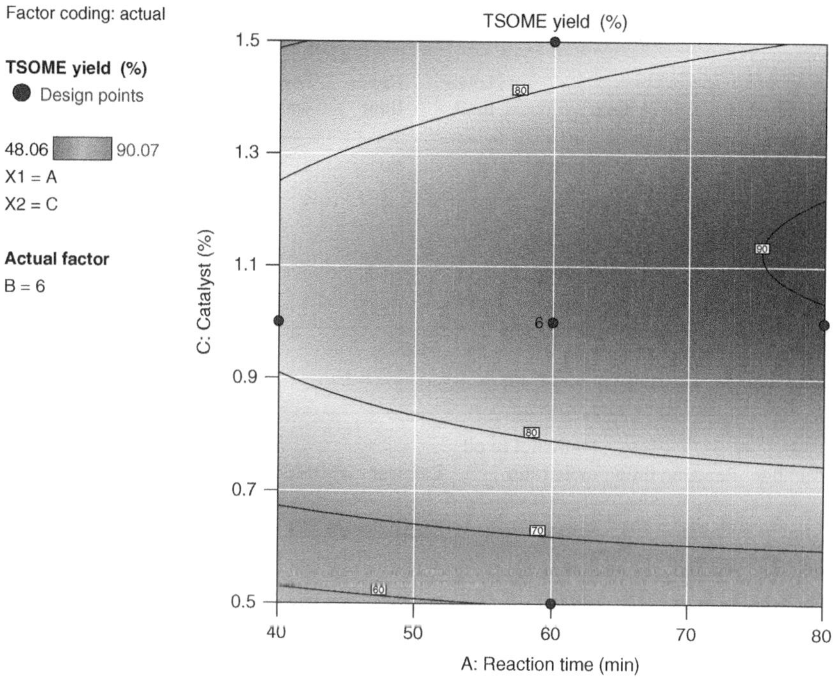

Figure 20.18 The contour plots of TSOME Yield against the catalyst and reaction time.

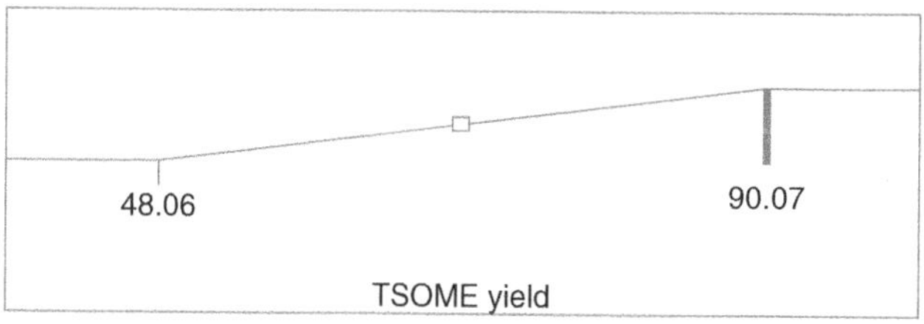

Figure 20.19 Range of TSOME yield within a specified period.

20.4.5 Evaluation Metrics for Model Performance Using Statistical Parameters for RSM and ANFIS Results

The efficacy of the formulated model was appraised using diverse performance evaluation metrics, which encompass mean square error (MSE), coefficient of determination (R^2), root mean square error (RMSE), and mean absolute percentage error (MAPE). MAPE gauges the average percentage difference between observed and predicted values, while RMSE quantifies the variability between predicted and actual data, serving as a measure of the model's accuracy. It is essential to note that RMSE is always positive. R^2 acts as an indicator of the

Table 20.12 Optimization constraints.

Name	Goal	Lower limit	Upper limit	Lower weight	Upper weight	Importance
A: Reaction time	Is in range	40	80	1	1	3
B: Methanol to oil mole ratio	Is in range	4	8	1	1	3
C: Catalyst	Is in range	0.5	1.5	1	1	3
TSOME yield	Maximize	48.06	90.07	1	1	3

Table 20.13 The optimal solutions.

Number	Reaction time	Methanol to oil mole ratio	Catalyst	TSOME yield	Desirability	
1	**77.918**	**5.904**	**1.125**	**90.399**	**1.000**	**Selected**
2	79.272	5.916	1.108	90.548	1.000	
3	76.436	5.827	1.136	90.191	1.000	
4	78.147	6.087	1.114	90.315	1.000	
5	78.254	5.978	1.061	90.100	1.000	
6	78.668	5.928	1.142	90.480	1.000	
7	76.185	5.972	1.147	90.091	1.000	
8	77.778	5.807	1.130	90.396	1.000	
9	79.613	6.187	1.166	90.299	1.000	
10	76.914	5.754	1.162	90.173	1.000	

strength of the linear association between predicted and measured values. Lower values of MAPE and RMSE indicate higher precision in predicting TSOME, with the optimal scenario being both metrics approaching zero. R^2 is a numeric value ranging from 0 to 1, where a value close to +1 signifies a perfect linear relationship. The expressions for R^2, MSE, RMSE, and MAPE (%) are articulated in the equations as presented:

$$R = \left(\frac{\sum_{m=1}^{n} \left(Y_{\text{Pred}.m} - y_{\text{Pred}}\right)\left(Y_{\text{exp}.m} - y_{\text{exp}}\right)}{\sqrt{\sum_{m=1}^{n} \left(Y_{\text{Pred}.m} - y_{\text{Pred}}\right)^2 \sum_{m=1}^{n} \left(Y_{\text{exp}.m} - y_{\text{exp}}\right)^2}} \right) \tag{20.5}$$

$$R^2 = 1 - \frac{\sum_{i=1}^{n} \left(Y_{i,P} - Y_{1,e}\right)^2}{\sum_{i=1}^{n} \left(Y_{i,P} - Y_{e,\text{ave}}\right)^2} \tag{20.6}$$

Table 20.14 The experimental runs matrix with predicted TSOME yield.

Run	Reaction time (min)	Methanol to oil mole ratio	Catalyst (%)	TSOME yield (%)	RSM TSOME yield (%)	ANFIS TSOME yield (%)	
1	40	4	0.5	48.06	48.45	48.12	48.42093
2	80	4	0.5	55.63	55.84	55.65	55.64193
3	40	8	0.5	56.52	55.06	55.98	56.51893
4	80	8	0.5	56.04	56.06	56.05	56.23493
5	40	4	1.5	62.97	62.46	62.68	62.26493
6	80	4	1.5	76.01	76.99	76.54	77.34093
7	40	8	1.5	65.13	64.44	64.98	64.60793
8	80	8	1.5	73.45	72.58	73.10	72.37893
9	40	6	1	79.38	81.65	80.21	81.88727
10	80	6	1	89.75	89.42	89.56	89.68327
11	60	4	1	81.15	80.09	80.96	79.95127
12	60	8	1	78.18	81.19	79.24	81.01927
13	60	6	0.5	59.71	60.55	59.98	60.18327
14	60	6	1.5	74.71	75.81	75.26	75.67727
15	60	6	1	90.07	86.43	88.06	86.99982
16	60	6	1	86.04	86.43	86.25	86.39982
17	60	6	1	86.04	86.43	86.18	86.19982
18	60	6	1	90.07	86.43	88.56	86.89982
19	60	6	1	85.13	86.43	85.98	85.39982
20	60	6	1	85.13	86.43	85.92	85.39982

$$\text{RMSE} = \sqrt{\frac{\sum_{i=1}^{n}\left(Y_{i,e} - Y_{i,p}\right)^2}{n}} \tag{20.7}$$

$$\text{MAE} = \sum_{i=1}^{n} \frac{\left(Y_{i,e} - Y_{i,p}\right)}{n} \tag{20.8}$$

$$\text{SEP} = \frac{\text{RMSE}}{Y_{e,\text{ave}}} \tag{20.9}$$

$$\text{AAD} = \frac{100}{n} \sum_{i=1}^{n} \frac{\left(Y_{i,e} - Y_{i,p}\right)}{Y_{i,e}} \tag{20.10}$$

To ascertain the superiority and predictive capability of the modeling techniques, statistical parameters including correlation coefficient (R), regression coefficient (R^2), RMSE, and others are utilized. The formulas for calculating the aforementioned statistical parameters are outlined in Eqs. (20.5–20.10). The analysis, as detailed in Table 20.14, presents the outcome obtained through classical RSM and the computational ANFIS. The comparison between experimental and RSM obtained result is presented in Figure 20.20, while the comparison between the experimental results and the ANFIS depicted result is obtained in Figure 20.21.

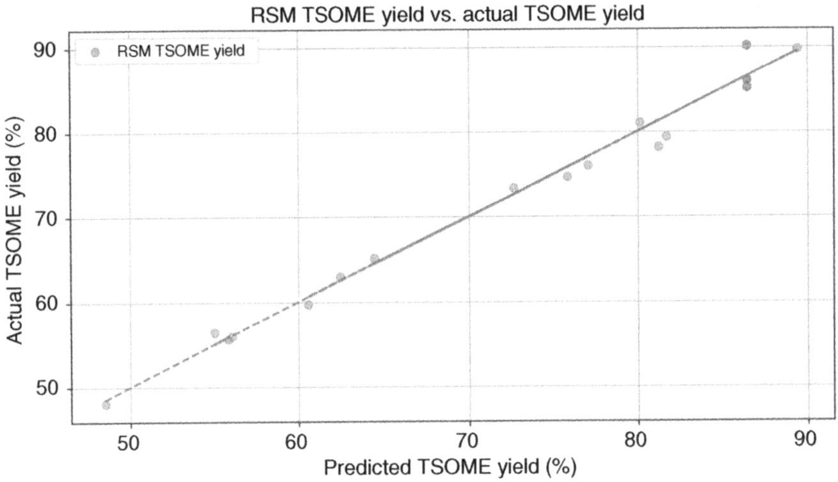

Figure 20.20 Predicted yield by RSM model (%).

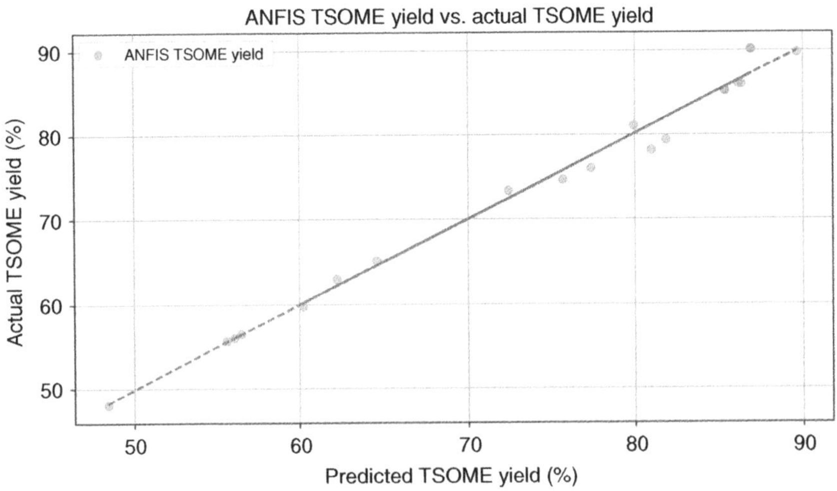

Figure 20.21 Predicted yield by ANFIS model (%).

The actual yield, as well as those predicted by the RSM and ANFIS models, have been effectively showcased. Figures demonstrate that the yield forecasted by the ANFIS model closely aligns with the actual TSOME Yield compared to the RSM model. Furthermore, the efficacy of these models is reaffirmed through plots depicting experimental values juxtaposed with model-predicted values (refer to Figures 20.20 and 20.21). These figures reveal that linear equations adequately capture the variations between RSM-predicted and actual TSOME Yields, as well as between ANFIS-predicted and actual TSOME Yields. The high regression coefficients (R^2) observed in both the RSM and ANFIS models indicate that over 90% of the data conforms to the respective models. Consequently, the ANFIS model demonstrates superior accuracy in yield prediction compared to the RSM model. This finding is consistent with similar reports from researchers elsewhere [22, 23].

The ANFIS was employed as a predictive model to estimate the yield of waste TSOME. In this context, three parameters, namely, reaction time, M:O, and catalyst amount, were selected as the model inputs in tandem, the TSOME Yield was identified as the response variable. To optimize the parameters of the power equation, the boundary conditions established for the RSM model were applied to both the inputs and outputs variables for ANFIS. The analyses were executed using ANFIS toolbox on MATLAB R2023 software. Table 20.14 outlines the optimal values determined by the ANFIS for achieving the highest TSOME Yield.

To assess the superiority and predictive capacity of the model techniques, statistical variables including the correlation coefficient (R), regression coefficient (R^2), RMSE, MAE, standard prediction error (SEP), and absolute average deviation (AAD) were well utilized. Observing the equations (Eqs. 20.5–20.10), Y_{pred} represents the predicted value of samples, Y_{exp} is the experimental value of samples, Y_i is the observed value of samples, Y_e is the estimated value of samples, and N represents the number of samples. The variables A–C denote independent variables, while β_{ij} signifies the coefficient of linear terms.

Figure 20.22 represent the statistical performance comparison of ANFIS and RSM TSOME Yield prediction. The plot comparing the results obtained from the ANFIS model and the RSM suggests that the ANFIS model performs slightly better than the RSM. Additionally, the ANFIS model has the capability to adjust the values of input and output variables as well as parameters, which allows it to achieve a more robust solution compared to RSM. Table 20.14 illustrates the actual yields alongside the predictions from the ANFIS and the RSM. Notably, the biodiesel yield model generated by ANFIS demonstrates closer proximity to the experimental yield in comparison to the RSM predictions across all runs. This observation aligns with findings reported by researchers in other studies [22, 23].

Figures 20.20 and 20.21 provide visual representations of the experimental yield vs. RSM-predicted yield and experimental yield vs ANFIS-predicted yield,

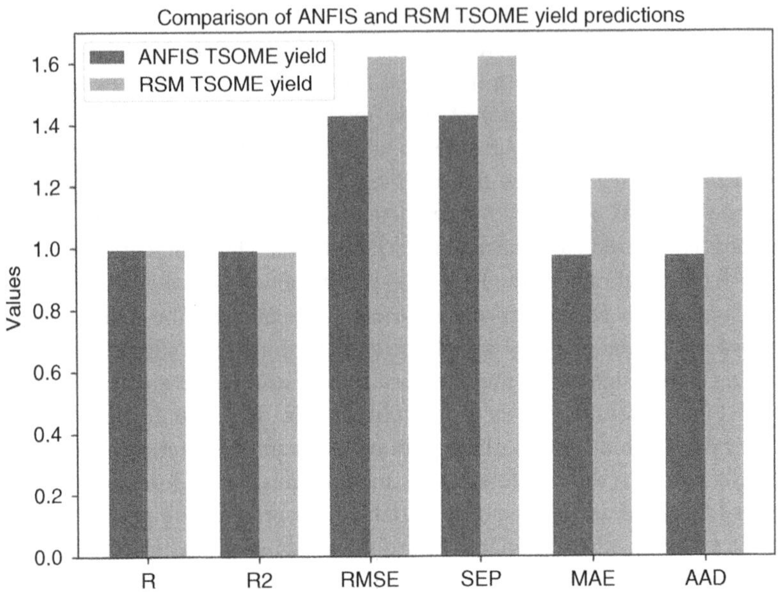

Figure 20.22 Comparison of ANFIS and RSM TSOME Yield predictions.

respectively. The values obtained are deemed suitable for capturing the variations in experimental and model-based yields, attributed to high regression coefficients (R^2) of 0.984 and 0.99415 for RSM and ANFIS, respectively. An R^2 of 0.9848 in the RSM model and 0.99415 in the ANFIS model indicates that over 98.4% and 99.4% of the data fit into the respective models. Consequently, the ANFIS model exhibits a higher accuracy in predicting yield compared to the RSM model. This corroborates similar findings reported by researchers elsewhere [24–26].

In Figures 20.20 and 20.21, estimated statistical indices, including correlation coefficient (R), regression coefficient (R^2), RMSE, SEP, MAE, and AAD are presented. The computed values for the RSM model are compared to those of the ANFIS model. The high R^2 values and low standard error indicate the model's robustness and accuracy in capturing the variability in biodiesel yield. These analytical results affirm the efficacy of the developed RSM model for biodiesel production from tobacco seed oil, providing valuable insights into the relationship between process parameters and TSOME Yield. However, these results underscore the superior performance of the ANFIS model over the RSM model, as evidenced by higher R and R^2 values and lower values for other statistical indices.

20.5 Economic and Environmental Implications of the Study

Economically, the findings suggest that making the most of underutilized tobacco seeds for biofuel production can potentially offer a sustainable and economically viable alternative to traditional fuel sources. By employing advanced modeling techniques such as RSM and ANFIS, the study demonstrates improved efficiency in biofuel production processes. This efficiency translates into reduced production costs and increased profitability for biofuel producers. Moreover, the use of underutilized resources like tobacco seeds can create new revenue streams for farmers and communities in tropical regions, contributing to local economic development and livelihood enhancement.

On the environmental front, the study indicates promising implications for sustainability and carbon footprint reduction. Biofuels derived from underutilized tobacco seeds offer a renewable and cleaner energy source compared to fossil fuels. Through the adoption of innovative modeling approaches, the study aims to optimize biofuel production processes, leading to decreased greenhouse gas emissions and mitigating environmental degradation associated with conventional fuel extraction and combustion. In addition, the utilization of underutilized resources promotes biodiversity conservation by reducing pressure on traditional agricultural lands and ecosystems.

In summary, the economic and environmental implications of the study highlight the potential of sustainable biofuel production from underutilized tobacco seeds to foster economic growth, mitigate environmental impacts, and promote a transition toward more sustainable energy systems in tropical regions and beyond.

20.5.1 Sustainability and Future Prospects

Sustainability and future prospects play a pivotal role in the study's implications for biofuel production from underutilized tobacco seeds in the tropics. From a sustainability perspective, the utilization of underutilized tobacco seeds offers a promising pathway toward addressing the dual challenges of energy security and environmental degradation. By harnessing these seeds for biofuel production, the study promotes resource efficiency and waste reduction in agricultural systems. Furthermore, the use of biofuels derived from renewable sources contributes to reducing greenhouse gas emissions and mitigating climate change impacts, aligning with global sustainability goals.

Looking ahead, the study's findings suggest encouraging prospects for the continued development and expansion of biofuel production from underutilized tobacco seeds. As advancements in modeling techniques such as RSM and ANFIS enhance process optimization and efficiency, the scalability and commercial viability of biofuel production are expected to improve. This could lead to broader adoption and integration of biofuels into existing energy infrastructures, diversifying energy sources and reducing dependency on finite fossil fuels.

Moreover, the study underscores the importance of ongoing research and innovation to address remaining challenges and unlock the full potential of biofuel production from underutilized resources. Continued investment in technology development, policy support, and stakeholder engagement will be crucial for overcoming barriers and maximizing the economic, environmental, and social benefits of sustainable biofuel production.

In summary, sustainability and future prospects are integral considerations in the study's exploration of biofuel production from underutilized tobacco seeds. By emphasizing resource efficiency, environmental stewardship, and technological innovation, the study contributes to shaping a more sustainable and resilient energy future for tropical regions and beyond.

20.6 Conclusion

In conclusion, this study showcases the application of RSM and ANFIS in sustainable biodiesel production from Tobacco seed oil. The study aimed to address the increasing demand for alternative and environmentally friendly sources of biofuels, and tobacco seed oil emerges as a promising and under-explored feedstock in this context. The transesterification parameters – reaction duration (T), M:O, and catalyst dosage – were well examined, and both ANFIS and RSM models were fine-tuned and validated. The statistical analysis revealed high accuracy for both models, with ANFIS exhibiting an R^2 value of 0.99415, RMSE is 1.42395, SEP is 1.42395, MAE is 0.972509, and AAD is 0.97. Similarly, RSM demonstrated robust performance with an R^2 value of 0.984822, RMSE of 1.6175, SEP of 1.618, and MAE of 1.22. From the results, ANFIS marginally outperforms RSM in terms of regression coefficient and error rates. This research emphasizes the effectiveness of the integrated computational system, providing valuable insights for sustainable energy in the biodiesel production domain. The outcomes aim to guide researchers, policymakers, and industry experts toward environmentally sustainable and economically feasible alternatives in biodiesel production.

20.A ANFIS Linguistic Rules

If (RT is low) and (methanol to oil ratio is low) and (catalyst is low) then (ANFIS predicted value is 2.output 1mf1).

If (RT is low) and (methanol to oil ratio is low) and (catalyst is medium) then (ANFIS predicted value is output 1mf2).

If (RT is low) and (methanol to oil ratio is low) and (catalyst is high) then (ANFIS predicted value is output 1mf3).

If (RT is low) and (methanol to oil ratio is medium) and (catalyst is low) then (ANFIS predicted value is output 1mf4).

If (RT is low) and (methanol to oil ratio is medium) and (catalyst is medium) then (ANFIS predicted value is output 1mf5).

If (RT is low) and (methanol to oil ratio is medium) and (catalyst is high) then (ANFIS predicted value is output 1mf6).

If (RT is low) and (methanol to oil ratio is high) and (catalyst is low) then (ANFIS predicted value is output 1mf7).

If (RT is low) and (methanol to oil ratio is high) and (catalyst is medium) then (ANFIS predicted value is output 1mf8).

If (RT is low) and (methanol to oil ratio is high) and (catalyst is high) then (ANFIS predicted value is output 1mf9).

If (RT is medium) and (methanol to oil ratio is low) and (catalyst is low) then (ANFIS predicted value is output 1mf9).

If (RT is medium) and (methanol to oil ratio is low) and (catalyst is low) then (ANFIS predicted value is output 1mf9).

If (RT is medium) and (methanol to oil ratio is low) and (catalyst is low) then (ANFIS predicted value is output 1mf9).

If (RT is medium) and (methanol to oil ratio is medium) and (catalyst is medium) then (ANFIS predicted value is output 1mf9).

If (RT is medium) and (methanol to oil ratio is medium) and (catalyst is medium) then (ANFIS predicted value is output 1mf9).

If (RT is medium) and (methanol to oil ratio is medium) and (catalyst is medium) then (ANFIS predicted value is output 1mf9).

If (RT is medium) and (methanol to oil ratio is high) and (catalyst is high) then (ANFIS predicted value is output 1mf9).

If (RT is medium) and (methanol to oil ratio is high) and (catalyst is high) then (ANFIS predicted value is output 1mf9).

If (RT is medium) and (methanol to oil ratio is high) and (catalyst is high) then (ANFIS predicted value is output 1mf9).

If (RT is high) and (methanol to oil ratio is low) and (catalyst is high) then (ANFIS predicted value is output 1mf9).

If (RT is high) and (methanol to oil ratio is medium) and (catalyst is high) then (ANFIS predicted value is output 1mf9).

If (RT is high) and (methanol to oil ratio is high) and (catalyst is high) then (ANFIS predicted value is output 1mf9).

If (RT is high) and (methanol to oil ratio is low) and (catalyst is high) then (ANFIS predicted value is output 1mf9).

If (RT is high) and (methanol to oil ratio is medium) and (catalyst is high) then (ANFIS predicted value is output 1mf9).

If (RT is high) and (methanol to oil ratio is high) and (catalyst is high) then (ANFIS predicted value is output 1mf9).

If (RT is high) and (methanol to oil ratio is low) and (catalyst is high) then (ANFIS predicted value is output 1mf9).

If (RT is high) and (methanol to oil ratio is medium) and (catalyst is high) then (ANFIS predicted value is output 1mf9).

If (RT is high) and (methanol to oil ratio is high) and (catalyst is high) then (ANFIS predicted value is output 1mf9).

References

1 Mbachu, V.M., Ovuworie, G.C., Okwu, M.O., and Tartibu, L.K. (2021). Modelling and sustainability of a demand-based biomass to biogas conversion system: a bio-mimicry feedstock inventory-based approach. *Biomass Convers. Biorefinery*. https://doi.org/10.1007/s13399-021-01581-z.

2 Okwu, M.O. and Adetunji, O. (2018). A comparative study of artificial neural network (ANN) and adaptive neuro-fuzzy inference system (ANFIS) models in distribution system with nondeterministic inputs. *Int. J. Eng. Bus. Manage.* 10: 1847979018768421. https://doi.org/10.1177/1847979018768421.

3 Samuel, O.D., Okwu, M.O., Tartibu, L.K. et al. (2021). Modelling of *Nicotiana tabacum* L. oil biodiesel production: comparison of ANN and ANFIS. *Front. Energy Res.* 8: 612165. https://doi.org/10.3389/fenrg.2020.612165.

4 Kumar, S. and Bansal, S. (2023). Performance evaluation of ANFIS and RSM in modeling biodiesel synthesis from soybean oil. *Biosens. Bioelectron.: X* 15: 100408.

5 Gupta, A. and Sharma, D.S.D. (2014). A survey on stock market prediction using various algorithms. *Int. J. Comput. Technol. Appl.* 5: 530–533.

6 Guo, J. and Baghban, A. (2017). Application of ANFIS strategy for prediction of biodiesel production using supercritical methanol. *Energy Sources Part A* 39 (17): 1862–1868.

7 Al-Shanableh, F., Bilin, M., Evcil, A., and Savas, M.A. (2023). Estimation of cold flow properties of biodiesel using ANFIS-based models. *Energy Sources Part A* 45 (2): 5440–5457.

8 Mostafaei, M. (2018). ANFIS models for prediction of biodiesel fuels cetane number using desirability function. *Fuel* 216: 665–672. https://doi.org/10.1016/j.fuel.2017.12.025.

9 Mostafaei, M. (2018). Prediction of biodiesel fuel properties from its fatty acids composition using ANFIS approach. *Fuel* 229: 22734. https://doi.org/10.1016/j.fuel.2018.04.148.

10 Aminian, A. and ZareNezhad, B. (2018). Accurate predicting the viscosity of biodiesels and blends using soft computing models. *Renew. Energy* 120: 488–500. https://doi.org/10.1016/j.renene.2017.12.038.

11 Mostafaei, M., Javadikia, H., and Naderloo, L. (2016). Modeling the effects of ultrasound power and reactor dimension on the biodiesel production yield: comparison of prediction abilities between response surface methodology (RSM) and adaptive neuro-fuzzy inference system (ANFIS). *Energy* 115: 626–636.

12 Naderloo, L., Javadikia, H., and Mostafaei, M. (2017). Modeling the energy ratio and productivity of biodiesel with different reactor dimensions and ultrasonic power using ANFIS. *Renew. Sust. Energ. Rev.* 70: 56–64.

13 Faizollahzadeh Ardabili, S., Mahmoudi, A., Mesri Gundoshmian, T., and Roshanianfard, A. (2016). Modeling and comparison of fuzzy and on/off controller in a mushroom growing hall. *Measurement* 90: 127–134.

14 Najafi, B., Faizollahzadeh Ardabili, S., Shamshirband, S. et al. (2018). Application of ANNs, ANFIS and RSM to estimating and optimizing the parameters that affect the yield and cost of biodiesel production. *Eng. Appl. Comput. Fluid Mech.* 12 (1): 611–624.

15 Kumar, S., Singh, J., Fore, V., and Kumar, A. (2023). Performance evaluation of ANFIS, ANN and RSM in biodiesel synthesis from Karanja oil with domestic microwave set up. *Multimed. Tools Appl.* 1–17.

16 Kikon, A., Dodamani, B.M., Barma, S.D., and Naganna, S.R. (2023). ANFIS-based soft computing models for forecasting effective drought index over an arid region of India. *AQUA Water Infrastruct. Ecosyst. Soc.* 72: 930–946.

17 Oladipo, S. and Sun, Y. (2023). Enhanced adaptive neuro-fuzzy inference system using genetic algorithm: a case study in predicting electricity consumption. *SN Appl. Sci.* 5 (7): 186.

18 Yazdani, A., Shanbehzadeh, M., and Kazemi-Arpanahi, H. (2023). Using an adaptive network-based fuzzy inference system for prediction of successful aging: a comparison with common machine learning algorithms. *BMC Med. Inf. Decis. Making* 23 (1): 229.

19 Suparta, W. and Samah, A.A. (2020). Rainfall prediction by using ANFIS times series technique in South Tangerang, Indonesia. *Geod. Geodyn.* 11 (6): 411–417.

20 Esmaili, M., Aliniaeifard, S., Mashal, M. et al. (2021). Assessment of adaptive neuro-fuzzy inference system (ANFIS) to predict production and water productivity of lettuce in response to different light intensities and CO_2 concentrations. *Agric. Water Manag.* 258: 107201.

21 Prasad, K., Gorai, A.K., and Goyal, P. (2016). Development of ANFIS models for air quality forecasting and input optimization for reducing the computational cost and time. *Atmos. Environ.* 128: 246–262.

22 Sodeifian, G., Ardestani, N.S.S.S.A., and Ghorbandoost, S. (2016). Application of supercritical carbondioxide to extract essential oil from Cleome coluteoides Boiss: experimental, response surface and grey wolf optimization methodology. *J. Supercrit. Fluids* 114: 55–63.

23 Betiku, E., Osunleke, A.S., Odude, V.O. et al. (2018). Performance evaluation of adaptive neuro-fuzzy inference system, artificial neural network and response surface methodology in modeling biodiesel synthesis from palm kernel oil by transesterification. *Biofuels* 1–16.

24 Samuel, O.D., Okwu, M.O., Amosun, S.T. et al. (2019). Production of fatty acid ethyl esters from rubber seed oil in hydrodynamic cavitation reactor: study of reaction parameters and some fuel properties. *Ind. Crop. Prod.* 141: 111658.

25 Encinar, J.M., Martínez, G., González, J.F., and Pardal, A.C. (2010). Transesterification of rapeseed oil with methanol in the presence of various co-solvents. *Proc. Venic* 1–17.

26 Samuel, O.D., Okwu, M.O., Oyejide, O.J. et al. (2020). Optimizing biodiesel production from abundant waste oils through empirical method and grey wolf optimizer. *Fuel* 281: 118701. https://doi.org/10.1016/j.fuel.2020.118701.

21

Sustainable Biodiesel Production

Navigating the Green Frontiers

21.1 Introduction to the Future of Sustainable Biodiesel

Having discussed in detail the comprehensive exploration into advancing sustainable biodiesel production with biodegradable feedstocks, it is imperative to focus on the future, envisioning a trajectory that holds promise for a greener and more sustainable energy frontier. The insights gained from this study not only shed light on the present state of biodiesel production but also pave the way for informed considerations about the future of this vital industry.

21.1.1 Innovation in Feedstock Selection

Innovation in feedstock selection is paramount for the advancement of biodiesel production. The future of this industry hinges on ongoing exploration and development in this field. The study's examination of a diverse range of feedstocks, including *chrysobalamus icaco*, water hyacinth, hemp, sunflower, tobacco seeds, and others, provides a exciting peek into the vast reservoir of potential sources available. However, the journey does not end here; rather, it marks the beginning of a continuous quest for even more innovative solutions.

The pursuit of novel, nonedible feedstocks stands as a cornerstone in this venture. By focusing on untapped resources, researchers can pave the way for a sustainable and environmentally friendly supply chain. This approach is vital to ensure that biodiesel production does not encroach upon essential food resources or exacerbate existing agricultural pressures. As the biodiesel industry evolves, the exploration of alternative feedstocks will remain a dynamic and essential aspect. Through ongoing research and experimentation, new possibilities can be unlocked, efficiency can be enhanced, and environmental

Feedstocks for Sustainable Biodiesel Production: Characterization, Selection, and Optimization, First Edition. Chinwe P. Okonkwo and Modestus O. Okwu.

impacts can be mitigated [1, 2]. By embracing innovation in feedstock selection, a course toward a more sustainable future for biodiesel production can be charted.

21.1.2 Optimization and Efficiency

The optimization and efficiency strategies elucidated in this study serve as the bedrock upon which future advancements in biodiesel production will be built. Fromfine-tuning transesterification techniques to maximizing the efficacy of catalysts and parametric studies facilitated by metaheuristic algorithms, each aspect contributes to a comprehensive framework aimed at enhancing overall productivity and sustainability [3]. Continued dedication to refining biodiesel production processes is imperative. By incessantly striving to improve efficiency, minimize waste, and mitigate environmental repercussions, the industry can solidify its position as a cornerstone of renewable energy.

Such ventures require a multifaceted approach, integrating cutting-edge technology, innovative methodologies, and a commitment to continuous improvement. Technological innovations play a pivotal role in this pursuit. Advancements in equipment design, automation, and process control systems offer promising avenues for optimizing production workflows. By harnessing the power of automation and data analytics, manufacturers can streamline operations, reduce resource consumption, and ultimately drive down costs. Furthermore, the strategic utilization of catalysts and additives can significantly augment process efficiency and product quality. Through consistent and methodical research and development, novel catalyst formulations and additives can be tailored to meet specific performance criteria while minimizing environmental impacts.

Ultimately, the quest for optimization and efficiency in biodiesel production is a dynamic and ongoing effort. By fostering collaboration between industry stakeholders, research institutions, and policymakers, the sector can be propelled collectively toward greater sustainability and resilience. With unwavering dedication and a commitment to innovation, the future of biodiesel production holds tremendous promise as a key player in the transition toward a more sustainable energy scenery.

21.1.3 Integration of Advanced Technologies

To meet the requirements for energy sustainability, manufacturers can employ data mining, artificial intelligence (AI), digital twin technology, and industrial simulation. These technologies enable the visualization and simulation of material flows, automation processes, and potential bottlenecks in manufacturing operations [4]. The integration of advanced technologies, such as AI, machine learning (ML), and precision farming, heralds a new era of innovation and efficiency within the biodiesel industry. Embracing these cutting-edge techniques presents a multitude of opportunities to revolutionize classical or traditional farming practices, amplify feedstock yields, and refine the biodiesel production process from start to finish. Precision farming, empowered by AI and ML algorithms, empowers farmers with real-time insights and granular data analytics thereby increasing the biodegradable wastes required for biodiesel production. By harnessing satellite imagery, drones, soil sensors, and Internet of Things (IoT) devices, farmers can optimize planting schedules, tailor irrigation and fertilization regimes, and mitigate crop stressors with unprecedented precision. This precise management not only maximizes feedstock yields but also minimizes resource inputs and environmental impacts. Moreover, the application of AI and ML extends beyond the field and into the production facility. These technologies enable predictive maintenance of equipment, proactive quality control measures, and real-time process optimization. By analyzing vast datasets generated throughout the production chain, AI algorithms can identify patterns, optimize parameters, and anticipate potential bottlenecks or inefficiencies, thus enhancing overall productivity and product quality.

The utilization of data-driven models such as artificial neural networks, response surface methodology, fuzzy logic grey wolf optimization, and other hybrid models, as exemplified in the optimization chapters, further underscores the transformative potential of advanced technologies in biodiesel production. By integrating historical data, sensor readings, and predictive analytics, decision-makers can make more informed choices regarding feedstock selection, process optimization, and resource allocation. This holistic approach not only maximizes efficiency but also enhances sustainability by minimizing waste and reducing environmental footprints. In essence, the integration of advanced technologies represents a paradigm shift in the biodiesel industry, unlocking untapped potential and driving exponential progress. By embracing innovation and collaboration, stakeholders can harness the full spectrum of benefits offered by AI, ML, and precision farming, propelling the sector toward a more sustainable, efficient, and resilient future.

21.1.4 Global Collaboration and Policy Support

Global collaboration and policy support are essential pillars upon which the future of biodiesel hinges. In an increasingly interconnected world, the collective efforts of governments, international organizations, and industry stakeholders are paramount to realizing the full potential of this renewable energy source and addressing pressing environmental challenges.

It is recommended that the Government exercise significant influence in shaping the trajectory of biodiesel adoption through the formulation and implementation of supportive policies. Incentives, such as tax credits, subsidies, and grants, can stimulate investment in biodiesel production infrastructure and encourage consumers to switch to biofuels. In addition, regulatory frameworks that mandate blending requirements or set emission standards can create a stable market demand for biodiesel, driving market growth and fostering innovation.

International collaboration is equally crucial, given the transnational nature of environmental issues. Multilateral agreements and initiatives facilitate knowledge exchange, technology transfer, and capacity-building efforts among countries. Furthermore, industry stakeholders, including producers, researchers, and advocacy groups, will play a vital role in shaping policy agendas and advocating for supportive measures. By engaging policymakers, raising awareness about the benefits of biodiesel, and providing evidence-based recommendations, these stakeholders can influence policy decisions and drive systemic change. Also, a campaign on reducing greenhouse gas emissions and transitioning toward sustainable energy sources like biodiesel is a way forward.

Moreover, fostering collaboration between the public and private sectors can unlock synergies and accelerate innovation in biodiesel production and utilization. Public–private partnerships can facilitate research and development initiatives, pilot projects, and technology demonstration programs, ultimately driving down costs and expanding market access for biodiesel. In summary, global collaboration and policy support are indispensable for realizing the full potential of biodiesel as a sustainable energy solution. By aligning incentives, regulations, and investments, governments and international organizations can create an enabling environment that fosters innovation, promotes adoption, and accelerates the transition toward a low-carbon future.

21.1.5 Public Awareness and Education

Public awareness and education are pivotal elements in molding the trajectory of biodiesel in the future. As societal consciousness regarding environmental issues continues to rise, there emerges an increasing appetite for sustainable and renewable energy alternatives. In this context, initiatives aimed at outreach, education, and awareness play a critical role in enlightening and engaging the public, ultimately fostering a collective dedication to embracing sustainable practices.

Outreach programs serve as invaluable platforms for disseminating information about the benefits of biodiesel and its role in mitigating climate change. These programs can take various forms, including workshops, seminars, community events, and public forums. By providing accessible and engaging avenues for learning, outreach efforts can empower individuals with the knowledge and understanding needed to make informed choices about energy consumption and support the transition to renewable fuels.

Educational initiatives targeting diverse audiencesfrom students in schools and universities to professionals in various industriesplay a crucial role in nurturing a culture of sustainability. Organizing knowledge-sharing workshops and incorporating topics related to biodiesel production, environmental conservation, and renewable energy into curricula equips future generations with the skills and mindset necessary to address pressing environmental challenges. Furthermore, vocational training programs and continuing education courses can provide professionals with the expertise needed to thrive in the evolving field of sustainable energy.

Awareness campaigns leverage various communication channels, including traditional media, social media platforms, and digital marketing, to reach broad audiences and catalyze behavioral change. By highlighting the environmental benefits, economic advantages, and societal implications of biodiesel use, these campaigns can inspire individuals to adopt more sustainable practices in their daily lives, whether through purchasing decisions, transportation choices, supply chains, or advocacy efforts.

Moreover, partnerships between governments, nongovernmental organizations (NGOs), educational institutions, and industry stakeholders can amplify the impact of public awareness and education initiatives. Collaborative efforts can pool resources, leverage expertise, and reach diverse demographics, fostering a holistic approach to raising awareness and driving positive change. In essence, public awareness and education serve as linchpins in the transition to a more sustainable energy future, with biodiesel playing a central role. By empowering individuals with knowledge, inspiring collective action, and fostering a culture of sustainability, these efforts lay the groundwork for a society that embraces renewable energy solutions and prioritizes environmental stewardship.

21.1.6 Emerging Feedstock Exploration

The horizon of biodiesel production brims with promise as exploration into emerging feedstocks gains momentum, unveiling a wealth of untapped potential. Through ongoing research efforts and innovative approaches, unconventional sources and hybrid feedstocks are poised to revolutionize the biodiesel industry, ushering in a new era of resilience and adaptability in the face of evolving environmental and economic dynamics.

Unconventional feedstocks, ranging from algae and microbial oils to waste oils and agricultural residues, present intriguing avenues for biodiesel production. These nontraditional sources offer unique advantages, such as rapid growth rates, high lipid content, and the ability to thrive in diverse environmental conditions. By harnessing these unconventional feedstocks, researchers can unlock previously unexplored pathways to sustainable biodiesel production, reducing reliance on finite resources and mitigating environmental impacts associated with conventional agriculture. Furthermore, the development of hybrid feedstocks represents a frontier of innovation in biodiesel research.

Leveraging advancements in biotechnology and genetic engineering, scientists can engineer crops with enhanced oil yield, nutrient uptake efficiency, and stress tolerance. Hybrid feedstocks offer the potential to optimize resource utilization, increase productivity, and mitigate risks associated with climate variability and resource constraints.

In addition, these crops can be positioned to thrive on marginal lands, minimizing competition with food crops and expanding the geographical scope of biodiesel production. The exploration of emerging feedstocks is not only driven by technological advancements but also by a growing recognition of the need for diversification and sustainability in the biodiesel industry. By expanding the feedstock portfolio, producers can mitigate supply chain risks, enhance resilience to market fluctuations, and meet evolving consumer preferences for environmentally friendly products. Moreover, the utilization of waste streams and by-products from other industries can create synergies and contribute to circular economy principles, further enhancing the sustainability of biodiesel production.

Collaboration between research institutions, industry stakeholders, and policymakers will be crucial in unlocking the full potential of emerging feedstocks. By fostering knowledge exchange, facilitating technology transfer, and providing regulatory support, governments can catalyze innovation and accelerate the adoption of new feedstock sources. In addition, investment in infrastructure and capacity-building initiatives can ensure the scalability and commercial viability of emerging feedstock-based biodiesel production systems. Finally, the exploration of emerging feedstocks represents a transformative opportunity for the biodiesel industry. By embracing innovation, collaboration, and sustainability, stakeholders can harness the diverse array of feedstock resources available, paving the way for a more resilient, adaptive, and sustainable future for biodiesel production.

In summary, the future of biodiesel is dynamic and holds immense promise. The journey toward sustainable horizons requires a collaborative effort from researchers, policymakers, industry stakeholders, and the public. The insights gained from this study serve as a foundation for future endeavors, providing a roadmap for the continued evolution of biodiesel production with a focus on biodegradable feedstocks. As we navigate toward a future marked by environmental stewardship and energy sustainability, the principles and findings presented in this comprehensive exploration will undoubtedly play a vital role in shaping the destiny of biodiesel on a global scale.

21.2 Biodiesel and Industry 4.0: Catalyzing Sustainable Transformation

The fourth industrial revolution (Industry 4.0) encompasses the utilization of digital technologies to enhance production efficiency and customization, the integration of connected physical assets with intelligent data processing, the increasing strategic significance of cognitive resources and decision-making processes, the emergence of intelligent machines including AI, ML, robotics and the effective utilization of online data for optimized decisions through analytics, smart decision making, and ML [4]. These technological advancements have the potential to enhance efficiency, productivity, and sustainability across various industries [4]. Concurrently, there is mounting of pressure on the renewable energy sector to transition from fossil fuels to cleaner, sustainable energy sources, and the sustainable materials sector aims to reduce the environmental impact associated with material production, usage, and disposal [5]. At the intersection of advanced technologies and sustainable energy solutions, the integration of biodiesel and Industry 4.0 emerges as a transformative force. Industry 4.0, characterized by the fusion of digital technologies with traditional industrial processes, presents a unique opportunity to revolutionize the biodiesel production setting. Within the renewable energy domain, Industry 4.0 technologies bolster the efficiency, dependability, and scalability of renewable energy set-ups [6].

Highlight of the intersection of Biodiesel and Industry 4.0.

21.2.1 Smart Farming Practices

Precision Agriculture: Industry 4.0 technologies enable precision farming practices, optimizing the cultivation of biodiesel feedstocks. Sensors, drones, and IoT devices can be employed to monitor crop health, assess soil conditions, and precisely apply fertilizers and pesticides. This leads to increased yields, reduced resource usage, and enhanced overall efficiency [7]. IoT sensors and data analytics oversee and enhance the performance of solar and wind energy systems, while AI algorithms anticipate and avert equipment malfunctions [8].

Smart farming practices, specifically precision agriculture, represent a transformative shift in the cultivation of a good number of biodiesel feedstocks, empowered by the integration of Industry 4.0 technologies. These advanced tools, including sensors, drones, and IoT devices, revolutionize traditional farming methods by providing real-time data and actionable insights for optimized decision-making and resource management. At the heart of precision agriculture lies the utilization of sensors and IoT devices deployed across fields to collect a wealth of environmental data [9, 10]. These sensors measure various parameters such as soil moisture, temperature, pH levels, and nutrient concentrations, offering farmers unprecedented visibility into crop health and soil conditions. By continuously monitoring these factors, farmers can identify potential issues early, proactively address challenges, and optimize inputs for maximum productivity.

Drones, equipped with high-resolution cameras and multi-spectral imaging sensors, serve as aerial scouts, capturing detailed images of fields from above. These images can be analyzed using advanced algorithms to detect anomalies, identify crop stressors, and assess overall crop health. By providing a bird's-eye view of the farm, drones enable farmers to make data-driven decisions regarding irrigation, pest management, and crop rotation, thereby optimizing yields and minimizing losses. Furthermore, precision agriculture facilitates the precise application of inputs such as fertilizers, pesticides, and herbicides, minimizing wastage and environmental impact. By leveraging data analytics and GPS technology, farmers can create customized prescription maps that delineate areas of varying crop needs within a field. This enables the targeted application of inputs, reducing over-application in areas with sufficient nutrient levels and ensuring optimal resource allocation across the farm.

The adoption of precision agriculture practices yields a multitude of benefits for biodiesel feedstock cultivation. Increased yields, driven by optimized resource management and proactive pest and disease control, translate into higher productivity and profitability for farmers. Moreover, reduced resource usage, including water, fertilizers, and agrochemicals, contributes to environmental sustainability and conservation efforts. In addition, by enhancing overall efficiency and resilience, precision agriculture helps farmers mitigate risks associated with climate variability and market fluctuations. In summary, precision agriculture represents a paradigm shift in the cultivation of biodiesel feedstocks, enabled by the convergence of advanced technologies and agricultural practices [11]. By embracing smart farming practices, farmers can unlock the full potential of their land, maximize productivity, and contribute to a more sustainable and resilient agricultural ecosystem.

21.2.2 Digital Supply Chain Management: Real-Time Monitoring

Digital supply chain management, empowered by Industry 4.0 technologies, revolutionizes the biodiesel industry through real-time monitoring of the entire supply chain [12]. By integrating IoT sensors and blockchain technology, stakeholders gain unprecedented visibility, transparency, and efficiency across all stages, from feedstock cultivation to product distribution. At the heart of this transformation are IoT sensors deployed throughout the supply chain, from fields and processing facilities to transportation vehicles and storage tanks. These sensors collect vast amounts of data in real time, monitoring key parameters such as temperature, humidity, location, and quality metrics. By continuously monitoring these variables, stakeholders can predictively identify issues, optimize processes, and ensure compliance with quality and sustainability standards. Blockchain technology serves as the backbone of digital supply chain management, providing a secure and immutable ledger for recording transactions and data exchanges. Each interaction within the supply chain, from the harvesting of feedstocks to the delivery of biodiesel to end-users, is cryptographically recorded on the blockchain. This transparent and tamper-proof record enables stakeholders to trace the journey of products, verify authenticity, and ensure compliance with regulatory requirements and sustainability certifications.

The integration of real-time monitoring and blockchain technology yields numerous benefits for the biodiesel supply chain. Enhanced transparency and traceability empower consumers to make informed decisions about the products they purchase, fostering trust and loyalty in biodiesel brands. Moreover, by capturing and analyzing data at every stage, stakeholders can identify inefficiencies, optimize workflows, and reduce waste, leading to cost savings and environmental benefits. Furthermore, real-time monitoring enables predictive maintenance of equipment and assets, minimizing downtime and ensuring continuous operation of production facilities and transportation fleets. By leveraging ML algorithms and predictive analytics, stakeholders can anticipate maintenance needs, optimize scheduling, and extend the lifespan of critical infrastructure, thereby maximizing uptime and productivity. In summary, digital supply chain management, with its focus on real-time monitoring and blockchain technology, represents a paradigm shift in the biodiesel industry. By embracing Industry 4.0 principles, stakeholders can unlock new levels of efficiency, transparency, and sustainability across the entire supply chain, driving growth and resilience in the transition toward a more sustainable energy future.

21.2.3 Automated Biodiesel Production

Smart Manufacturing: The implementation of smart manufacturing processes within biodiesel production plants enhances efficiency and reduces operational costs. By integrating automated systems, robotics, and IoT-enabled equipment, biodiesel plants can streamline operations, enhance process control, and optimize resource utilization throughout the production process. At the core of smart manufacturing in biodiesel production plants are automated systems that orchestrate and coordinate various production activities seamlessly. These systems leverage advanced sensors and actuators to monitor and control critical parameters such as temperature, pressure, flow rates, and catalyst ratios in real time. By continuously collecting and analyzing data, automated systems can adjust process parameters dynamically to maintain optimal operating conditions, minimize waste, and ensure consistent product quality.

Robotics plays a pivotal role in automating labor-intensive tasks and handling hazardous materials, thereby improving workplace safety and efficiency. Automated robotic arms can perform tasks such as loading feedstocks, transferring materials between process units, and packaging finished products with precision and reliability. By freeing up human operators from repetitive or dangerous tasks, robotics not only enhance productivity but also create a safer and more ergonomic work environment.

IoT-enabled equipment further enhances the capabilities of automated biodiesel production plants by enabling connectivity and data exchange between different components and systems. Sensors embedded in equipment and machinery capture real-time data on performance metrics, equipment health, and energy consumption, providing valuable insights for optimization and predictive maintenance. In addition, IoT platforms facilitate remote monitoring and control, allowing operators to oversee production processes from anywhere, optimize resource allocation, and respond swiftly to changing conditions or alarms.

The implementation of smart manufacturing principles in biodiesel production plants yields numerous benefits for producers and consumers alike. Enhanced efficiency and productivity translate into cost savings and competitive advantages for producers while ensuring a consistent and high-quality product for consumers. Moreover, by optimizing resource utilization and minimizing environmental impacts, smart manufacturing contributes to the sustainability of biodiesel production and reduces its carbon footprint.

In summary, automated biodiesel production, driven by smart manufacturing technologies, represents a significant step forward in the evolution of the biodiesel industry. By embracing automation, robotics, and IoT-enabled equipment, biodiesel plants can achieve unprecedented levels of efficiency, reliability, and sustainability, paving the way for a more resilient and environmentally friendly energy future.

21.2.4 Data-Driven Decision-Making

Data-driven decision-making, powered by advanced analytics and big data, stands as a cornerstone of Industry 4.0 within the biodiesel production settings. By harnessing the vast amount of data generated throughout the production process, stakeholders can extract valuable insights, optimize operations, and drive continuous improvement across all facets of production.

At the heart of data-driven decision-making in biodiesel production is advanced analytics, encompassing a diverse array of techniques such as predictive modeling, ML, and statistical analysis. These methodologies enable producers to uncover patterns, correlations, and trends within their data, facilitating informed decision-making and proactive problem-solving.

Predictive analytics plays a particularly crucial role in optimizing production schedules, maintenance routines, and overall process efficiency. By analyzing historical data and real-time sensor readings, predictive models can forecast equipment failures, identify potential bottlenecks, and optimize resource allocation to maximize throughput and minimize downtime. This proactive approach not only enhances operational efficiency but also reduces maintenance costs and extends the lifespan of critical assets.

Moreover, advanced analytics can inform strategic decision-making at every stage of the biodiesel production process. From feedstock selection and blending ratios to reaction kinetics and product purification, data-driven insights enable producers to fine-tune processes, improve product quality, and meet regulatory requirements with precision. Furthermore, big data analytics enable producers to leverage external data sources, such as market trends, weather forecasts, and commodity prices, to inform strategic decision-making and risk management. By integrating external data into their analytics framework, producers can anticipate market fluctuations, optimize procurement strategies, and mitigate supply chain risks, thereby enhancing competitiveness and resilience in a dynamic marketplace.

The adoption of data-driven decision-making in biodiesel production yields tangible benefits for producers, including improved yields, reduced costs, and enhanced competitiveness. By leveraging advanced analytics and big data, producers can unlock new levels of efficiency, agility, and innovation, positioning themselves for success in an increasingly data-driven world.

In summary, data-driven decision-making, underpinned by advanced analytics and big data, represents a transformative opportunity for the biodiesel industry. By harnessing the power of data to inform decision-making, producers can optimize operations, improve sustainability, and drive continuous improvement, ultimately shaping more efficient, resilient, and competitive biodiesel production systems.

21.2.5 Energy Management and Optimization

Energy Efficiency: Industry 4.0 technologies contribute to the efficient management of energy resources in biodiesel production. By leveraging smart grids, energy monitoring systems, and AI-driven algorithms, producers can optimize energy utilization throughout the production process, minimizing environmental impact and operational costs while maximizing efficiency.

At the forefront of energy management in biodiesel production are smart grids, which enable dynamic monitoring, control, and optimization of energy distribution and consumption. By integrating renewable energy sources such as solar, wind, and biomass with traditional grid infrastructure, smart grids can balance supply and demand in real time, optimizing energy usage and reducing reliance on fossil fuels. Furthermore, smart grid technologies enable producers to participate in demand response programs, adjusting energy consumption in response to market signals and grid conditions to maximize cost savings and grid stability.

Energy monitoring systems play a crucial role in providing real-time insights into energy consumption patterns and identifying opportunities for optimization. These systems utilize sensors and IoT devices to collect data on energy usage, equipment performance, and environmental conditions throughout the production facility. By analyzing this data using advanced analytics and AI-driven algorithms, producers can identify inefficiencies, optimize equipment operation, and implement energy-saving measures to reduce overall energy consumption and associated costs.

Moreover, AI-driven algorithms offer predictive capabilities that enable proactive energy management and optimization. By analyzing historical data and real-time sensor readings, AI algorithms can forecast energy demand, optimize production schedules, and dynamically adjust energy usage to minimize peak loads and maximize efficiency. In addition, AI-based predictive maintenance algorithms can anticipate equipment failures, optimizing maintenance schedules and reducing downtime, further enhancing energy efficiency and operational reliability.

The adoption of energy management and optimization strategies in biodiesel production yields numerous benefits for producers and the environment. By reducing energy consumption and reliance on fossil fuels, producers can lower greenhouse gas emissions, mitigate climate change impacts, and enhance overall sustainability. In addition, optimizing energy usage helps reduce operational costs, improve competitiveness, and enhance resilience to energy price fluctuations and supply chain disruptions. In summary, energy management and optimization, empowered by Industry 4.0 technologies, play a crucial role in driving sustainability and cost-effectiveness in biodiesel production. By leveraging smart grids, energy monitoring systems, and AI-driven algorithms, producers can maximize energy efficiency, minimize environmental impact, and ensure long-term viability in an increasingly energy-conscious world.

21.2.6 Quality Control and Assurance: Real-Time Monitoring

Quality control and assurance are paramount in biodiesel production, and Industry 4.0 technologies enable real-time monitoring of biodiesel properties to ensure adherence to stringent quality standards. By leveraging sensors, analytical instruments, and a centralized control system, producers can continuously monitor key parameters throughout the production process, promptly identify deviations, and implement corrective actions to maintain product quality and consistency. At the heart of real-time monitoring in biodiesel production are sensors and analytical instruments strategically deployed at critical points along the production line. These sensors measure various parameters such as viscosity, density, acidity, moisture content, and methanol content, providing real-time data on the quality and composition of the biodiesel being produced. In addition, spectroscopic techniques such as infrared spectroscopy and chromatography can be utilized to analyze biodiesel samples with high precision and accuracy.

The data collected by sensors and analytical instruments are transmitted to a centralized control system, where they are analyzed in realtime to assess compliance with quality standards and specifications. By comparing the measured values to predefined thresholds and specifications, the control system can detect deviations from the desired quality parameters and trigger immediate corrective actions. These actions may include adjusting process parameters, adding corrective additives, or diverting the production stream for further analysis or treatment.

Furthermore, Industry 4.0 technologies enable the integration of advanced data analytics and ML algorithms into the quality control process. By analyzing historical data and patterns, these algorithms can identify trends, predict potential quality issues, and optimize process parameters to prevent deviations before they occur. In addition, ML algorithms can continuously improve over time as they are exposed to more data, enhancing their predictive capabilities and refining the quality control process.

The implementation of real-time monitoring for biodiesel properties offers several benefits for producers and consumers alike. Producers can ensure consistent product quality, minimize waste and rework, and reduce the risk of noncompliance with regulatory requirements. In addition, real-time monitoring enhances transparency and trust with consumers by providing assurance of product quality and safety. In summary, Industry 4.0 technologies enable real-time monitoring of biodiesel properties, enhancing quality control and assurance in the production process. By leveraging sensors, analytical instruments, and advanced analytics, producers can maintain product quality and consistency, meet regulatory requirements, and ensure customer satisfaction in the competitive biodiesel market.

21.2.7 Digital Twin Technology

Digital twin technology, heralded as a game-changer in various industries, is gaining traction in biodiesel production, offering a virtual replication of physical processes. This innovative approach allows producers to create digital replicas of their production facilities, enabling simulation, testing, and optimization of processes in a virtual environment before implementing changes in the actual production line. This not only reduces risks and costs but also enhances efficiency and innovation in biodiesel production.

At the core of digital twin technology in biodiesel production are sophisticated modeling and simulation tools that replicate the behavior and performance of physical assets and processes. These digital twins are created by integrating data from various sources, including sensors, historical records, and engineering models, to accurately represent the production facility and its components. Through advanced algorithms and physics-based models, digital twins can simulate complex interactions between equipment, materials, and environmental factors, providing insights into process dynamics and performance.

Digital twins enable producers to conduct virtual experiments and scenario analyses to optimize process parameters, identify bottlenecks, and explore alternative production strategies. By simulating different operating conditions and scenarios, producers can evaluate the impact of changes in real time, assess potential risks, and make informed decisions to improve process efficiency and productivity.

Furthermore, digital twins facilitate predictive maintenance by continuously monitoring equipment health and performance in real time. By analyzing sensor data and performance metrics, digital twins can detect anomalies and predict potential failures before they occur, enabling proactive maintenance interventions to prevent downtime and costly repairs.

The implementation of digital twin technology offers several advantages for biodiesel producers. First, it reduces the need for costly and time-consuming physical experiments and trials, accelerating the development and optimization of production processes. Second, it enhances operational efficiency by enabling continuous monitoring and optimization of equipment and processes in realtime. Third, it enhances innovation by providing a platform for experimentation and exploration of new ideas and technologies.

In conclusion, digital twin technology represents a paradigm shift in biodiesel production, offering a powerful tool for simulation, testing, and optimization of processes in a virtual environment. By leveraging digital twins, producers can minimize risks, reduce costs, and drive innovation, ultimately improving efficiency and sustainability in the biodiesel industry.

21.2.8 Cyber-Physical Systems (CPS): Integration of Cyber and Physical Elements

Cyber-physical systems (CPS), at the forefront of Industry 4.0, represent a pivotal advancement in the field of biodiesel production by seamlessly integrating computational and physical processes. This integration fosters a dynamic environment where machines, processes, and human operators communicate and collaborate in real time, enabling heightened responsiveness and adaptability within the production system [17]. At its essence, CPS in biodiesel production involves the interconnection of physical components such as machinery, sensors, and production units – with digital systems, including data analytics platforms, control systems, and human–machine interfaces [18]. This interconnectedness facilitates the exchange of data, commands, and feedback loops between the physical and digital fields, allowing for comprehensive monitoring, control, and optimization of production processes.

One of the key benefits of CPS in biodiesel production is enhanced operational efficiency and productivity. By leveraging real-time data from sensors embedded within production equipment, CPS can optimize process parameters, anticipate equipment failures, and streamline workflows to maximize throughput and minimize downtime [19]. In addition, CPS enables adaptive control strategies that adjust production operations in response to changing conditions, such as variations in feedstock composition or market demand, ensuring optimal performance under dynamic circumstances.

Moreover, CPS facilitates improved safety and risk management within biodiesel production facilities. By integrating sensors and monitoring systems, CPS can detect anomalies, hazardous conditions, or equipment malfunctions in real time, triggering immediate alerts and shutdown procedures to prevent accidents or mitigate potential hazards. Furthermore, CPS enables predictive maintenance strategies that anticipate equipment failures before they occur, reducing the risk of unplanned downtime and costly repairs.

In addition, CPS fosters greater transparency, collaboration, and decision-making capabilities among stakeholders within the biodiesel production ecosystem. Through digital dashboards, visualization tools, and collaborative platforms, CPS provides real-time insights into production performance, resource utilization, and quality metrics, empowering operators, engineers, and managers to make data-driven decisions and coordinate efforts effectively.

In summary, CPSrepresents a transformative paradigm in biodiesel production, enabling seamless integration of computational and physical elements to create responsive, adaptive, and efficient production environments. By harnessing the power of CPS, biodiesel producers can optimize operations, enhance safety, and drive innovation, ultimately contributing to a more sustainable and resilient biofuel industry.

The synergy between Biodiesel and Industry 4.0 represents a paradigm shift in the way biodiesel is produced, monitored, and managed. This convergence not only enhances efficiency and sustainability but also positions biodiesel as a key player in the broader context of smart and sustainable industrial practices. As we embrace the era of Industry 4.0, the biodiesel industry stands poised for a dynamic and technologically advanced future, contributing significantly to the global transition toward greener and more sustainable energy solutions.

21.3 Unveiling the Green Frontier of Biodiesel Production

In the culmination of the exploration within the pages of the book *Advancing Sustainable Biodiesel Production with Biodegradable Feedstocks: Characterization, Selection, and Optimization,* we stand at the threshold of a new era for biodiesel. This book has been designed to unveil the intricacies and potentials within the biodiesel industry; it serves as a testament to the transformative power embedded in sustainable practices and innovative feedstocks.

The journey commenced with a recognition of the urgent need for alternatives to conventional fossil fuels. As global concerns loom about population growth, escalating energy demands, and pressing environmental issues, the focus is directed toward biodiesel as a symbol of hope in the 21st century. Its affordability, renewability, and minimal ecological impact make it a guiding light toward a greener energy future.

Crucial to the narrative is the spotlight on biodegradable feedstocks, a diverse array ranging from *Chrysobalamus icao*, sunflower, hemp, water hyacinth, and others [13–16]. By emphasizing nonedible sources, a course is charted that ensures sustainability without compromising vital agricultural resources. Each chapter unfurls the story of these novel feedstocks, from cultivation to the production processes, which require transforming their oils into high-quality biodiesel. Optimization processes underscore the commitment to enhancing efficiency and effectiveness in biodiesel production. Various aspects have been considered in this study, from sourcing and production to parametric studies and environmental impacts; a comprehensive view of the biodiesel production system has emerged.

This resource, intentionally designed as an open text, welcomes a diverse audience of undergraduate and graduate students, research fellows, and professionals across an array of academic domains. It stands as a guide, nurturing a deeper understanding of biodiesel production using innovative feedstocks. Whether one's focus is on scientific intricacies, economic considerations, or societal implications, this book provides a foundation for exploration. Looking toward the future, the integration of Industry 4.0 technologies emerges as a catalyzing force, propelling biodiesel into a field of smart and sustainable industrial practices. The fusion of digital technologies with traditional processes opens doors to precision agriculture, real-time monitoring, automated production, and data-driven decision-making.

21.4 Conclusion

In the final observation, it is affirmed that biodegradable feedstocks are not merely components of biodiesel production but agents of change, aligning perfectly with globally recognized standards. As the intricacies of each feedstock unfold, the collective exploration serves as both a guide and inspiration. It empowers students, researchers, practitioners, and enthusiasts to focus more on the promise and potential of biodiesel production, fostering a deeper understanding of sustainable energy practices. In conclusion, this book not only serves as a comprehensive guide but also as a catalyst for further exploration, innovation, and commitment toward a biodiesel future that is not only sustainable but integral to the broader discourse on environmental stewardship. There is no doubt that the revelations within the pages of this book will propel readers toward a future where biodiesel, shaped by biodegradable feedstocks and propelled by Industry 4.0, will take center stage in steering humanity toward a more sustainable and harmonious energy frontier.

References

1 OtanochaO., OkwuM.O., L.K.Tartibu (2021) Modified biogas tank for production of gas from decomposable organic waste. *Biomass Convers. Biorefin.* https://doi.org/10.1007/s13399-020-01195-x.

2 Mbachu, V.M., Okwu, M.O., Chiabuotu, C.C., and Tartibu, L.K. (2022). Biogas for electricity generationin Nigeria: a systematic review of the prospect, efforts and contemporary challenges. In: *Advances in Biofeed Stocks and Biofuels*. Scrivener-Wiley. https://doi.org/10.1002/9781119785842.ch4.

3 Samuel, O.D., Okwu, M.O., Oyejide, O.J. et al. (2020). Optimizing biodiesel production from abundant waste oils through the empirical method and grey wolf optimizer. *Fuel* 281: 118701.

4 Okwu, M.O., Tartibu, L.K., Maware, C. et al. (2023). Emerging technologies of Industry 4.0: challenges and opportunities. *IEEE, icABCD* 1–13. https://doi.org/10.1109/icABCD54961.2022.9856002.

5 Deng, C., Guo, R., Liu, C. et al. (2018). Data cleansing for energy-saving: a case of cyber-physical machine tools health monitoring system. *Int. J. Prod. Res.* 56 (1–2): 16. https://doi.org/10.1080/00207543.2017.1394596.

6 Onu, P. and Mbohwa, C. (2019). Renewable energy technologies in brief. *Int. J. Sci. Technol. Res.* 8 (10): 1283–1289.

7 Jena, M.C., Mishra, S.K., and Moharana, H.S. (2020). Application of Industry 4.0 to enhance sustainable manufacturing. *Environ. Prog. Sustain. Energy* 39 (1): 13360.

8 Pease, S.G., Trueman, R., Davies, C. et al. (2018). An intelligent real-time cyber-physical toolset for energy and process prediction and optimisation in the future industrial Internet of Things. *Future Gener. Comput. Syst.* 79: 815–829.

9 Arshad, R., Zahoor, S., Shah, M.A. et al. (2017). Green IoT: an investigation on energy saving practices for 2020 and beyond. *IEEE Access*.

10 Beier, G., Ullrich, A., Niehoff, S. et al. (2020). Industry 4.0: how it is defined from a sociotechnical perspective and how much sustainability it includes – a literature review. *J. Clean. Prod.*

11 Onu, P. and Mbohwa, C. (2021). Nascent technologies in resources conservation and sustainable agricultural development. *Agric. Waste Diver. Sustain. Issues.*

12 Esmaeilian, B., Sarkis, J., Lewis, K., and Behdad, S. (2020). Blockchain for the future of sustainable supply chain management in Industry 4.0. *Resour. Conserv. Recycl.*

13 Okonkwo, C.P., Ajiwe, V.I.E., Obiadi, M.C. et al. (2023). Production of biodiesel from the novel non-edible seed of Chrysobalanus icaco using natural heterogeneous catalyst: Modeling and prediction using Artificial Neural Network. *J. Clean. Prod.* 385: 135631.

14 Okonkwo, C.P., Ajiwe, V.I., Ikeuba, A.I. et al. (2023). Production and performance evaluation of biodiesel from *Elaeis guineensis* using natural snail shell-based heterogeneous catalyst: kinetics, modeling and optimisation by artificial neural network. *RSC Adv.* 13 (28): 19495–19507.

15 Okonkwo, C.P., Ajiwe, V.I.E., Obiadi, M.C. et al. (2021). Kinetics and transesterification of the oil obtained from *Cussonia bateri* (Jansa seed) as a Stepin biodiesel production using natural heterogeneous catalyst. *Am. J. Appl. Chem.* 9 (5): 164–171.

16 Okonkwo, C.P., Ajiwe, V.I.E., Obiadi, M.C., and Okwu, M. (2021). A systematic review of biodiesel production with sustainable feedstock using assorted catalyst. *Am. J. Appl. Chem.* 9 (5): 154–163.

17 Onu, P. and Mbohwa, C. (2021). Reimagining the future: techno innovation advancement in manufacturing. *Mater. Today Proc.* 44 (1): 1953–1959.

18 Ghobakhloo, M. (2020). Industry 4.0, digitization, and opportunities for sustainability. *J. Clean. Prod.*

19 Ma, S., Zhang, Y., Lv, J. et al. (2019). Energy-cyber-physical system enabled management for energy-intensive manufacturing industries. *J. Clean. Prod.*

Conclusion

In the pursuit of a greener and more sustainable future, the exploration and utilization of biodegradable feedstocks for biodiesel production stand as a symbol of hope and innovation. Through the comprehensive journey undertaken in this book, *Feedstocks for Sustainable Biodiesel Production: Characterization, Selection, and Optimization*, a profound understanding has been reached regarding the pivotal role these feedstocks play in reshaping the biodiesel industry. From the outset, the imperative need for sustainable alternatives to conventional fossil fuels has been underscored against the backdrop of escalating energy demands and pressing environmental concerns. Biodiesel, derived from both edible and nonedible sources, emerges as a promising solution due to its renewability, affordability, ecological friendliness, and minimal toxicity. Crucially, this book highlights the immense potential lying within nonedible feedstocks, offering sustainable alternatives without compromising valuable agricultural resources.

Throughout the pages, this book has exposed the intricate processes involved in biodiesel production using a diverse array of biodegradable feedstocks. From the cultivation and processing of these feedstocks to the transformation of their oils into high-quality biodiesel, each chapter has provided invaluable insights into optimization techniques, economic and technological considerations, and environmental impacts. The wealth of knowledge encapsulated within this resource serves as a guiding light for students, industry professionals, researchers and scientists, policy makers and regulators, farmers and agricultural professionals, environmentalists and sustainability advocates, investors and entrepreneurs, educators and trainers, consultants non-governmental organizations (NGO), general public and enthusiasts, fostering a deeper understanding of sustainable energy practices and biodiesel production. Whether one's interests lie in scientific intricacies, economic analyses, or societal implications, this book serves as a comprehensive

Feedstocks for Sustainable Biodiesel Production: Characterization, Selection, and Optimization, First Edition. Chinwe P. Okonkwo and Modestus O. Okwu.

and accessible guide, empowering individuals to navigate the multifaceted field of biodiesel production with confidence, assurance and deeper insight.

Positioned at the leading edge of a crucial historical moment marked by increasing awareness and the pressing demand for sustainable energy solutions, this book serves as a catalyst for change. It inspires and empowers individuals to embrace the promise and potential inherent in advancing sustainable biodiesel production with biodegradable feedstocks. By harnessing the transformative power of innovation and collaboration, it becomes possible to pave the way toward a greener, more harmonious future for generations to come.

In summary, achieving sustainability is not a destination but rather an ongoing journey demanding unwavering dedication, ingenuity, and resilience. The insights gleaned from the pages of this book serve as more than mere knowledge, they represent actionable inspiration that can be translated into tangible solutions for renewable energy challenges. By applying these insights, the opportunity to drive progress and expand the horizons of what can be achieved in the quest for a more sustainable future becomes possible. Thus, the insights obtained within the pages of this book can be implemented to address renewable energy challenges and expand the horizons of what is achievable in the quest for a brighter future.

Index

Feedstocks for Sustainable Biodiesel Production: Characterization, Selection, and Optimization, First Edition. Chinwe P. Okonkwo and Modestus O. Okwu.

d

m

n

Printed and bound by CPI Group (UK) Ltd, Croydon, CR0 4YY

17/09/2026

14964331-0006